Advances in Computer Vision and Pattern Recognition

More information about this series at http://www.springer.com/series/4205

Hà Quang Minh · Vittorio Murino
Editors

Algorithmic Advances in Riemannian Geometry and Applications

For Machine Learning, Computer Vision, Statistics, and Optimization

Springer

Editors
Hà Quang Minh
Pattern Analysis and Computer Vision
Istituto Italiano di Tecnologia
Genoa
Italy

Vittorio Murino
Pattern Analysis and Computer Vision
Istituto Italiano di Tecnologia
Genoa
Italy

ISSN 2191-6586 ISSN 2191-6594 (electronic)
Advances in Computer Vision and Pattern Recognition
ISBN 978-3-319-45025-4 ISBN 978-3-319-45026-1 (eBook)
DOI 10.1007/978-3-319-45026-1

Library of Congress Control Number: 2016948260

Printed on acid-free paper

This Springer imprint is published by Springer Nature
The registered company is Springer International Publishing AG
The registered company address is: Gewerbestrasse 11, 6330 Cham, Switzerland

Preface

Overview and Goals

The theme of this volume is the application of the rich and powerful theories and techniques of Riemannian geometry to the problems in machine learning, statistics, optimization, computer vision, and related fields.

Traditional machine learning and data analysis methods often assume that the input data can be represented by vectors in Euclidean space. While this assumption has worked well for many applications, researchers have increasingly realized that if the data is intrinsically non-Euclidean, ignoring this geometrical structure can lead to suboptimal results.

In the existing literature, there are two common approaches for exploiting data geometry when the data is assumed to lie on a Riemannian manifold.

In the first direction, often referred to as manifold learning, the data is assumed to lie on an unknown Riemannian manifold and the structure of this manifold is exploited through the training data, either labeled or unlabeled. Examples of manifold learning techniques include manifold regularization via the graph Laplacian, locally linear embedding, and isometric mapping.

In the second direction, which is gaining increasing importance and success, the Riemannian manifold representing the input data is assumed to be known explicitly. Some manifolds that have been widely used for data representation are the manifold of symmetric, positive definite matrices, the Grassmannian manifold of subspaces of a vector space, and the Kendall manifold of shapes. When the manifold is known, the full power of the mathematical theory of Riemannian geometry can be exploited in both the formulation of algorithms as well as their theoretical analysis. Successful applications of this approach are numerous and range from brain imaging, kernel learning, and low rank matrix completion, to computer vision tasks such as object detection and tracking.

This volume focuses on the latter research direction. The forthcoming chapters were written by researchers currently active in the fields. Overall, the book describes some of the latest algorithmic advances using Riemannian geometry, both theoretically and computationally, with potentially large impact on many research areas in these fields.

The volume targets a broad audience, consisting of Ph.D. students and researchers in machine learning, statistics, optimization, computer vision, and related fields.

Acknowledgments

We wish to thank all the authors for contributing some of their latest works to the volume. We also wish to thank Pierre-Antoine Absil, Gregory Chirikjian, Mark Girolami, Pavan Turaga, Bart Vandereycken, and Baba Vemuri, for their help in reviewing the manuscript. Finally, we wish to thank Simon Rees and the Springer editing team for helping us bring this volume to fruition.

Genoa, Italy
June 2016

Hà Quang Minh
Vittorio Murino

Contents

Contributors

Anoop Cherian ARC Centre of Excellence for Robotic Vision, Australian National University, Canberra, Australia

Adam Duncan Department of Statistics, Florida State University, Tallahassee, FL, USA

P. Thomas Fletcher University of Utah, Salt Lake City, UT, USA

Mehrtash Harandi College of Engineering and Computer Science, Australian National University, Canberra, ACT, Australia

Richard Hartley College of Engineering and Computer Science, Australian National University, Canberra, ACT, Australia

Reshad Hosseini School of ECE, College of Engineering, University of Tehran, Tehran, Iran

Shiwei Lan Department of Statistics, University of Warwick, Coventry, UK

Hà Quang Minh Pattern Analysis and Computer Vision (PAVIS), Istituto Italiano di Tecnologia (IIT), Genoa, Italy

Vittorio Murino Pattern Analysis and Computer Vision (PAVIS), Istituto Italiano di Tecnologia (IIT), Genoa, Italy

Fatih Porikli Australian National Univeristy, Canberra, Australia; Data61/CSIRO, Eveleigh, Australia

Mathieu Salzmann CVLab, EPFL, Lausanne, Switzerland

Babak Shahbaba Department of Statistics and Department of Computer Science, University of California, Irvine, CA, USA

Suvrit Sra Laboratory for Information & Decision Systems (LIDS), Massachusetts Institute of Technology, Cambridge, MA, USA

Anuj Srivastava Department of Statistics, Florida State University, Tallahassee, FL, USA

Jochen Trumpf College of Engineering and Computer Science, Australian National University, Canberra, ACT, Australia

Miaomiao Zhang Massachusetts Institute of Technology, Cambridge, MA, USA

Zhengwu Zhang SAMSI, Research Triangle Park, Durham, NC, USA

Introduction

Themes of the Volume

The aim of this book is to present some of the most recent algorithmic advances in Riemannian geometry in the context of machine learning, statistics, optimization, computer vision, and related fields. The unifying theme of the different chapters in the book is the exploitation of the geometry of data using the mathematical machinery of Riemannian geometry. As demonstrated by all the subsequent chapters, when the data is intrinsically non-Euclidean, the utilization of this geometrical information can lead to better algorithms that can capture more accurately the structures inherent in the data, leading ultimately to better empirical performance.

This book is not intended to be an encyclopedic compilation of the applications of Riemannian geometry. Instead, it focuses on several important research directions that are currently actively pursued by researchers in the field. These include statistical modeling and analysis on manifolds, optimization on manifolds, Riemannian manifolds and kernel methods, and dictionary learning and sparse coding on manifolds. We now describe these topics in more detail, noting that a particular chapter in the book may cover more than one of them.

1. **Statistical modeling and analysis on manifolds**. This research direction seeks to develop theories and techniques for statistical modeling and analysis on Riemannian manifolds by utilizing the intrinsic geometry of the underlying manifolds. This theme is covered by several chapters in the book. First, it is considered in the chapter by Lan and Shahbaba, which develops Hamiltonian Monte Carlo on the sphere, using spherical geometry, for sampling constrained probability distributions. Second, it is covered in the chapter by Zhang and Fletcher, which develops Bayesian models for shape analysis, with shape variability being represented as random variables on the manifold of diffeomorphic transformations. Statistical shape analysis is also the theme of the chapter by Duncan et al., which presents a Riemannian framework for curves in Euclidean and Hilbert spaces and for tree-like structures. Finally, the chapter by

Porikli treats regression on the matrix Lie group of affine transformations, with applications in computer vision.

2. **Optimization on manifolds**. This research direction is concerned with the generalization of the theories and algorithms for optimization in Euclidean space to the manifold setting. This is the theme of the chapter by Sra and Hosseini, which deals with optimization on the manifold of symmetric, positive definite (SPD) matrices by exploiting its geometric structure. From an application perspective, this theme is considered in the chapter by Cherian and Sra for the problem of Riemannian Dictionary Learning and Sparse Coding.
3. **Riemannian manifolds and kernel methods**. Kernel methods are among the most powerful paradigms in machine learning and its applications. However, most of the kernels employed in the literature are defined on Euclidean space and applying them directly to data that lies on a nonlinear manifold generally leads to suboptimal results. In order to capture the manifold structure in the data, it is necessary to define *kernels on the manifold* itself, using a notion of distance that captures its intrinsic geometry. This theme is pursued in the chapter by Harandi et al, which considers kernels defined on Grassmann manifolds, and the chapter by Minh and Murino, which discusses kernels defined on the manifold of SPD matrices. Moreover, the interplay between kernels and Riemannian manifolds can also go in the direction of *kernels giving rise to manifolds*. In the chapter by Minh and Murino, it is shown how a positive definite kernel gives rise to covariance operators, which lie on the infinite-dimensional manifold of positive definite operators, on which another kernel can be defined, leading to a two-layer kernel machine.
4. **Dictionary learning and sparse coding on manifolds**. This research direction seeks to generalize the algorithms for dictionary learning and sparse coding on Euclidean space to the Riemannian manifold setting by utilizing the intrinsic geometry of the underlying manifold. This is the theme of the chapter by Cherian and Sra, which considers dictionary learning and sparse coding on the manifold of SPD matrices, and the chapter by Harandi et al, which considers this problem on the Grassmann manifolds.

Organization of the Volume

We now give a summary of the chapters in the book, in order of appearance.

The chapter by Zhang and Fletcher is titled *Bayesian Statistical Shape Analysis on the Manifold of Diffeomorphisms*. In this chapter, the authors present two recently introduced Bayesian models for the statistical analysis of anatomical shapes through diffeomorphic transformations on the image domain. The first model, namely *Bayesian diffeomorphic atlas building*, is a probabilistic formulation for computing an atlas, or template image, that is most representative of a set of input images. In this model, the distance between images is measured via an energy function, whose regularization term is defined using a Riemannian metric on the manifold of diffeomorphisms. The model parameters are estimated via Monte Carlo expectation maximization (EM) algorithm, with the E step carried out via Hamiltonian Monte Carlo sampling on the manifold of diffeomorphisms. The mathematical formulation is accompanied by numerical examples involving atlas building for 3D images. The second model, namely *Bayesian principal geodesic analysis* (BPGA), generalizes the Bayesian formulation of principal component analysis (PCA) to the manifold of diffeomorphisms. Using experimental results on the task of reconstructing 3D brain MRI, the authors demonstrate that BPGA results in a much more compact representation compared with both linear PCA and tangent PCA.

The chapter by Lan and Shahbaba is titled *Sampling Constrained Probability Distributions using Spherical Augmentation*. In this chapter, the authors present their recently introduced approach, namely spherical augmentation, for sampling from constrained probability distributions. In this approach, the constrained domain, defined by norm or functional constraints, is mapped to a sphere in an augmented space. Sampling algorithms then generate new proposals on the sphere, using the spherical metric, which satisfy the required constraints when mapped back to the original space. The authors use this approach to obtain several novel Monte Carlo sampling algorithms, namely spherical Hamiltonian Monte Carlo and spherical Lagrangian Monte Carlo. The mathematical formulation is accompanied by many numerical examples, including Bayesian Lasso, Bayesian bridge regression, and latent Dirichlet allocation (LDA) for topic modeling, tested in particular on the Wikipedia corpus, among others.

The chapter by Sra and Hosseini is titled *Geometric Optimization in Machine Learning*. In this chapter, the authors report some of the most recent algorithmic developments in solving optimization problems on the manifold of positive definite matrices. Two key mathematical concepts involved are *geodesic convexity*, which is the generalization of Euclidean convexity to the Riemannian manifold setting, and *Thompson nonexpansivity*, for a class of nonconvex functions that are not necessarily geodesically convex. Together, these concepts enable the global optimization of many functions that are nonconvex in the Euclidean sense. In particular, the authors exploit geodesic convexity in the problem of fitting Gaussian mixture models (GMMs), leading to an algorithm with substantial improvement in performance compared to the classical expectation minimization (EM) algorithm.

The chapter by Cherian and Sra is titled *Positive Definite Matrices: Data Representation and Applications to Computer Vision*. In this chapter, the authors consider positive definite matrices in the form of covariance descriptors, a powerful data representation paradigm in computer vision. In particular, the authors present their recent approach on Riemannian dictionary learning and sparse coding on the manifold of positive definite matrices, using the affine-invariant Riemannian metric and the Riemannian conjugate gradient algorithm. Using experimental results involving face recognition, person re-identification, and 3D object recognition, the authors demonstrate that the Riemannian approach performs substantially better than its Euclidean counterpart.

The chapter by Minh and Murino is titled *From Covariance Matrices to Covariance Operators: Data Representation from Finite to Infinite-Dimensional Settings*. In this chapter, the authors report on the recent generalization of the covariance matrix representation for images to the infinite-dimensional setting, using covariance operators in reproducing kernel Hilbert spaces (RKHS). In particular, the authors describe the generalizations of the affine-invariant Riemannian and Log-Euclidean distances between positive definite matrices to the infinite-dimensional manifold of positive definite operators. In the case of RKHS covariance operators, these distances admit closed form expressions via the Gram matrices corresponding to the reproducing kernels. The mathematical formulation is accompanied by numerical examples in image classification, demonstrating that the infinite-dimensional framework substantially outperforms its finite-dimensional counterpart.

The chapter by Porikli is titled *Regression on Lie Groups and Its Application to Affine Motion Tracking*. In this chapter, the author treats regression on matrix Lie groups, which are used in most of the transformations in computer vision, such as affine motions and rotations. The proposed formulation goes beyond the typical Euclidean approximation in the literature by providing a solution consistent with the underlying topology. The mathematical formulation is accompanied by numerical examples in a fundamental computer vision task, namely affine motion tracking.

The chapter by Harandi, Hartley, Salzmann, and Trumpf, is titled *Dictionary Learning on Grassmann Manifolds*. In this chapter, the authors present their recent work on dictionary learning and sparse coding on the Grassmann manifolds of subspaces of Euclidean space. In particular, the authors propose to embed Grassmann manifolds into reproducing kernel Hilbert spaces (RKHS) by defining a family of positive definite kernels on these manifolds. Thus, the problems of dictionary learning and sparse coding on the Grassmann manifolds are transformed into the corresponding ones in kernel spaces, which can be solved efficiently. The mathematical formulation is accompanied by numerical examples in action recognition in videos.

The chapter by Duncan, Zhang, and Srivastava is titled *An Elastic Riemannian Framework for Shape Analysis of Curves and Tree-like Structures*. In this chapter, the authors present a Riemannian framework for shape analysis that is invariant under the action of re-parametrization for three different types of objects. The chapter begins with the elastic Riemannian metric framework for shape analysis of Euclidean curves using square-root velocity functions. This framework is then extended to trajectories in Hilbert spaces and tree-like structures, which are treated as composite trajectories in Hilbert spaces. The mathematical formulation is accompanied by numerical examples involving planar shapes and neuron morphology, using digital 3D neuronal reconstructions.

Hà Quang Minh
Vittorio Murino

Chapter 1
Bayesian Statistical Shape Analysis on the Manifold of Diffeomorphisms

Miaomiao Zhang and P. Thomas Fletcher

Abstract In this chapter, we present Bayesian models for diffeomorphic shape variability in populations of images. The first model is a probabilistic formulation of the image *atlas* construction problem, which seeks to compute an atlas image most representative of a set of input images. The second model adds diffeomorphic modes of shape variation, or *principal geodesics*. Both of these models represent shape variability as random variables on the manifold of diffeomorphic transformations. We define a Gaussian prior distribution for diffeomorphic transformations using the inner product in the tangent space to the diffeomorphism group. We develop a Monte Carlo Expectation Maximization (MCEM) algorithm for the Bayesian inference, due to the lack of closed-form solutions, where the expectation step is approximated via Hamiltonian Monte Carlo (HMC) sampling of diffeomorphisms. The resulting inference produces estimates of the image atlas, principal geodesic modes of variation, and model parameters. We show that the advantage of the Bayesian formulation is that it provides a principled way to estimate both the regularization parameter of the diffeomorphic transformations and the intrinsic dimensionality of the input data.

1.1 Introduction

The key philosophy in computational anatomy is to quantify anatomical shapes in images through diffeomorphic transformations of the image domain. These diffeomorphic transformations are smooth, bijective mappings with smooth inverses, which guarantee topological consistency between images, e.g., no folding or tearing. An elegant mathematical formulation for estimating diffeomorphism transformations between images is *Large Deformation Diffeomorphic Metric Mapping (LDDMM)*,

M. Zhang (✉)
Massachusetts Institute of Technology, Cambridge, MA, USA
e-mail: miao86@mit.edu

P.T. Fletcher
University of Utah, Salt Lake City, UT, USA
e-mail: fletcher@sci.utah.edu

H.Q. Minh and V. Murino (eds.), *Algorithmic Advances in Riemannian Geometry and Applications*, Advances in Computer Vision and Pattern Recognition, DOI 10.1007/978-3-319-45026-1_1

proposed by Beg et al. [6]. In this setting, the space of diffeomorphisms of an image domain forms an infinite-dimensional Lie group equipped with a right-invariant Riemannian metric. This gives rise to a variational principle that expresses the optimal registration between two images as a geodesic flow.

Computing a representative image for a population, or atlas, is an important first step in shape analysis, with the goal of finding a common coordinate system for comparisons across individuals. Principal modes of variation, represented in the tangent space of the diffeomorphism group, provide second-order statistics of shape variability. Many diffeomorphic shape models of images are set up as optimization problems, rather than as statistical inference problems. These approaches (1) perform poorly in the presence of noises or outliers, (2) have free parameters that are difficult and time consuming to select, even for experienced users, and (3) do not provide probabilistic conclusions about the data. To address these problems, this chapter presents a Bayesian framework of statistical shape analysis including diffeomorphic atlas building and principal geodesic analysis of diffeomorphisms.

Previous methods often formulate diffeomorphic atlas building as a maximum a posterior (MAP) optimization problem [15, 16, 29] with an image matching likelihood of the atlas and each input image, as well as a prior arising from a metric on an infinite-dimensional manifold of diffeomorphisms that encourages smooth deformations. The regularity of smoothness is typically controlled by parameters describing the metric on the tangent space of the diffeomorphism group. However, Allassonniére et al. [1] pointed out that the common mode approximation scheme performs poorly under image noise, even for a simple 1D template estimation problem where the transformations are discrete shifts. Besides this, the regularity parameters were specified in an ad hoc manner rather than being estimated directly from data, due to the fact that the log posterior of the metric parameters does not have a closed form and is computationally problematic to solve using direct optimization. To overcome these disadvantages, several works [14, 20, 21, 28] have then developed Bayesian formulations of the atlas estimation problem in a small deformation setting. They estimated the atlas by marginalizing over the posterior distribution of image transformations using a Monte Carlo sampling procedure. Furthermore inference of the level of regularization in nonrigid registration by a Bayesian model were developed in [3, 23, 24].

Image atlas is a mean point estimation and it does not encode the group shape variability. Extracting an efficient low-dimensional, second-order statistics from high-dimensional diffeomorphisms is critical to improve the power and interpretability of further statical analysis. A standard way to analyze manifold data variability is principal geodesic analysis (PGA) [11, 12], which generalizes principal component analysis (PCA) in Euclidean space to nonlinear manifolds. PGA estimates lower dimensional geodesic subspaces by minimizing the sum-of-squared geodesic distance to the data. Based on this work, exact solutions to PGA were developed in [22, 26]. To give a probabilistic interpretation of these geodesic analysis, Zhang and Fletcher [33] developed a latent variable model for PGA that provides a probabilistic framework for factor analysis on finite-dimensional manifolds called PPGA. Previous methods [13, 19, 27] performed the dimensionality reduction after the fact, i.e., as a PCA of diffeomorphisms in the tangent space as a second stage after the estimation step. All these models have two major disadvantages. First, they do not explicitly optimize the fit of the principal modes to the data intrinsically in the space of diffeomorphisms, which results in a suboptimal fit to the data. Second, the number of dimensions to use is selected manually rather than inferred directly from the data. A related Bayesian model of PCA (BPCA) [7] for Euclidean data was proposed to automatically learn the dimension of the latent space from data by including a sparsity-inducing prior on each component of the factor matrix. This linear factor analysis model, however, is not applicable to nonlinear diffeomorphic transformations. The main difficulty of generalizing the model definition of BPCA to generic manifolds is the lack of an explicit formulation for the normalizing constant of distributions on manifolds.

The following sections present an integrated framework for two recently introduced statistical shape models, Bayesian diffeomorphic atlas building [36] and Bayesian principal geodesic analysis (BPGA) of diffeomorphisms [34, 35]. We first introduce a posterior distribution of diffeomorphisms, followed by methods for MCMC sampling of diffeomorphisms, as well as parameter estimation via marginalization of the diffeomorphic transformations. There has been some work on stochastic flows of diffeomorphisms [8], which are Brownian motions, i.e., small perturbations integrated along a time-dependent flow. A similar idea of using Gaussian Process prior on the stochastic velocity field to generate random transformations is developed in [31]. This chapter focuses on a different prior distribution on the tangent space of initial velocity fields rather than on the entire time-dependent flow, which leads to random geodesics in the space of diffeomorphisms. Other prior distributions can also be adapted to this Bayesian framework such as control points parameterization of diffeomorphisms [2]. We next describe the BPGA model, which treats the dimensionality reduction step as a probabilistic inference problem on discrete images, and an inference procedure that jointly estimates the image atlas and principal geodesic modes of variation. This model goes beyond the PPGA algorithm by introducing automatic dimensionality reduction and extending from finite-dimensional manifolds to the infinite-dimensional case of diffeomorphic image registration.

1.2 Mathematical Background

In this section, we briefly review the basic concepts of diffeomorphisms and their applications in image analysis.

1.2.1 Space of Diffeomorphisms

Consider a d-dimensional torus $\Omega = \mathbb{R}^d / \mathbb{Z}^d$ as the domain in which we define images. A diffeomorphism is a bijective smooth invertible mapping $\phi : \Omega \to \Omega$

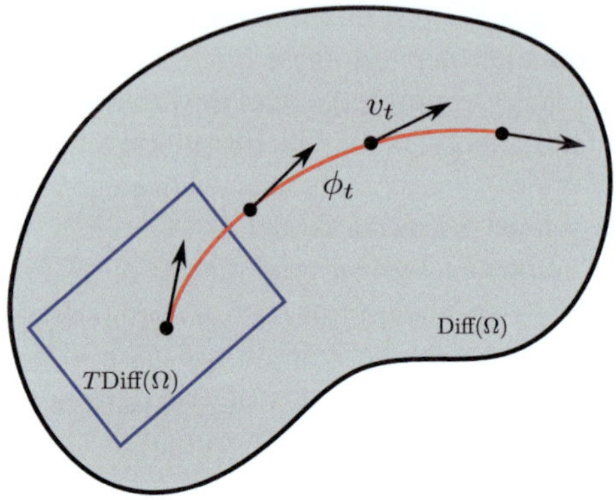

Fig. 1.1 Space of diffeomorphisms

with its smooth inverse ϕ^{-1}. We denote by $\text{Diff}(\Omega)$ the space of diffeomorphisms whose derivatives of all orders exist and are square-integrable. This space forms an infinite-dimensional *Lie group*. The *Lie algebra* of $\text{Diff}(\Omega)$ is the tangent space at the identity transform, $V = T_{\text{id}}\text{Diff}(\Omega)$, and consists of all vector fields equipped with the *Lie bracket* operation. Most of the computations w.r.t. the diffeomorphism group are done in the space of the Lie algebra because it is a linear vector space.

Given a flow of time-varying velocity field, $v_t : [0, 1] \to V$, we generate a flow of diffeomorphisms, $t \mapsto \phi_t \in \text{Diff}(\Omega)$ (see Fig. 1.1), as a solution to the ordinary differential equation,

$$\frac{d\phi_t}{dt}(x) = v_t \circ \phi_t(x). \tag{1.1}$$

Note that we use subscripts for the time variable, i.e., $v_t(x) = v(t, x)$, and $\phi_t(x) = \phi(t, x)$.

1.2.2 Metrics on Diffeomorphisms

A distance metric provides a way to measure the difference between diffeomorphisms, which forms the mathematical foundation for statistical analysis of diffeomorphisms such as the Fréchet mean, variability quantification, and regression. We first define an inner product on the tangent space of diffeomorphisms at identity, $V = T_e\text{Diff}(\Omega)$, where $e \in \text{Diff}(\Omega)$ is the identity transformation. This inner product is of the form

$$\langle v, w \rangle_V = \int_\Omega (Lv(x), w(x))dx,$$

for $v, w \in V$, and a symmetric, positive-definite differential operator $L : V \to V^*$, mapping to the dual space, V^*. The dual to the vector v is a momentum, $m \in V^*$, such that $m = Lv$ or $v = Km$, where K is the inverse of L. In this chapter, we use

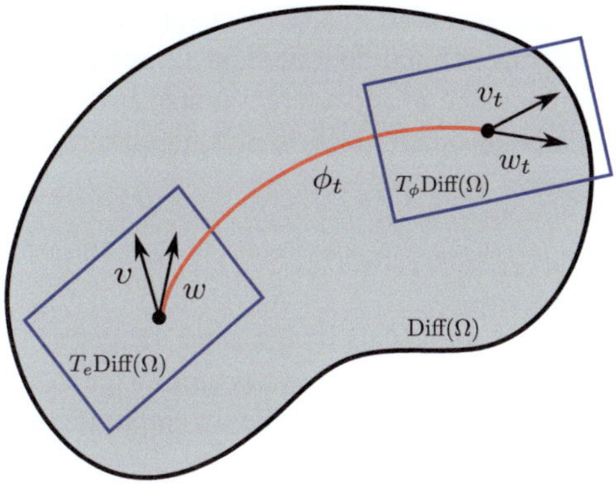

Fig. 1.2 Metrics on diffeomorphisms

$L = -\alpha\Delta + \beta$, where Δ is the discrete Laplacian, and α and β are positive numbers. A major advantage of using Laplacian is that it is a diagonal matrix in the Fourier space, thus makes the inverse of L much easier to compute.

Next we define a right-invariant metric as an inner product at any other point $\phi \in \mathrm{Diff}(\Omega)$, by pulling back the velocities at ϕ to the identity by right composition. For $v, w \in T_\phi\mathrm{Diff}(\Omega)$, the right-invariant metric is given by

$$\langle v, w\rangle_{T_\phi \mathrm{Diff}(\Omega)} = \langle v \circ \phi^{-1}, w \circ \phi^{-1}\rangle_V.$$

A geodesic curve $\{\phi_t\} \in \mathrm{Diff}(\Omega)$, illustrated in Fig. 1.2, is a flow of diffeomorphisms that minimizes the energy

$$E(\phi_t) = \int_0^1 \left\| \frac{d\phi_t}{dt} \circ \phi_t^{-1} \right\|_V^2 dt,$$

and it is characterized by the Euler–Poincaré equations (EPDiff) [5, 17],

$$\frac{\partial v}{\partial t} = -K\left[(Dv)^T m + Dm\, v + m \,\mathrm{div}\, v\right], \tag{1.2}$$

where D denotes the Jacobian matrix and div is the divergence operator.

Given an initial velocity, $v_0 \in V$, at $t = 0$, the EPDiff equation (1.2) can be integrated forward in time, resulting in a time-varying velocity $v_t : [0, 1] \to V$, which itself is subsequently integrated in time by the rule (1.1) to arrive at the geodesic path, $\phi_t \in \mathrm{Diff}(\Omega)$. This process is known as *geodesic shooting*.

1.2.3 Diffeomorphic Atlas Building with LDDMM

Before introducing the diffeomorphic atlas building problem in the setting of LDDMM [6] with geodesic shooting [25, 30, 32], we first review a distance metric between pairwise images. Consider images $I_0, I_1 \in L^2(\Omega, \mathbb{R})$ as square-integrable functions defined on a domain Ω, we compute the diffeomorphic matching from a source image I_0 and a target image I_1 by minimizing an energy function of sum-of-squared distance function plus regularization term as

$$E(v_0, I_0, I_1) = \frac{1}{2\sigma^2}\|I_0 \circ \phi_1^{-1} - I_1\|_{L^2}^2 + \frac{1}{2}\|v_0\|_V^2, \tag{1.3}$$

where σ^2 represents image noise variance.

When the energy above is minimized over all initial velocities, it yields a squared distance metric between the two input images, i.e.,

$$d(I_0, I_1)^2 = \min_{v_0 \in V} E(v_0, I_0, I_1).$$

Using this distance metric between images, we find the Fréchet mean of a group of images $J^1, \ldots, J^N \in L^2(\Omega, \mathbb{R})$ and the initial velocities $\{v_0^k \in L^2([0, 1], V)\}_{k=1\ldots N}$ that minimize the distances function, i.e.,

$$\arg\min_{I, v_0^k} \frac{1}{N} \sum_{k=1}^{N} d(I, J^k)^2. \tag{1.4}$$

Putting (1.3) and (1.4) together, we have

$$E(v_0^k, I) = \sum_{k=1}^{N} \frac{1}{2\sigma^2} \left\| I \circ (\phi_1^k)^{-1} - J^k \right\|_{L^2}^2 + \frac{1}{2}\|v_0^k\|_V^2, \tag{1.5}$$

where the deformation ϕ_1^k is defined in (1.1) as the integral flow of v_0^k with $\phi_0^k = Id$. Because the distance function between images is itself a minimization problem, the atlas estimation is typically done by alternating between the minimization to find the optimal v_0^k and the update of the atlas, I. Note that for notation simplicity, we denote v_0^k as v^k and ϕ_1^k as ϕ^k in the following sections.

1.3 A Bayesian Model for Atlas Building

For a continuous domain $\Omega \subset \mathbb{R}^n$, direct interpretation of (1.3) as a negative log posterior is problematic, as the image match term would be akin to isotropic Gaussian noise in the infinite-dimensional Hilbert space $L^2(\Omega, \mathbb{R})$. This is not a well-defined probability distribution as it has an infinite measure. More appropriately, we can instead consider our input images, J^k, and our atlas image, I, to be measured on a discretized grid, $\Omega \subset \mathbb{Z}^n$. That is, images are elements of the finite-dimensional Euclidean space $l^2(\Omega, \mathbb{R})$. We will also consider velocity fields v^k and the resulting diffeomorphisms ϕ^k to be defined on the discrete grid, Ω. Now our noise model is i.i.d. Gaussian noise at each image voxel, with the likelihood given by

$$p(J^k \mid v^k, I) = \frac{1}{(2\pi)^{M/2}\sigma^M} \exp\left(-\frac{\|I \circ (\phi^k)^{-1} - J^k\|^2}{2\sigma^2}\right), \tag{1.6}$$

where M is the number of voxels, and the norm inside the exponent is the Euclidean norm of $l^2(\Omega, \mathbb{R})$.

The negative log prior on the v^k is a discretized version of the squared Hilbert space norm above. Now consider L to be a discrete, self-adjoint, positive-definite differential operator on the domain Ω. The prior on each v^k is given by a multivariate Gaussian,

$$p(v^k) = \frac{1}{(2\pi)^{\frac{M}{2}} |L^{-1}|^{\frac{1}{2}}} \exp\left(-\frac{(Lv^k, v^k)}{2}\right), \tag{1.7}$$

where $|L|$ is the determinant of L. In the sequel, we could put noninformative priors on $\theta = (\alpha, \sigma^2, I)$ and jointly marginalize them out with v^k. Instead, we simply treat θ to be parameters that we wish to estimate. We fix β to a small number to ensure that the L operator is nonsingular. Putting together the likelihood (1.6) and prior (1.7), we arrive at the log joint posterior for the diffeomorphisms, via initial velocities, v^k, as

$$\log \prod_{k=1}^{N} p\left(v^k \mid J^k; \theta\right) = \frac{N}{2} \log |L| - \frac{1}{2} \sum_{k=1}^{N} (Lv^k, v^k) - \frac{MN}{2} \log \sigma$$
$$- \frac{1}{2\sigma^2} \sum_{k=1}^{N} \|I \circ (\phi^k)^{-1} - J^k\|^2 + \text{const.} \tag{1.8}$$

1.4 Estimation of Model Parameters

We now present an algorithm for estimating the parameters, θ, of the probabilistic image atlas model specified in the previous section. These parameters include the image atlas, I, the smoothness level, or metric parameter, α, and the variance of the image noise, σ^2. We treat the v^k, i.e., the initial velocities of the image diffeomorphisms, as latent random variables with log posterior given by (1.8). This requires integration over the latent variables, which is intractable in closed form. We thus develop a Hamiltonian Monte Carlo procedure for sampling v^k from the posterior and use this in a Monte Carlo Expectation Maximization algorithm to estimate θ. It consists of two main steps:

1. E-step

We draw a sample of size S from the posterior distribution (1.8) using HMC with the current estimate of the parameters, $\theta^{(i)}$. Let $v^{kj}, j = 1, \ldots, S$, denote the jth point in this sample for the kth velocity field. The sample mean is taken to approximate the Q function,

$$\begin{aligned} Q(\theta \mid \theta^{(i)}) &= E_{v^k \mid J^k; \theta^{(i)}} \left[\sum_{k=1}^{N} \log p \left(v^k \mid J^k; \theta \right) \right] \\ &\approx \frac{1}{S} \sum_{j=1}^{S} \sum_{k=1}^{N} \log p \left(v^{kj} \mid J^k; \theta \right). \end{aligned} \tag{1.9}$$

2. M-step

Update the parameters by maximizing $Q(\theta \mid \theta^{(i)})$. The maximization is closed form in I and σ^2, and a one-dimensional gradient ascent in α.

Image Matching Gradient

In our HMC sampling procedure, we will need to compute gradients, with respect to initial momenta, of the diffeomorphic image matching problem in (1.3), for matching the atlas I to an input image J^k.

Following the optimal control theory approach in [30], we add Lagrange multipliers to constrain the diffeomorphism $\phi^k(t)$ to be a geodesic path. This is done by introducing time-dependent adjoint variables, $\hat{m}$, $\hat{I}$ and $\hat{v}$, and writing the augmented energy,

$$\begin{aligned} \tilde{E}(m_0) =& E(Km_0, I, J^k) + \\ & \int_0^1 \langle \hat{m}, \dot{m} + \mathrm{ad}_v^* m \rangle dt + \int_0^1 \langle \hat{I}, \dot{I} + \nabla I \cdot v \rangle dt + \int_0^1 \langle \hat{v}, m - Lv \rangle dt, \end{aligned}$$

where E is the diffeomorphic image matching energy from (1.3), and the other terms correspond to Lagrange multipliers enforcing: (a) the geodesic constraint, which comes from the EPDiff equation (1.2), (b) the image transport equation, $\dot{I} = -\nabla I \cdot v$, and c) the constraint that $m = Lv$, respectively.

The optimality conditions for m, I, v are given by the following time-dependent system of ODEs, termed the *adjoint equations*:

$$-\dot{\hat{m}} + \mathrm{ad}_v \hat{m} + \hat{v} = 0, \qquad -\dot{\hat{I}} - \nabla \cdot (\hat{I} v) = 0, \qquad -\mathrm{ad}_{\hat{m}}^* m + \hat{I} \nabla I - L\hat{v} = 0,$$

subject to initial conditions

$$\hat{m}(1) = 0, \qquad \hat{I}(1) = \frac{1}{\sigma^2}(I(1) - J^k).$$

Finally, after integrating these adjoint equations backwards in time to $t = 0$, the gradient of $\tilde{E}$ with respect to the initial momenta is

$$\nabla_{m_0}\tilde{E} = Km_0 - \hat{m}_0. \tag{1.10}$$

1.4.1 Hamiltonian Monte Carlo (HMC) Sampling

Hamiltonian Monte Carlo [9] is a powerful MCMC sampling methodology that is applicable to a wide array of continuous probability distributions. It utilizes Hamiltonian dynamics as a Markov transition probability and efficiently explores the space of a target distribution. The integration through state space results in more efficient, global moves, while it also uses gradient information of the log probability density to sample from higher probability regions. In this section, we derive a HMC sampling method to draw a random sample from the posterior distribution of our latent variables, v^k, the initial velocities defining the diffeomorphic image transformations from the atlas to the data.

To sample from a pdf $f(x)$ using HMC, one first sets up a Hamiltonian $H(x, \mu) = U(x) + V(\mu)$, consisting of a "potential energy," $U(x) = -\log f(x)$, and a "kinetic energy", $V(\mu) = -\log g(\mu)$. Here $g(\mu)$ is some proposal distribution (typically isotropic Gaussian) on an auxiliary momentum variable, μ. An initial random momentum μ is drawn from the density $g(\mu)$. Starting from the current point x and initial random momentum μ, the Hamiltonian system is integrated forward in time to produce a candidate point, $\tilde{x}$, along with the corresponding forward-integrated momentum, $\tilde{\mu}$. The candidate point $\tilde{x}$ is accepted as a new point in the sample with probability

$$P(\text{accept}) = \min(1, \exp(-U(\tilde{x}) - V(\tilde{\mu}) + U(x) + V(\mu)).$$

This acceptance–rejection method is guaranteed to converge to the desired density $f(x)$ under fairly general regularity assumptions on f and g.

In our model, to sample v^k from the posterior in (1.8), we equivalently sample m^k from the dual momenta, using $v^k = Km^k$, so we define our potential energy as $U(m^k) = -\log p(m^k | J^k; \theta)$. We use the prior distribution on the dual momenta as our proposal density, in other words, we use $p(K\mu)$ defined as in (1.7), taking care to include the appropriate change-of-variables. As shown in [18], the form of the kinetic energy can be chosen to enforce certain conditions in the sampling. In our work, we define $V(\mu) = (\mu, K\mu)$, which helps to enforce that the velocity samples be smooth vector fields via application of the low-pass filter K. This gives us the following Hamiltonian system to integrate in the HMC:

$$\begin{aligned}\frac{dm^k}{dt} &= \frac{\partial H}{\partial \mu} &&= K\mu, \\ \frac{d\mu}{dt} &= -\frac{\partial H}{\partial m^k} &&= -\nabla_{m^k}\tilde{E},\end{aligned}$$

where the last term comes from the gradient defined in (1.10). As is standard practice in HMC, we use a "leap-frog" integration scheme, which better conserves the Hamiltonian and results in high acceptance rates.

1.4.2 The Maximization Step

We now derive the M-step for updating the parameters $\theta = (\alpha, \sigma^2, I)$ by maximizing the HMC approximation of the Q function, which is given in (1.9). This turns out to be a closed-form update for the noise variance σ^2 and the atlas I, and a simple one-dimensional gradient ascent for α.

From (1.9), it is easy to derive the closed-form update for σ^2 as

$$\sigma^2 = \frac{1}{MNS}\sum_{j=1}^{S}\sum_{k=1}^{N}\|I \circ (\phi^{kj})^{-1} - J^k\|^2. \tag{1.11}$$

For updating the atlas image I, we set the derivative of the Q function approximation which with respect to I to zero. The solution for I gives a closed-form update,

$$I = \frac{\sum_{j=1}^{S}\sum_{k=1}^{N} J^k \circ \phi^{kj}|D\phi^{kj}|}{\sum_{j=1}^{S}\sum_{k=1}^{N} |D\phi^{kj}|}.$$

The gradient ascent over α requires that we take the derivative of the metric $L = -\alpha\Delta + \beta I$, with respect to α. We do this in the Fourier domain, where the discrete Laplacian is a diagonal operator. For a 3D grid, the coefficients A_{xyz} of the discrete Laplacian at coordinate (x, y, z) in the Fourier domain is

$$A_{xyz} = -2\left(\cos\frac{2\pi x}{W-1} + \cos\frac{2\pi y}{H-1} + \cos\frac{2\pi z}{D-1}\right) + 6,$$

where W, H, D are the dimension of each direction. Hence, the determinant of the operator L is

$$|L| = \prod_{x,y,z} A_{xyz}\alpha + \beta.$$

The gradient of the HMC approximated Q function, with respect to α, is

$$\nabla_\alpha Q(\theta \,|\, \theta^{(i)}) \approx \frac{1}{2}\sum_{j=1}^{S}\sum_{k=1}^{N}\left[\sum_{x,y,z}\frac{A_{xyz}}{A_{xyz}\alpha+\beta} - \left\langle -\Delta v^{kj}, v^{kj}\right\rangle\right].$$

1.5 Bayesian Principal Geodesic Analysis

Before introducing our BPGA model for diffeomorphisms, we first review BPCA [7] for Euclidean data. The main idea of BPCA is to formulate a generative latent variable model for PCA that automatically selects the appropriate dimensionality of the model. Note that since our main goal is to quantify the data variability in this section, we fix the regularity parameter α estimated as described in Sect. 1.3. Consider a set y of n-dimensional Euclidean random variables $\{y_j\}_{j=1,\ldots,N} \in \mathbb{R}^n$, the relationship between each variable y_j and its corresponding q-dimensional ($q < n$) latent variable x_j is

$$y_j = \mu + Bx_j + \varepsilon, \tag{1.12}$$

where μ is the mean of dataset $\{y_j\}$, x_j is conventionally defined as a random variable generated from $N(0, I)$, B is an $n \times q$ factor matrix that relates x_j and y_j, and $\varepsilon \sim N(0, \sigma^2 I)$ represents error. This definition gives a data likelihood as

$$p(y \,|\, x; B, \mu, \sigma^2) \propto \prod_{j=1}^{N} \exp\left(-\frac{\left\|y_j - \mu - Bx_j\right\|^2}{2\sigma^2}\right).$$

To automatically select the principal components from data, BPCA includes a Gaussian prior over each column of B, which is known as an automatic relevance determination (ARD) prior. Each such Gaussian has an independent variance associated with a precision hyperparameter γ_i, so that

$$p(B \,|\, \gamma) = \prod_{i=1}^{q}\left(\frac{\gamma_i}{2\pi}\right)^{n/2} \exp\left(-\frac{1}{2}\gamma_i B_i^T B_i\right),$$

where B_i denotes the ith column of B.

The inference of BPCA is an EM algorithm that iteratively estimates the model parameters. At each iteration the value of γ_i is approximated by $\gamma_i = \frac{n}{\|B_i\|^2}$, using the current estimate of B_i. This induces sparsity by driving the corresponding component B_i to zero. More specifically, if γ_i is large, B_i will be effectively removed in the latent space. This arises naturally because the larger γ_i is, the lower the probability of B_i will be. Notice that the columns of B define the principal subspace of standard PCA, therefore, inducing sparsity on B has the same effect as removing irrelevant dimensions in the principal subspace.

1.5.1 Probability Model

We formulate the random initial velocity for the kth individual as $v^k = Wx^k$, where W is a matrix with q columns of principal initial velocities, and $x^k \in \mathbb{R}^q$ is a latent variable that lies in a low-dimensional space, with

$$p(x^k \mid W) \propto \exp\left(-\frac{1}{2} \|Wx^k\|_V^2\right). \tag{1.13}$$

Compared to BPCA, the difference of this latent variable prior is incorporating W as a conditional probability, which guarantees smoothness of the geodesic shooting path.

Our noise model is based on the assumption of i.i.d. Gaussian at each image voxel, much like [16, 19, 36]. This can be varied under different conditions, for instance, spatially dependent model for highly correlated noise data. In this chapter, we will focus on the commonly used and simple Gaussian noise model, with the likelihood given by

$$p(J^k \mid I, \sigma^2, x^k) = \frac{1}{(2\pi)^{M/2}\sigma^M} \exp\left(-\frac{\|I \circ (\phi^k)^{-1} - J^k\|_{L^2}^2}{2\sigma^2}\right). \tag{1.14}$$

The prior on W is a sparsity prior that suppresses the small principal initial velocity to zero. This prior is analogous to the hierarchical sparsity prior proposed by [10], with the difference that we use the natural Hilbert space norm for the velocity. The prior is based on Laplacian distribution, a widely used and exploited way to achieve sparse estimation. It presses the irrelevant or redundant components exactly to zero. As first introduced by [4], the Laplace distribution is equivalent to the marginal distribution of a hierarchical-Bayes model: a Gaussian prior with zero mean and exponentially distributed variances. Let i denote the ith principal component of W. We define each component W_i as a random variable with the hierarchical model distribution

$$p(W_i \mid \tau_i) \sim N(0, \tau_i),$$
$$p(\tau_i \mid \gamma_i) \sim \text{Exp}\left(\frac{\gamma_i}{2}\right),$$

After integrating out τ_i, we have the marginalized distribution as

$$p(W_i \mid \gamma_i) = \int_0^\infty p(W_i \mid \tau_i) p(\tau_i \mid \gamma_i) d\tau_i = \frac{\sqrt{\gamma_i}}{2} \exp\left(-\sqrt{\gamma_i}\, \|W_i\|_1\right),$$

which is a Laplacian distribution with scale parameter $\gamma_i/2$. The degree of sparsity is controlled by the hyperparameter γ_i on the l_1 penalty. However, the sparsity parameter is specified in an ad hoc manner. As in [10], an effective model is proposed to remove

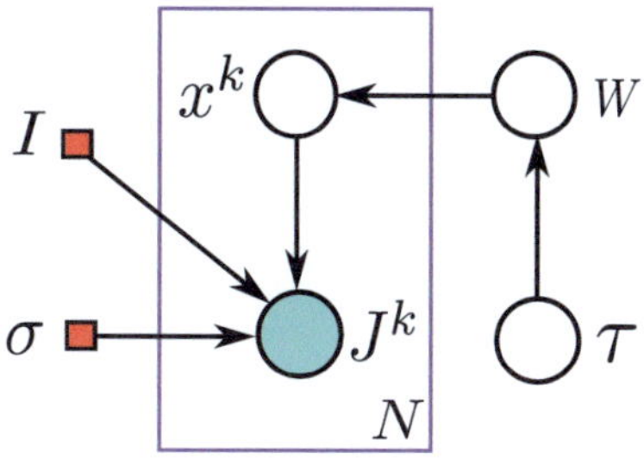

Fig. 1.3 Graphical representation of BPGA for the kth subject J^k

γ_i by adopting a Jeffreys' noninformative hyperprior as $p(\tau_i) \sim 1/\tau_i$. This has the advantages that (1) the improper hyperprior is scale-invariant, and (2) the model is parameter-free. Using this hierarchical sparsity prior on the columns of W for the automatic mode selection, we formulate the problem as

$$p(W, x \mid \tau) \propto \exp\left(-\frac{1}{2} \sum_{k=1}^{N} \| W x^k \|_V^2 - \sum_{i=1}^{q} \frac{\| W_i \|_V^2}{2\tau_i} \right), \tag{1.15}$$

$$p(\tau) \propto \frac{1}{\tau},$$

where $x = [x^1, \ldots, x^k]$, $\tau = [\tau_1, \ldots, \tau_q]$. We will later integrate out the latent variable τ using expectation maximization.

We can express our model for the kth subject using the graphical representation shown in Fig. 1.3.

1.5.2 Inference

Due to the high computational cost of treating $\theta = \{I, \sigma^2\}$ as random variables and sampling them, we use MAP estimation to determine θ as model parameters. After defining the likelihood (1.14) and prior (1.15) in the previous section, we now arrive at the joint posterior for BPGA as

$$\prod_{k=1}^{N} p\left(W, x, \tau \mid J^k; \theta\right) \propto \left[\prod_{k=1}^{N} p(J^k \mid x^k, \theta)\, p(x^k \mid W) \right] p(W \mid \tau)\, p(\tau). \tag{1.16}$$

In order to treat the W, x^k and τ as latent random variables with the log posterior given by (1.16), we would ideally integrate out the latent variables, which are intractable in closed form for W, x^k. Instead, we develop an expectation maximization algorithm to compute a closed-form solution to integrate out τ first, and then use a mode approximation for W, x^k to the posterior distribution. It contains two alternating steps:

E-step

Using the current estimate of the parameters $\hat{\theta}$, we compute the expectation Q of the complete log posterior of (1.16) with respect to the latent variables τ as

$$\begin{aligned} Q(W, x^k, \theta \mid \hat{\theta}, \hat{W}) \propto & -\frac{1}{2\sigma^2} \sum_{k=1}^{N} \left\| I \circ (\phi^k)^{-1} - J^k \right\|_{L^2}^2 - \frac{MN}{2} \log \sigma \\ & - \frac{1}{2} \sum_{k=1}^{N} \left\| W x^k \right\|_V^2 - \sum_{i=1}^{q} \frac{\| W_i \|_V^2}{2 \| \hat{W}_i \|_V^2}. \end{aligned} \tag{1.17}$$

Note that we use the same approach to integrate out τ in [10]. Details are in Appendix A.

M-step: Gradient Ascent for W, x^k

We introduce a gradient ascent scheme to estimate W, x^k, and $\theta = (I, \sigma^2)$ simultaneously. We compute the gradient with respect to the initial momentum m^k of the diffeomorphic image matching problem in (1.5), and then apply the chain rule to obtain the gradient term w.r.t. W and x^k. Following the same derivation in (1.10), we obtain the gradient of Q with respect to W after applying the chain rule as

$$\nabla_W Q = -\sum_{k=1}^{N} K(m^k - K\hat{m}^k)(x^k)^T - W\Lambda,$$

where Λ is a diagonal matrix with diagonal element $\frac{1}{\left\| \hat{W}_i \right\|_V^2}$. The gradient with respect to x^k is

$$\nabla_{x^k} Q = -W^T K(m^k - K\hat{m}^k).$$

Closed-Form Solution for θ

We now derive the maximization for updating the parameters θ. This turns out to be a closed-form update for the atlas I, noise variance σ^2. For updating I and σ^2, we set the derivative of the expectation with respect to I, σ^2 to zero. The solution for I, σ^2 gives an update

$$I = \frac{\sum_{k=1}^{N} J^k \circ \phi^k |D\phi^k|}{\sum_{k=1}^{N} |D\phi^k|}, \quad \sigma^2 = \frac{1}{MN} \sum_{k=1}^{N} \left\| I \circ (\phi^k)^{-1} - J^k \right\|_{L^2}^2.$$

1.6 Results

In this section, we first demonstrate the effectiveness of our proposed model and MCEM estimation routine using both 2D synthetic data and real 3D MRI brain data. Because we have a generative model, we can forward simulate a random sample of images from a distribution with known parameters $\theta = (\alpha, \sigma^2, I)$. Then, in the next subsection, we test if we can recover those parameters using our MCEM algorithm. Figure 1.4 illustrates this process. We simulated a 2D synthetic dataset starting from an atlas image, I, of a binary circle with resolution 100×100. We then generated 20 smooth initial velocity fields from the prior distribution, $p(v^k)$, defined in (1.15), setting $\alpha = 0.025$ and $\beta = 0.001$. Deformed circle images were constructed by shooting the initial velocities by the EPDiff equations and transforming the atlas by the resulting diffeomorphisms, ϕ^k. Finally, we added i.i.d. Gaussian noise according to our likelihood model (1.14). We used a standard deviation of $\sigma = 0.05$, which corresponds to an SNR of 20 (which is more noise than typical structural MRI).

Parameter Estimation on Synthetic Data

In our estimation procedure, we initialized α with 0.002 for noise free, and 0.01 for noise corrupted images. The step size of 0.005 for leap-frog integration is used in HMC with 10 units of time discretization in integration of EPDiff equations.

Figure 1.5 compares the true atlas and estimated atlases in the clean and noisy case. Figure 1.6 shows the convergence graph for α and σ estimation by using 100 samples. It shows that our method recovers the model parameters fairly well. However, the iterative mode approximation algorithm does not recover the α parameter as nicely as our method. In the noisy case, the mode approximation algorithm estimates α as

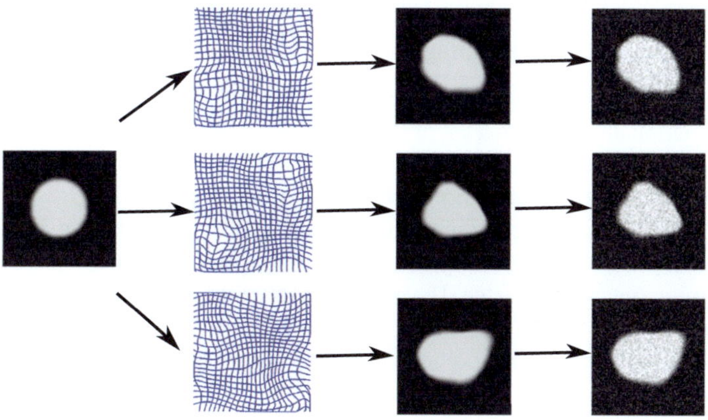

Fig. 1.4 Simulating synthetic 2D data from the generative diffeomorphism model. From *left* to *right* the ground truth template image, random diffeomorphisms from the prior model, deformed images, and final noise corrupted images

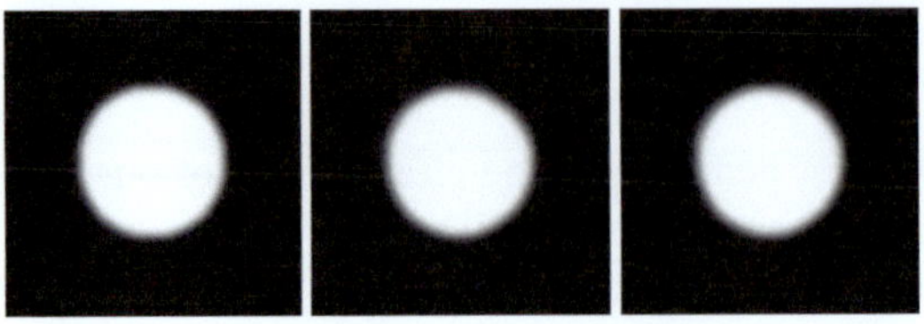

Fig. 1.5 Atlas estimation results. *Left* ground truth template. *Center* estimated template from noise-free dataset. *Right* estimated template from noise-corrupted dataset

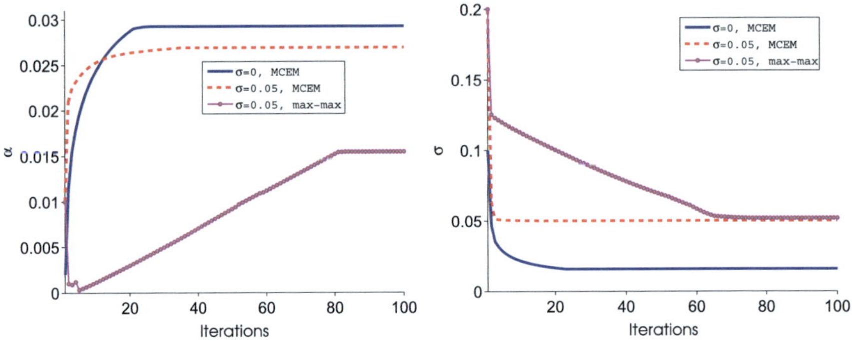

Fig. 1.6 Estimation of α, σ. *Left* α estimation. *Right* σ estimation. In our MCEM method, final estimated α and σ for noise free data are 0.028, 0.01, and for noise data are 0.026, 0.0501. Compared with max-max method, for the noise data, estimated α and σ are 0.0152, 0.052

0.0152, which is far from the ground truth value of 0.025. This is compared with our estimation of 0.026. In addition, in the noise-free example, the mode approximation algorithm blows up due to the σ dropping close to 0, thus making the image match term numerically too high and the geodesic shooting unstable.

Atlas Building on 3D Brain Images

To demonstrate the effectiveness of our method on the real data, we apply our MCEM atlas estimation algorithm to a set of brain MRI from ten healthy subjects. The MRI have resolution $108 \times 128 \times 108$ and are skull stripped, intensity normalized, and co-registered with rigid transforms. We set the initial $\alpha = 0.01$, $\beta = 0.001$ with 15 time steps. Due to the massive computational cost of sampling in the high-dimensional image space, we implement a message passing interface (MPI) parallel programming on a GPU cluster. The entire inference costs approximately a week to complete.

The left side (the first five columns) of Fig. 1.7 shows coronal and axial view of the 3D MRI used as input. The right side shows the initialization (greyscale average of the input images), followed by the final atlas estimated by our method. The final atlas estimate correctly aligns the anatomy of the input images, producing a sharper average image. The algorithm also jointly estimated the smoothness parameter to be $\alpha = 0.028$ and the image noise standard deviation to be $\sigma = 0.031$.

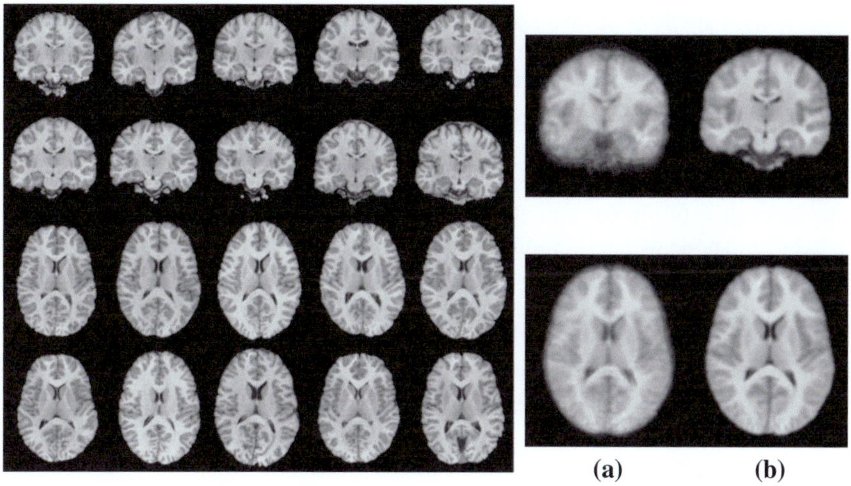

Fig. 1.7 *Left* (*the first five columns*) coronal and axial view of the input 3D MRIs. *Right* (*a*) initial greyscale average of the input images. *Right* (*b*) final atlas estimated by our MCEM estimation procedure

Image Matching Accuracy

Finally, we demonstrate that another benefit of our HMC sampling methodology is improved performance in the standard image registration problem under large deformation shooting. Rather than using a direct gradient descent to solve the image registration problem, we instead can find the posterior mean of the model (1.16), where for image matching we fix the "atlas," I, as the source image and have just one target image, I_1. The stochastic behavior in the sampling helps to get out of local minima, where the direct gradient descent can get stuck. We compared our proposed method with direct gradient descent image registration by geodesic shooting from [30]. We used the authors' uTIlzReg package for geodesic shooting, which is available freely online. For the comparison, we registered the image pair shown in the first two panels of Fig. 1.8, which requires a large deformation. The source and

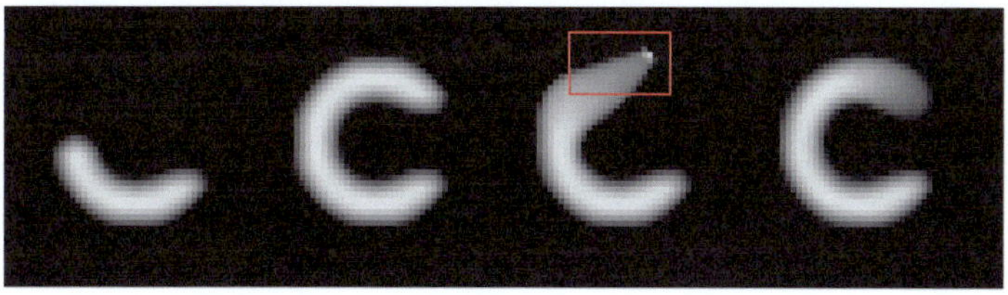

Fig. 1.8 The first two images from left to right are the source and target image, respectively. Third is the matched image obtained by geodesic shooting method using [30]. Last image is the matched image from our MCEM method

target images are 50×50. We used $\alpha = 0.02$, $\beta = 0.001$ for smoothing kernel, and $h = 40$ time steps between $t = 0$ and $t = 1$. Note that we only want to compare the image matching here, so we fix the α and σ parameters.

Figure 1.8 demonstrates the results of the direct geodesic shooting registration with our HMC posterior mean. It shows that the geodesic shooting method gets stuck in a local minima and cannot make it to the target image even with a large number of time steps ($h = 60$) in the time discretization (we tried several time discretizations up to 60, and none worked). Though our method did not match perfectly in the tip of the "C," it still recovers the full shape while retaining a diffeomorphic transformation.

Principal Geodesics Estimation on OASIS Brain Data

We then demonstrate the effectiveness of BPGA model by applying it to a set of brain magnetic resonance images (MRI) from the 3D OASIS brain database. The data consists of MRI from 130 subjects between the age of 60 and 95. The MRI have a resolution of $128 \times 128 \times 128$ with an image spacing of $1.0 \times 1.0 \times 1.0\,\text{mm}^3$ and are skull stripped, intensity normalized, and co-registered with rigid transforms. To set the parameters in L operator, we did initial step of estimating $\alpha = 0.01$ using the Bayesian atlas building procedure introduced in Sect. 1.3. We used 15 time steps in geodesic shooting and initialize the template I as the average of image intensities, with W as the matrix of principal components from tangent PCA (TPCA) [19].

The proposed BPGA model automatically determined that the latent dimensionality of the data was 15. Figure 1.9 displays the automatically estimated modes, $i = 1, 2$, of the brain MRI variation. We forward shoot the constructed atlas, I, by the estimated principal momentum $a_i W_i$ along the geodesics. For the purpose of visualization, we demonstrate the brain variation from the atlas by $a_i = -3, -1.5, 0, 1.5, 3$. We also show the log determinant of Jacobians at $a_i = 3$, with red representing regions of expansion and blue representing regions of contraction. The first mode of variation clearly shows that ventricle size change is a dominant source of variability in brain shape. Our algorithm also jointly estimated the image noise standard deviation parameter as $\sigma = 0.04$.

Image Reconstruction Accuracy

We validated the ability of our BPGA model to compactly represent the space of brain variation by testing how well it can reconstruct unseen images. After estimating the principal initial velocity and parameters from the training subjects above, we used these estimates to reconstruct another 20 testing subjects from the same OASIS database that were not included in the training. We then measured the discrepancy between the reconstructed images and the testing images. Note that our reconstruction only used the first 15 principal modes, which were automatically selected by our algorithm.

We use the first fifteen dimensions to compare our model with linear PCA (LPCA) on image intensities and TPCA. Examples of the reconstructed images and their error maps from these models are shown in Figs. 1.10 and 1.11. Table 1.1 shows the comparison of the reconstruction accuracy as measured by the average and standard

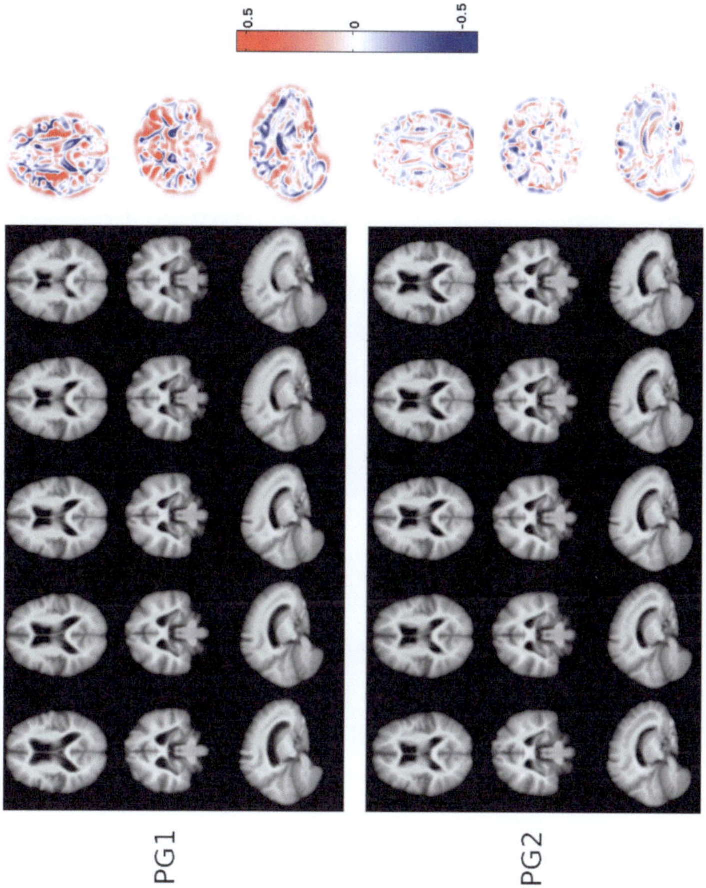

Fig. 1.9 *Top* to *bottom* axial, coronal, and sagittal views of shooting the atlas by the first and second principal modes. *Left* to *right* BPGA model of image variation evaluated at $a_i = -3, -1.5, 0, 1.5, 3$, and log determinant of Jacobians at $a_i = 3$

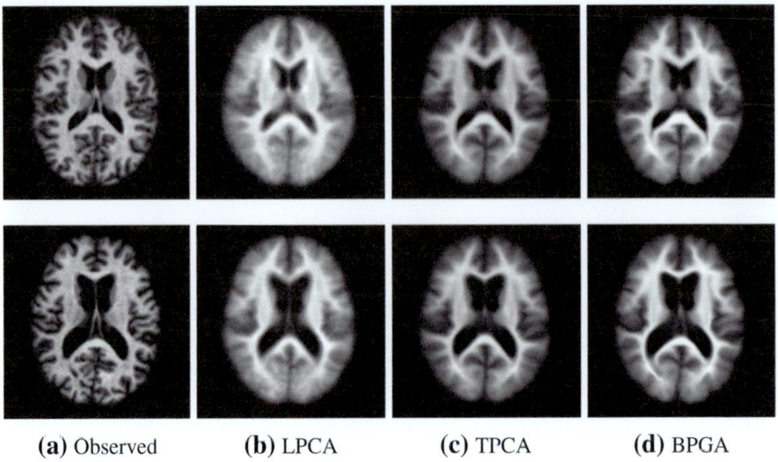

Fig. 1.10 *Left* to *right* original data, reconstruction by LPCA, TPCA, and BPGA

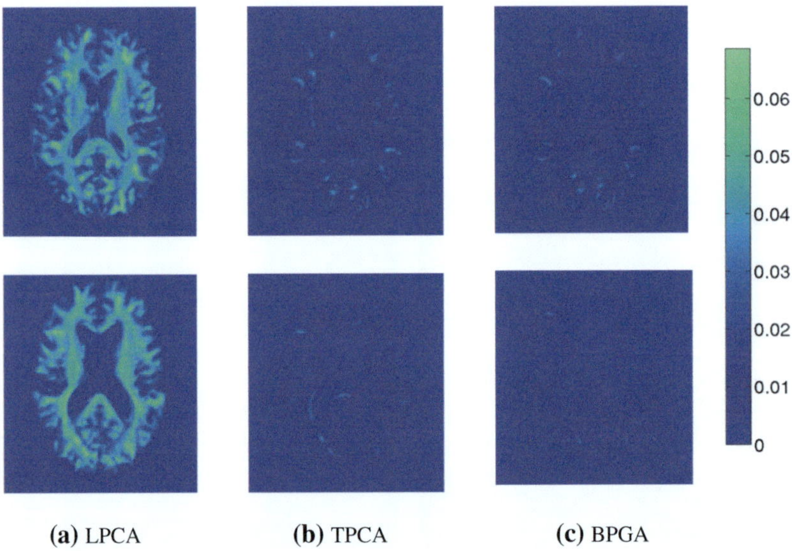

Fig. 1.11 *Left* to *right* absolute value of reconstruction error map by LPCA, TPCA, and BPGA

Table 1.1 Comparison of mean squared reconstruction error between LPCA, TPCA, and BPGA models. Average and standard deviation over 20 test images

	LPCA	TPCA	BPGA
Average MSE	4.2×10^{-2}	3.4×10^{-2}	2.8×10^{-2}
Std of MSE	1.25×10^{-2}	4.8×10^{-3}	4.2×10^{-3}

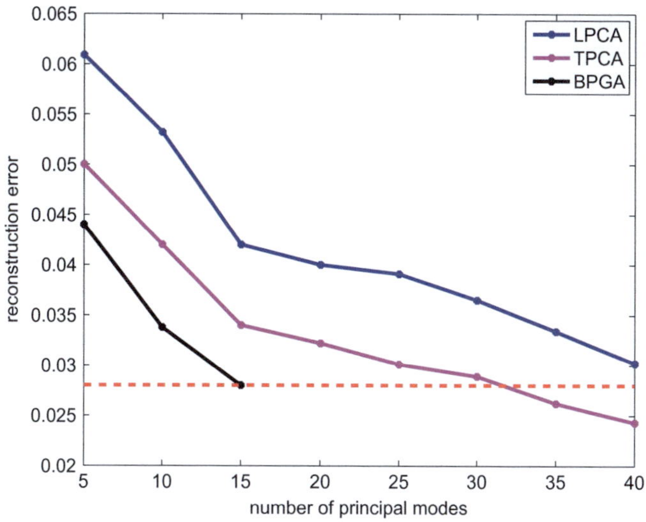

Fig. 1.12 Averaged mean squared reconstruction error with different number of principal modes by LPCA, TPCA, and BPGA over 20 test images

deviation of the mean squared error (MSE). The table indicates that our model outperforms both LPCA and TPCA in the diffeomorphic setting. We also display the reconstruction error with increasing number of principal modes. Figure 1.12 shows that TPCA requires approximately 32 principal modes, more than twice as much as our model does, to achieve the same level of reconstruction accuracy. LPCA cannot match the BPGA reconstruction accuracy with even 40 principal modes. This reflects that our model BPGA gains a more compact representation than TPCA and LPCA.

Acknowledgments This work was supported by NIH Grant R01 MH084795, NIH Grant 5R01EB007688, and NSF CAREER Grant 1054057.

References

1. S. Allassonnière, Y. Amit, A. Trouvé, Toward a coherent statistical framework for dense deformable template estimation. J. R. Stat. Soc. Ser. B **69**, 3–29 (2007)
2. S. Allassonniere, S. Durrleman, E. Kuhn, Bayesian mixed effect atlas estimation with a diffeomorphic deformation model. SIAM J. Imaging Sci. **8**(3), 1367–1395 (2015)
3. S. Allassonnière, E. Kuhn, Stochastic algorithm for parameter estimation for dense deformable template mixture model. ESAIM-PS **14**, 382–408 (2010)
4. D.F. Andrews, C.L. Mallows, Scale mixtures of normal distributions. J. R. Stat. Soc. Ser. B (Methodological) **36**, 99–102 (1974)
5. V.I. Arnol'd, Sur la géométrie différentielle des groupes de Lie de dimension infinie et ses applications à l'hydrodynamique des fluides parfaits. Ann. Inst. Fourier **16**, 319–361 (1966)
6. M.F. Beg, M.I. Miller, A. Trouvé, L. Younes, Computing large deformation metric mappings via geodesic flows of diffeomorphisms. Int. J. Comput. Vis. **61**(2), 139–157 (2005)

7. C.M. Bishop, Bayesian PCA, in *Advances in neural information processing systems* (MIT press, Cambridge, 1999), pp. 382–388
8. A. Budhiraja, P. Dupuis, V. Maroulas, Large deviations for stochastic flows of diffeomorphisms. Bernoulli **16**, 234–257 (2010)
9. S. Duane, A. Kennedy, B. Pendleton, D. Roweth, Hybrid Monte Carlo. Phys. Lett. B **195**, 216–222 (1987)
10. M.A.T. Figueiredo, Adaptive sparseness for supervised learning. IEEE Trans. Pattern Anal. Mach. Intell. **25**(9), 1150–1159 (2003)
11. P.T. Fletcher, C. Lu, S. Joshi, Statistics of shape via principal geodesic analysis on Lie groups, in *Computer Vision and Pattern Recognition* (IEEE Computer Society, Washington, DC, 2003), pp. 95–101
12. P.T. Fletcher, C. Lu, S.M. Pizer, S. Joshi, Principal geodesic analysis for the study of nonlinear statistics of shape. IEEE Trans. Med. Imaging **23**(8), 995–1005 (2004)
13. P. Gori, O. Colliot, Y. Worbe, L. Marrakchi-Kacem, S. Lecomte, C. Poupon, A. Hartmann, N. Ayache, S. Durrleman, Bayesian atlas estimation for the variability analysis of shape complexes, in *Medical Image Computing and Computer-Assisted Intervention*, vol. 8149 (Springer, Heidelberg, 2013). pp. 267–274
14. J.E. Iglesias, M.R. Sabuncu, K. Van Leemput, ADNI, Incorporating parameter uncertainty in Bayesian segmentation models: application to hippocampal subfield volumetry, in *MICCAI*, (Springer, Heidelberg, 2012)
15. S. Joshi, B. Davis, M. Jomier, G. Gerig, Unbiased diffeomorphic atlas construction for computational anatomy. NeuroImage **23**(Supplement 1), 151–160 (2004)
16. J. Ma, M.I. Miller, A. Trouvé, L. Younes, Bayesian template estimation in computational anatomy. NeuroImage **42**, 252–261 (2008)
17. M.I. Miller, A. Trouvé, L. Younes, Geodesic shooting for computational anatomy. J. Math. Imaging Vis. **24**(2), 209–228 (2006)
18. R.M. Neal, MCMC using Hamiltonian dynamics. Handb. Markov Chain Monte Carlo **2**, 113–162 (2011)
19. A. Qiu, L. Younes, M.I. Miller, Principal component based diffeomorphic surface mapping. IEEE Trans. Med. Imaging **31**(2), 302–311 (2012)
20. P. Risholm, S. Pieper, E. Samset, W.M. Wells, Summarizing and visualizing uncertainty in non-rigid registration, in *MICCAI* (Springer, Heidelberg, 2010)
21. P. Risholm, E. Samset, W.M. Wells, Bayesian estimation of deformation and elastic parameters in non-rigid registration, in *WBIR* (Springer, Heidelberg, 2010)
22. S. Said, N. Courty, N. Le Bihan, S.J. Sangwine, Exact principal geodesic analysis for data on SO(3), in *Proceedings of the 15th European Signal Processing Conference* (2007). pp. 1700–1705
23. I.J.A. Simpson, M.J. Cardoso, M. Modat, D.M. Cash, M.W. Woolrich, J.L.R. Andersson, J.A. Schnabel, S. Ourselin, Alzheimers Disease Neuroimaging Initiative et al., Probabilistic nonlinear registration with spatially adaptive regularisation. Med. Image Anal. **26**(1), 203–216 (2015)
24. I.J.A. Simpson, A.S. Julia, R.G. Adrian, L.R.A. Jesper, W.W. Mark, Probabilistic inference of regularisation in non-rigid registration. NeuroImage **59**, 2438–2451 (2012)
25. N. Singh, J. Hinkle, S. Joshi, P. Thomas Fletcher, A vector momenta formulation of diffeomorphisms for improved geodesic regression and atlas construction, in *International Symposium on Biomedial Imaging (ISBI)*, April 2013
26. S. Sommer, F. Lauze, S. Hauberg, M. Nielsen, Manifold valued statistics, exact principal geodesic analysis and the effect of linear approximations, in *Proceedings of the European Conference on Computer Vision* (2010). pp. 43–56
27. M. Vaillant, M.I. Miller, L. Younes, A. Trouvé, Statistics on diffeomorphisms via tangent space representations. NeuroImage **23**, S161–S169 (2004)
28. K. Van Leemput, Encoding probabilistic brain atlases using Bayesian inference. IEEE Trans. Med. Imaging **28**, 822–837 (2009)

29. F.-X. Vialard, L. Risser, D. Holm, D. Rueckert, Diffeomorphic atlas estimation using Kärcher mean and geodesic shooting on volumetric images, in *MIUA* (2011)
30. F.-X. Vialard, L. Risser, D. Rueckert, C.J. Cotter, Diffeomorphic 3d image registration via geodesic shooting using an efficient adjoint calculation. Int. J. Comput. Vis. **97**, 229–241 (2012)
31. D. Wassermann, M. Toews, M. Niethammer, W. Wells III, Probabilistic diffeomorphic registration: representing uncertainty, in *Biomedical Image Registration* (Springer, Switzerland, 2014). pp. 72–82
32. L. Younes, F. Arrate, M.I. Miller, Evolutions equations in computational anatomy. NeuroImage **45**(1S1), 40–50 (2009)
33. M. Zhang, P.T. Fletcher, Probabilistic principal geodesic analysis, in *Advances in Neural Information Processing Systems* (2013). pp. 1178–1186
34. M. Zhang, P.T. Fletcher, Bayesian principal geodesic analysis in diffeomorphic image registration, in *Medical Image Computing and Computer-Assisted Intervention–MICCAI 2014* (Springer, Heidelberg, 2014). pp. 121–128
35. M. Zhang, P.T. Fletcher, Bayesian principal geodesic analysis for estimating intrinsic diffeomorphic image variability. Med. Image Anal. **25**, 37–44 (2015)
36. M. Zhang, N. Singh, P.T. Fletcher, Bayesian estimation of regularization and atlas building in diffeomorphic image registration, in *Information Processing in Medical Imaging* (Springer, Heidelberg, 2013). pp. 37–48

Chapter 2
Sampling Constrained Probability Distributions Using Spherical Augmentation

Shiwei Lan and Babak Shahbaba

Abstract Statistical models with constrained probability distributions are abundant in machine learning. Some examples include regression models with norm constraints (e.g., Lasso), probit, many copula models, and latent Dirichlet allocation (LDA). Bayesian inference involving probability distributions confined to constrained domains could be quite challenging for commonly used sampling algorithms. In this work, we propose a novel augmentation technique that handles a wide range of constraints by mapping the constrained domain to a sphere in the augmented space. By moving freely on the surface of this sphere, sampling algorithms handle constraints implicitly and generate proposals that remain within boundaries when mapped back to the original space. Our proposed method, called Spherical Augmentation, provides a mathematically natural and computationally efficient framework for sampling from constrained probability distributions. We show the advantages of our method over state-of-the-art sampling algorithms, such as exact Hamiltonian Monte Carlo, using several examples including truncated Gaussian distributions, Bayesian Lasso, Bayesian bridge regression, reconstruction of quantized stationary Gaussian process, and LDA for topic modeling.

2.1 Introduction

Many commonly used statistical models in Bayesian analysis involve high dimensional probability distributions confined to constrained domains. Some examples include regression models with norm constraints (e.g., Lasso), probit, many copula models, and latent Dirichlet allocation (LDA). Very often, the resulting models are

S. Lan (✉)
Department of Statistics, University of Warwick, Coventry CV4 7AL, UK
e-mail: s.lan@warwick.ac.uk

B. Shahbaba
Department of Statistics and Department of Computer Science,
University of California, Irvine, CA 92697, USA
e-mail: babaks@uci.edu

H.Q. Minh and V. Murino (eds.), *Algorithmic Advances in Riemannian Geometry and Applications*, Advances in Computer Vision and Pattern Recognition, DOI 10.1007/978-3-319-45026-1_2

intractable and the imposed constraints add another layer of complexity. Therefore, in these problems simulating samples for Monte Carlo estimation is quite challenging [12, 40, 41, 47, 57]. Although the literature on improving the efficiency of computational methods for Bayesian inference is quite extensive see, for example, [1–3, 5, 7, 9, 11, 14, 15, 18, 20, 22, 25–28, 32, 33, 35, 37, 39, 42–46, 51–53, 56, 63–65, 67], these methods do not directly address the complications due to constrained target distributions. When dealing with such distributions, MCMC algorithms typically evaluate each proposal to ensure it is within the boundaries imposed by the constraints. Computationally, this is quite inefficient, especially in high dimensional problems where proposals are very likely to miss the constrained domain. Alternatively, one could map the original domain to the entire Euclidean space to remove the boundaries [49]. This approach too is computationally inefficient since the sampler needs to explore a much larger space than needed.

In this chapter, we study a novel technique, *Spherical Augmentation (SA)*, for a family of MCMC algorithms to handle a specific class of constraints. SA was recently proposed by [34] for Hamiltonian Monte Carlo (HMC) [23, 46] to handle constraints involving norm inequalities (Fig. 2.1). SA method augments the parameter space and maps the constrained domain to a sphere in the augmented space. The sampling algorithm explores the surface of this sphere. This way, it handles constraints implicitly and generates proposals that remain within boundaries when mapped back to the original space. Here, we generalize the work of [34] to handle a broader class of constraints, which are still convertible to norm constraints. We will also discuss an improved version of Spherical HMC [34] designed specifically for box-type constraints. Additionally, we will show how SA can be applied to Lagrangian Monte Carlo [35] with application to LDA problems and sampling from probability simplex.

There have been some related works recently. Reference [46] discusses modifying standard HMC such that the sampler bounces off the boundaries by letting the poten-

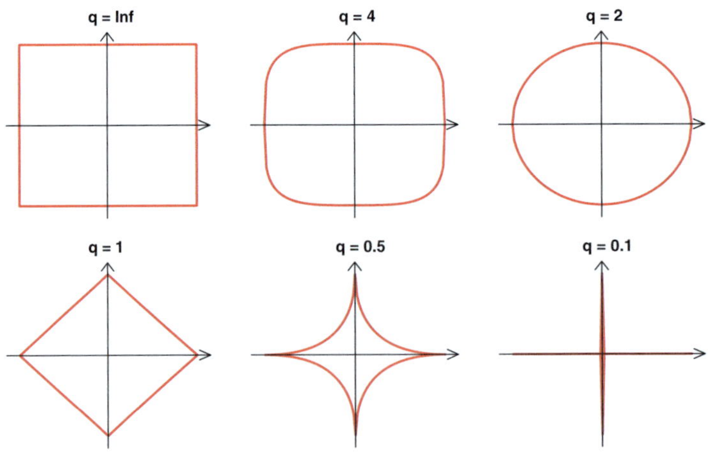

Fig. 2.1 q-Norm constraints

tial energy go to infinity for parameter values that violate the constraints. This creates "energy walls" at boundaries. This approach, henceforth called *Wall HMC*, has limited applications and tends to be computationally inefficient, because the frequency of hitting and bouncing increases exponentially as dimension grows. Reference [47] follow the idea of Wall HMC and propose an exact HMC algorithm specifically for truncated Gaussian distributions with constraints. Reference [12] on the other hand propose a modified version of HMC for handling holonomic constraint $c(\theta) = 0$ by using Lagrange multipliers. Reference [13] discuss an alternative approach for situations where constrained domains can be identified as sub-manifolds. In particular, Spherical HMC [34] is motivated by [13] but removes the requirement of embedding, thus it is applicable to more general settings. All these methods provide interesting solutions for specific types of constraints. In contrast, our proposed method offers a general and efficient framework applicable to a wide range of problems.

The chapter is structured as follows. Before presenting our methods, in Sect. 2.2 we provide a brief overview of HMC and one of its variants, namely, Lagrangian Monte Carlo (LMC) [35]. We then present the underlying idea of *Spherical Augmentation*, first for two simple cases, ball type (Sect. 2.3.1) and box type (Sect. 2.3.2) constraints, then for more general q-norm type constraints (Sect. 2.3.3), as well as some functional constraints (Sect. 2.3.4). In Sect. 2.4, we apply the SA technique to HMC (Sect. 2.4.2) and LMC (Sect. 2.4.3) for sampling from constrained target distributions. We evaluate our proposed methods using simulated and real data in Sect. 2.5. Finally, Sect. 2.6 is devoted to discussion and future directions.

2.2 Preliminaries

2.2.1 Hamiltonian Monte Carlo

HMC improves upon random walk Metropolis (RWM) by proposing states that are distant from the current state, but nevertheless accepted with high probability. These distant proposals are found by numerically simulating Hamiltonian dynamics, whose state space consists of its *position*, denoted by the vector $\boldsymbol{\theta}$, and its *momentum*, denoted by the vector $\mathbf{p}$. Our objective is to sample from the continuous probability distribution of $\boldsymbol{\theta}$ with the density function $f(\boldsymbol{\theta})$. It is common to assume that the fictitious momentum variable $\mathbf{p} \sim \mathcal{N}(\mathbf{0}, \mathbf{M})$, where $\mathbf{M}$ is a symmetric, positive-definite matrix known as the *mass matrix*, often set to the identity matrix $\mathbf{I}$ for convenience.

In this Hamiltonian dynamics, the *potential energy*, $U(\boldsymbol{\theta})$, is defined as minus the log density of $\boldsymbol{\theta}$ (plus any constant), that is $U(\boldsymbol{\theta}) := -\log f(\boldsymbol{\theta})$; the *kinetic energy*, $K(\mathbf{p})$ for the auxiliary momentum variable $\mathbf{p}$ is set to be minus the log density of $\mathbf{p}$ (plus any constant). Then the total energy of the system, *Hamiltonian* function, is defined as their sum,

$$H(\boldsymbol{\theta}, \mathbf{p}) = U(\boldsymbol{\theta}) + K(\mathbf{p}) \tag{2.1}$$

Given the Hamiltonian $H(\boldsymbol{\theta}, \mathbf{p})$, the system of $(\boldsymbol{\theta}, \mathbf{p})$ evolves according to the following *Hamilton's equations*

$$\begin{aligned} \dot{\boldsymbol{\theta}} &= \nabla_{\mathbf{p}} H(\boldsymbol{\theta}, \mathbf{p}) &&= \mathbf{M}^{-1}\mathbf{p} \\ \dot{\mathbf{p}} &= -\nabla_{\boldsymbol{\theta}} H(\boldsymbol{\theta}, \mathbf{p}) &&= -\nabla_{\boldsymbol{\theta}} U(\boldsymbol{\theta}) \end{aligned} \tag{2.2}$$

In practice when the analytical solution to Hamilton's equations is not available, we need to numerically solve these equations by discretizing them, using some small time step ε. For the sake of accuracy and stability, a numerical method called *leapfrog* is commonly used to approximate the Hamilton's equations [46]. We usually solve the system for L steps, with some step size, ε, to propose a new state in the Metropolis algorithm, and accept or reject it according to the Metropolis acceptance probability. [See [46], for more discussions].

2.2.2 *Lagrangian Monte Carlo*

Although HMC explores the target distribution more efficiently than RWM, it does not fully exploit the geometric properties of the parameter space. Reference [28] propose Riemannian HMC (RHMC), which adapts to the local Riemannian geometry of the target distribution by using a position-specific mass matrix $\mathbf{M} = \mathbf{G}(\boldsymbol{\theta})$. More specifically, they set $\mathbf{G}(\boldsymbol{\theta})$ to the Fisher information matrix. In this chapter, we mainly use the *spherical metric* instead, to serve the purpose of constraint handling. The proposed method can be viewed as an extension to this approach since it explores the geometry of sphere.

Following the argument of [4, 28] define Hamiltonian dynamics on the Riemannian manifold endowed with metric $\mathbf{G}(\boldsymbol{\theta})$. With the non-flat metic, the momentum vector becomes $\mathbf{p}|\boldsymbol{\theta} \sim \mathscr{N}(\mathbf{0}, \mathbf{G}(\boldsymbol{\theta}))$ and the Hamiltonian is therefore defined as follows:

$$H(\boldsymbol{\theta}, \mathbf{p}) = \phi(\boldsymbol{\theta}) + \frac{1}{2}\mathbf{p}^{\mathsf{T}}\mathbf{G}(\boldsymbol{\theta})^{-1}\mathbf{p}, \quad \phi(\boldsymbol{\theta}) := U(\boldsymbol{\theta}) + \frac{1}{2}\log\det\mathbf{G}(\boldsymbol{\theta}) \tag{2.3}$$

Unfortunately the resulting Riemannian manifold Hamiltonian dynamics becomes nonseparable since it contains products of $\boldsymbol{\theta}$ and $\mathbf{p}$, and the numerical integrator, *generalized leapfrog*, is an *implicit* scheme that involves time-consuming fixed-point iterations.

Reference [35] proposes to change the variables $\mathbf{p} \mapsto \mathbf{v} := \mathbf{G}(\boldsymbol{\theta})^{-1}\mathbf{p}$ and define an *explicit* integrator for RHMC by using the following equivalent *Lagrangian* dynamics:

$$\begin{aligned} \dot{\boldsymbol{\theta}} &= \mathbf{v} \\ \dot{\mathbf{v}} &= -\mathbf{v}^{\mathsf{T}}\boldsymbol{\Gamma}(\boldsymbol{\theta})\mathbf{v} - \mathbf{G}(\boldsymbol{\theta})^{-1}\nabla_{\boldsymbol{\theta}}\phi(\boldsymbol{\theta}) \end{aligned} \tag{2.4}$$

where the *velocity* $\mathbf{v}|\boldsymbol{\theta} \sim \mathcal{N}(\mathbf{0}, \mathbf{G}(\boldsymbol{\theta})^{-1})$. Here, $\boldsymbol{\Gamma}(\boldsymbol{\theta})$ is the Christoffel symbols derived from $\mathbf{G}(\boldsymbol{\theta})$.

The proposed *explicit* integrator is time reversible but not volume preserving. Based on the change of variables theorem, one can adjust the acceptance probability with Jacobian determinant to satisfy the detailed balance condition. The resulting algorithm, *Lagrangian Monte Carlo (LMC)*, is shown to be computationally more efficient than RHMC [See [35] for more details].

Throughout this chapter, we express the kinetic energy K in terms of velocity, $\mathbf{v}$, instead of momentum, $\mathbf{p}$ [9, 35].

2.3 Spherical Augmentation

In this section, we introduce the *Spherical Augmentation* technique for handling norm constraints implicitly. We start with two simple constraints: ball type (2-norm) and box type (∞-norm). Then, we generalize the methodology to arbitrary q-norm type constraints for $q > 0$. Finally, we discuss some functional constraints that can be reduced to norm constraints.

2.3.1 *Ball Type Constraints*

Consider probability distributions confined to the D-dimensional unit ball $\mathscr{B}_{\mathbf{0}}^D(1) := \{\boldsymbol{\theta} \in \mathbb{R}^D : \|\boldsymbol{\theta}\|_2 = \sqrt{\sum_{i=1}^D \theta_i^2} \leq 1\}$. The constraint is given by restricting the 2-norm of parameters: $\|\boldsymbol{\theta}\|_2 \leq 1$.

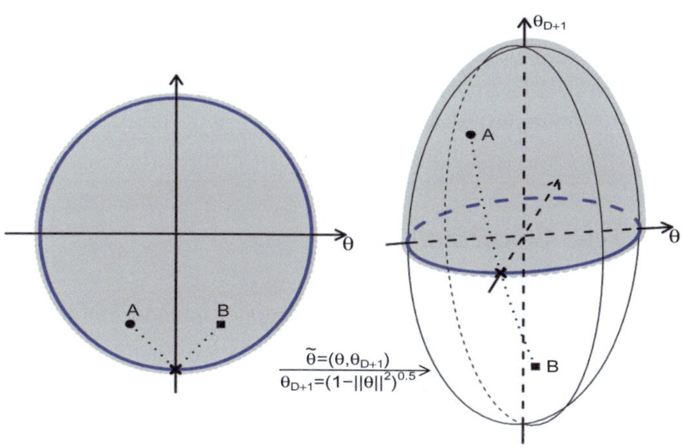

Fig. 2.2 Transforming the unit ball $\mathscr{B}_{\mathbf{0}}^D(1)$ to the sphere $\mathscr{S}^D$

The idea of SA is to augment the original D-dimensional manifold of unit ball $\mathscr{B}_{\mathbf{0}}^D(1)$ to a hypersphere $\mathscr{S}^D := \{\tilde{\boldsymbol{\theta}} \in \mathbb{R}^{D+1} : \|\tilde{\boldsymbol{\theta}}\|_2 = 1\}$ in a $(D+1)$-dimensional space. This can be done by adding an auxiliary variable θ_{D+1} to the original parameter $\boldsymbol{\theta} \in \mathscr{B}_{\mathbf{0}}^D(1)$ to form an extended parameter $\tilde{\boldsymbol{\theta}} = (\boldsymbol{\theta}, \theta_{D+1})$ such that $\theta_{D+1} = \sqrt{1 - \|\boldsymbol{\theta}\|_2^2}$. Next, we identify the lower hemisphere $\mathscr{S}_-^D$ with the upper hemisphere $\mathscr{S}_+^D$ by ignoring the sign of θ_{D+1}. This way, the domain of the target distribution is changed from the unit ball $\mathscr{B}_{\mathbf{0}}^D(1)$ to the D-dimensional sphere, $\mathscr{S}^D := \{\tilde{\boldsymbol{\theta}} \in \mathbb{R}^{D+1} : \|\tilde{\boldsymbol{\theta}}\|_2 = 1\}$, through the following transformation:

$$T_{\mathscr{B}\to\mathscr{S}} : \mathscr{B}_{\mathbf{0}}^D(1) \longrightarrow \mathscr{S}_{\pm}^D, \quad \boldsymbol{\theta} \mapsto \tilde{\boldsymbol{\theta}} = \left(\boldsymbol{\theta}, \pm\sqrt{1 - \|\boldsymbol{\theta}\|_2^2}\right) \tag{2.5}$$

which can also be recognized as the coordinate map from the Euclidean coordinate chart $\{\boldsymbol{\theta}, \mathscr{B}_{\mathbf{0}}^D(1)\}$ to the manifold $\mathscr{S}^D$.

After collecting samples $\{\tilde{\boldsymbol{\theta}}\}$ using a sampling algorithm (e.g., HMC) defined on the sphere, $\mathscr{S}^D$, we discard the last component θ_{D+1} and obtain the samples $\{\boldsymbol{\theta}\}$ that automatically satisfy the constraint $\|\boldsymbol{\theta}\|_2 \leq 1$. Note that the sign of θ_{D+1} does not affect our Monte Carlo estimates. However, after applying the above transformation, we need to adjust our estimates according to the change of variables theorem as follows: [58]

$$\int_{\mathscr{B}_{\mathbf{0}}^D(1)} f(\boldsymbol{\theta}) d\boldsymbol{\theta}_{\mathscr{B}} = \int_{\mathscr{S}_+^D} f(\tilde{\boldsymbol{\theta}}) \left| \frac{d\boldsymbol{\theta}_{\mathscr{B}}}{d\boldsymbol{\theta}_{\mathscr{S}_c}} \right| d\boldsymbol{\theta}_{\mathscr{S}_c} \tag{2.6}$$

where $\left| \frac{d\boldsymbol{\theta}_{\mathscr{B}}}{d\boldsymbol{\theta}_{\mathscr{S}_c}} \right| = |\theta_{D+1}|$ as shown in Corollary 2.1 in Appendix "Canonical Metric in Cartesian Coordinates". Here, $d\boldsymbol{\theta}_{\mathscr{B}}$ and $d\boldsymbol{\theta}_{\mathscr{S}_c}$ are volume elements under the Euclidean metric and the *canonical spherical metric*, respectively.

With the above transformation (2.5), the resulting sampler is defined and moves freely on $\mathscr{S}^D$ while implicitly handling the constraints imposed on the original parameters. As illustrated in Fig. 2.2, the boundary of the constraint, i.e., $\|\boldsymbol{\theta}\|_2 = 1$, corresponds to the equator on the sphere $\mathscr{S}^D$. Therefore, as the sampler moves on the sphere, e.g., from A to B, passing across the equator from one hemisphere to the other translates to "bouncing back" off the boundary in the original parameter space.

2.3.2 *Box-Type Constraints*

Many constraints are given by both lower and upper bounds. Here we focus on a special case that defines a hyper-rectangle $\mathscr{R}_{\mathbf{0}}^D := [0, \pi]^{D-1} \times [0, 2\pi)$; other *box* type constraints can be transformed to this hyper-rectangle. This constrained domain can be mapped to the unit ball $\mathscr{B}_{\mathbf{0}}^D(1)$ and thus reduces to the ball type constraint discussed in Sect. 2.3.1. However, a more natural approach is to use *spherical* coordinates, which directly maps the hyperrectangle $\mathscr{R}_{\mathbf{0}}^D$ to the sphere $\mathscr{S}^D$,

$$T_{\mathscr{R}_0 \to \mathscr{S}} : \mathscr{R}_{\mathbf{0}}^D \longrightarrow \mathscr{S}^D, \quad \boldsymbol{\theta} \mapsto \mathbf{x}, \; x_d = \begin{cases} \cos(\theta_d)\prod_{i=1}^{d-1} \sin(\theta_i), & d < D+1 \\ \prod_{i=1}^{D} \sin(\theta_i), & d = D+1 \end{cases} \tag{2.7}$$

Therefore, we use $\{\boldsymbol{\theta}, \mathscr{R}_{\mathbf{0}}^D\}$ as the spherical coordinate chart for the manifold $\mathscr{S}^D$. Instead of being appended with an extra dimension as in Sect. 2.3.1, here $\boldsymbol{\theta} \in \mathbb{R}^D$ is treated as the spherical coordinates of the point $\mathbf{x} \in \mathbb{R}^{D+1}$ with $\|\mathbf{x}\|_2 = 1$.

After obtaining samples $\{\mathbf{x}\}$ on the sphere $\mathscr{S}^D$, we transform them back to $\{\boldsymbol{\theta}\}$ in the original constrained domain $\mathscr{R}_{\mathbf{0}}^D$ using the following inverse mapping of (2.7):

$$T_{\mathscr{S} \to \mathscr{R}_0} : \mathscr{S}^D \longrightarrow \mathscr{R}_{\mathbf{0}}^D, \; \mathbf{x} \mapsto \boldsymbol{\theta}, \; \theta_d = \begin{cases} \operatorname{arccot} \frac{x_d}{\sqrt{1-\sum_{i=1}^{d} x_i^2}}, & d < D \\ 2 \operatorname{arccot} \frac{x_D + \sqrt{x_{D+1}^2 + x_D^2}}{x_{D+1}}, & d = D \end{cases} \tag{2.8}$$

Similarly, we need to adjust the estimates based on the following change of variables formula:

$$\int_{\mathscr{R}_{\mathbf{0}}^D} f(\boldsymbol{\theta}) d\boldsymbol{\theta}_{\mathscr{R}_0} = \int_{\mathscr{S}^D} f(\boldsymbol{\theta}) \left| \frac{d\boldsymbol{\theta}_{\mathscr{R}_0}}{d\boldsymbol{\theta}_{\mathscr{S}_r}} \right| d\boldsymbol{\theta}_{\mathscr{S}_r} \tag{2.9}$$

where $\left| \frac{d\boldsymbol{\theta}_{\mathscr{R}_0}}{d\boldsymbol{\theta}_{\mathscr{S}_r}} \right| = \prod_{d=1}^{D-1} \sin^{-(D-d)}(\theta_d)$ as shown Proposition 2.7.3 in Appendix "Round Metric in the Spherical Coordinates". Here, $d\boldsymbol{\theta}_{\mathscr{R}_0}$ and $d\boldsymbol{\theta}_{\mathscr{S}_r}$ are volume elements under the Euclidean metric and the *round spherical metric*, respectively.

With the above transformation (2.7), we can derive sampling methods on the sphere to implicitly handle box-type constraints. As illustrated in Fig. 2.3, the red vertical boundary of $\mathscr{R}_{\mathbf{0}}^D$ collapses to the north pole of $\mathscr{S}^D$, while the green vertical boundary collapses to the South pole. Two blue horizontal boundaries are mapped to the same prime meridian of $\mathscr{S}^D$ shown in blue color. As the sampler moves freely on the sphere $\mathscr{S}^D$, the resulting samples automatically satisfy the original constraint after being transformed back to the original domain.

2.3.3 General *q*-Norm Constraints

The ball and box-type constraints discussed in previous sections are in fact special cases of more general q-norm constraints with q set to 2 and ∞ respectively. In general, these constraints are expressed in terms of q-norm of the parameter vector $\boldsymbol{\beta} \in \mathbb{R}^D$,

$$\|\boldsymbol{\beta}\|_q = \begin{cases} (\sum_{i=1}^{D} |\beta_i|^q)^{1/q}, & q \in (0, +\infty) \\ \max_{1 \le i \le D} |\beta_i|, & q = +\infty \end{cases} \tag{2.10}$$

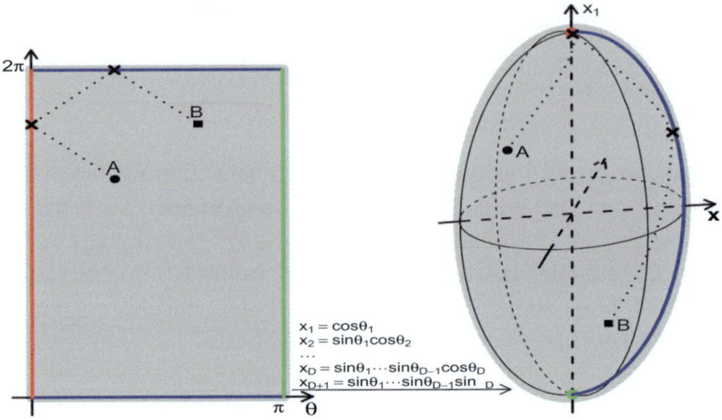

Fig. 2.3 Transforming the hyperrectangle $\mathscr{R}_{\mathbf{0}}^D$ to the sphere $\mathscr{S}^D$

This class of constraints is very common in statistics and machine learning. For example, when $\boldsymbol{\beta}$ are regression parameters, $q = 2$ corresponds to the ridge regression and $q = 1$ corresponds to Lasso [61].

Denote the domain constrained by general q-norm as $\mathscr{Q}^D := \{\boldsymbol{\beta} \in \mathbb{R}^D : \|\boldsymbol{\beta}\|_q \leq 1\}$. It could be quite challenging to sample probability distributions defined on $\mathscr{Q}^D$ (see Fig. 2.1). To address this issue, we propose to transform $\mathscr{Q}^D$ to the unit ball $\mathscr{B}_{\mathbf{0}}^D(1)$ so that the method discussed in Sect. 2.3.1 can be applied. As before, sampling methods defined on the sphere $\mathscr{S}^D$ generate samples that automatically fall within $\mathscr{B}_{\mathbf{0}}^D(1)$. Then we transform those samples back to the q-norm domain, $\mathscr{Q}^D$, and adjust the estimates with the following change of variables formula:

$$\int_{\mathscr{Q}^D} f(\boldsymbol{\beta}) d\boldsymbol{\beta}_{\mathscr{Q}} = \int_{\mathscr{S}_+^D} f(\tilde{\boldsymbol{\theta}}) \left| \frac{d\boldsymbol{\beta}_{\mathscr{Q}}}{d\boldsymbol{\theta}_{\mathscr{S}_c}} \right| d\boldsymbol{\theta}_{\mathscr{S}_c} \tag{2.11}$$

where $\left| \frac{d\boldsymbol{\beta}_{\mathscr{Q}}}{d\boldsymbol{\theta}_{\mathscr{S}_c}} \right| = \left| \frac{d\boldsymbol{\beta}_{\mathscr{Q}}}{d\boldsymbol{\theta}_{\mathscr{B}}^{\mathsf{T}}} \right| \left| \frac{d\boldsymbol{\theta}_{\mathscr{B}}}{d\boldsymbol{\theta}_{\mathscr{S}_c}} \right| = \left| \frac{d\boldsymbol{\beta}_{\mathscr{Q}}}{d\boldsymbol{\theta}_{\mathscr{B}}^{\mathsf{T}}} \right| |\theta_{D+1}|$. In the following, we introduce the bijective mappings between $\mathscr{Q}^D$ and $\mathscr{B}_{\mathbf{0}}^D(1)$ and specify the associated Jacobian determinants $\left| \frac{d\boldsymbol{\beta}_{\mathscr{Q}}}{d\boldsymbol{\theta}_{\mathscr{B}}^{\mathsf{T}}} \right|$.

2.3.3.1 Norm Constraints with $q \in (0, +\infty)$

For $q \in (0, +\infty)$, q-norm domain $\mathscr{Q}^D$ can be transformed to the unit ball $\mathscr{B}_0^D(1)$ bijectively via the following map (illustrated by the left panel of Fig. 2.4):

$$T_{\mathscr{Q} \to \mathscr{B}} : \mathscr{Q}^D \to \mathscr{B}_0^D(1), \quad \beta_i \mapsto \theta_i = \text{sgn}(\beta_i)|\beta_i|^{q/2} \tag{2.12}$$

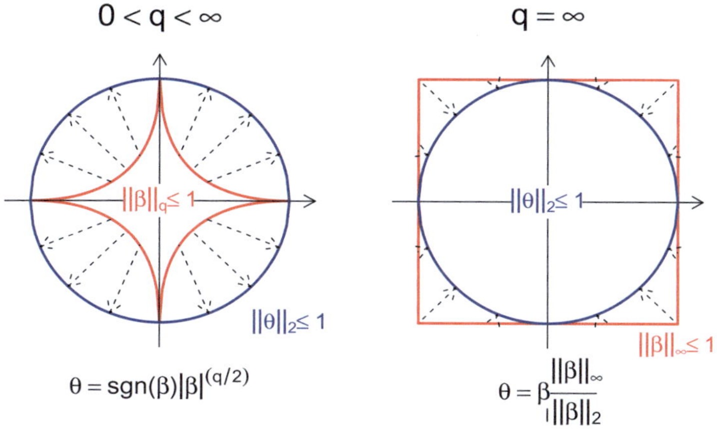

Fig. 2.4 Transforming q-norm constrained domain to unit ball. *Left* from unit cube $\mathscr{C}^D$ to unit ball $\mathscr{B}_0^D(1)$; *Right* from general q-norm domain $\mathscr{Q}^D$ to unit ball $\mathscr{B}_0^D(1)$

The Jacobian determinant of $T_{\mathscr{B}\to\mathscr{Q}}$ is $\left|\frac{d\boldsymbol{\beta}_{\mathscr{Q}}}{d\boldsymbol{\theta}_{\mathscr{B}}^{\mathsf{T}}}\right| = \left(\frac{2}{q}\right)^D \left(\prod_{i=1}^D |\theta_i|\right)^{2/q-1}$. See Appendix "Jacobian of the transformation between q-norm domains" for more details.

2.3.3.2 Norm Constraints with $q = +\infty$

When $q = +\infty$, the norm inequality defines a unit *hypercube*, $\mathscr{C}^D := [-1, 1]^D = \{\boldsymbol{\beta} \in \mathbb{R}^D : \|\boldsymbol{\beta}\|_\infty \le 1\}$, from which the more general form, *hyper-rectangle*, $\mathscr{R}^D := \{\boldsymbol{\beta} \in \mathbb{R}^D : \mathbf{l} \le \boldsymbol{\beta} \le \mathbf{u}\}$, can be obtained by proper shifting and scaling. The unit hypercube $\mathscr{C}^D$ can be transformed to its inscribed unit ball $\mathscr{B}_{\mathbf{0}}^D(1)$ through the following map (illustrated by the right panel of Fig. 2.4):

$$T_{\mathscr{C}\to\mathscr{B}} : [-1, 1]^D \to \mathscr{B}_0^D(1), \quad \boldsymbol{\beta} \mapsto \boldsymbol{\theta} = \boldsymbol{\beta}\frac{\|\boldsymbol{\beta}\|_\infty}{\|\boldsymbol{\beta}\|_2} \tag{2.13}$$

The Jacobian determinant of $T_{\mathscr{B}\to\mathscr{R}}$ is $\left|\frac{d\boldsymbol{\beta}_{\mathscr{R}}}{d\boldsymbol{\theta}_{\mathscr{B}}^{\mathsf{T}}}\right| = \frac{\|\boldsymbol{\theta}\|_2^D}{\|\boldsymbol{\theta}\|_\infty^D} \prod_{i=1}^D \frac{u_i - l_i}{2}$. More details can be found in Appendix "Jacobian of the transformation between q-norm domains".

2.3.4 *Functional Constraints*

Many statistical problems involve functional constraints. For example, [47] discusses linear and quadratic constraints for multivariate Gaussian distributions. Since the target distribution is truncated Gaussian, Hamiltonian dynamics can be exactly simulated and the boundary-hitting time can be analytically obtained. However, finding the hitting time and reflection trajectory is computationally expensive. Some constraints of this type can be handled by the spherical augmentation method more efficiently. Further, our method can be used for sampling from a wide range of distributions beyond Gaussian.

2.3.4.1 Linear Constraints

In general, M linear constraints can be written as $\mathbf{l} \leq \mathbf{A}\boldsymbol{\beta} \leq \mathbf{u}$, where $\mathbf{A}$ is $M \times D$ matrix, $\boldsymbol{\beta}$ is a D-vector, and the boundaries $\mathbf{l}$ and $\mathbf{u}$ are both M-vectors. Here, we assume $M = D$ and $\mathbf{A}_{D\times D}$ is invertible. (Note that we generally do not have $\mathbf{A}^{-1}\mathbf{l} \leq \boldsymbol{\beta} \leq \mathbf{A}^{-1}\mathbf{u}$.) Instead of sampling $\boldsymbol{\beta}$ directly, we sample $\boldsymbol{\eta} := \mathbf{A}\boldsymbol{\beta}$ with the box-type constraint: $\mathbf{l} \leq \boldsymbol{\eta} \leq \mathbf{u}$. Now we can apply our proposed method to sample $\boldsymbol{\eta}$ and transform it back to $\boldsymbol{\beta} = \mathbf{A}^{-1}\boldsymbol{\eta}$. In this process, we use the following change of variables formula:

$$\int_{\mathbf{l}\leq\mathbf{A}\boldsymbol{\beta}\leq\mathbf{u}} f(\boldsymbol{\beta})d\boldsymbol{\beta} = \int_{\mathbf{l}\leq\boldsymbol{\eta}\leq\mathbf{u}} f(\boldsymbol{\eta}) \left|\frac{d\boldsymbol{\beta}}{d\boldsymbol{\eta}}\right| d\boldsymbol{\eta} \tag{2.14}$$

where $\left|\frac{d\boldsymbol{\beta}}{d\boldsymbol{\eta}}\right| = |\mathbf{A}|^{-1}$.

Figure 2.5 illustrates that both exact HMC [47] and HMC with spherical augmentation can handle linear constraints, here $\mathbf{l} = \mathbf{0}$, $\mathbf{A} = \begin{bmatrix} -0.5 & 1 \\ 1 & 1 \end{bmatrix}$ and $\mathbf{u} = \mathbf{2}$, imposed on a 2D Gaussian distribution $\mathcal{N}(\boldsymbol{\mu}, \boldsymbol{\Sigma})$ with $\boldsymbol{\mu} = \begin{bmatrix} 0 \\ 1 \end{bmatrix}$ and $\boldsymbol{\Sigma} = \begin{bmatrix} 1 & 0.5 \\ 0.5 & 1 \end{bmatrix}$ (first row). They all yield good estimates (colored heatmaps) that match the truth (solid contour curves) very well. However, the exact HMC is not applicable to more complicated distributions such as the damped sine wave distribution (second row in Fig. 2.5) with the following density:

$$f(\boldsymbol{\beta}) \propto \frac{\sin^2 Q(\boldsymbol{\beta})}{Q(\boldsymbol{\beta})}, \quad Q(\boldsymbol{\beta}) = \frac{1}{2}(\boldsymbol{\beta} - \boldsymbol{\mu})^{\mathsf{T}}\boldsymbol{\Sigma}^{-1}(\boldsymbol{\beta} - \boldsymbol{\mu}) \tag{2.15}$$

Therefore, there is no colored density estimation plot in the second row and second column subfigure. However, it is worth noting that for truncated Gaussian distributions, since the exact HMC method of [47] does not require $\mathbf{A}$ to be square invertible, it can handle a wider range of linear constraints compared to our method.

2.3.4.2 Quadratic Constraints

General quadratic constraints can be written as $l \leq \boldsymbol{\beta}^{\mathsf{T}}\mathbf{A}\boldsymbol{\beta} + \mathbf{b}^{\mathsf{T}}\boldsymbol{\beta} \leq u$, where $l, u > 0$ are scalars. We assume $\mathbf{A}_{D\times D}$ symmetric and positive-definite. By the spectral theorem, we have the decomposition $\mathbf{A} = \mathbf{Q}\boldsymbol{\Sigma}\mathbf{Q}^{\mathsf{T}}$, where $\mathbf{Q}$ is an orthogonal matrix

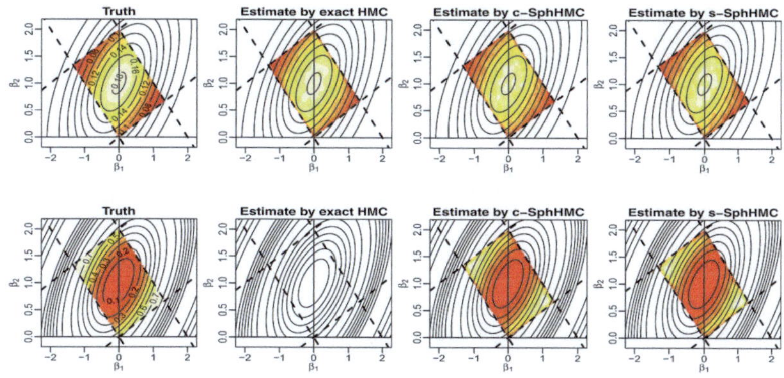

Fig. 2.5 Sampling from a Gaussian distribution (*first row*) and a damped sine wave distribution (*second row*) with linear constraints. *Solid elliptical curves* always show true unconstrained probability density contours. *Dashed lines* define linear constrained domains. Colored heatmaps indicate constrained probability density based on truth or estimation from MCMC samples. First column shows the true distributions. The results of exact HMC method [47] are shown in the second column (not applicable in the *second row*). The *last two columns* show the results of our proposed methods

and $\mathbf{\Sigma}$ is a diagonal matrix of eigenvalues of $\mathbf{A}$. By shifting and scaling, $\boldsymbol{\beta} \mapsto \boldsymbol{\beta}^* = \sqrt{\mathbf{\Sigma}}\mathbf{Q}^\mathsf{T}(\boldsymbol{\beta} + \frac{1}{2}\mathbf{A}^{-1}\mathbf{b})$, we only need to consider the *ring* type constraints for $\boldsymbol{\beta}^*$,

$$\circledcirc : l^* \leq \|\boldsymbol{\beta}^*\|_2^2 = (\boldsymbol{\beta}^*)^\mathsf{T}\boldsymbol{\beta}^* \leq u^*, \quad l^* = l + \frac{1}{4}\mathbf{b}^\mathsf{T}\mathbf{A}^{-1}\mathbf{b},\ u^* = u + \frac{1}{4}\mathbf{b}^\mathsf{T}\mathbf{A}^{-1}\mathbf{b} \tag{2.16}$$

which can be mapped to the unit ball as follows:

$$T_{\circledcirc \to \mathscr{B}} : \mathscr{B}_{\mathbf{0}}^D(\sqrt{u^*}) \backslash \mathscr{B}_{\mathbf{0}}^D(\sqrt{l^*}) \longrightarrow \mathscr{B}_{\mathbf{0}}^D(1), \quad \boldsymbol{\beta}^* \mapsto \boldsymbol{\theta} = \frac{\boldsymbol{\beta}^*}{\|\boldsymbol{\beta}^*\|_2} \frac{\|\boldsymbol{\beta}^*\|_2 - \sqrt{l^*}}{\sqrt{u^*} - \sqrt{l^*}} \tag{2.17}$$

We can then apply our proposed method in Sect. 2.3.1 to obtain samples $\{\boldsymbol{\theta}\}$ in $\mathscr{B}_{\mathbf{0}}^D(1)$ and transform them back to the original domain with the following inverse operation of (2.17):

$$T_{\mathscr{B} \to \circledcirc} : \mathscr{B}_{\mathbf{0}}^D(1) \longrightarrow \mathscr{B}_{\mathbf{0}}^D(\sqrt{u^*}) \backslash \mathscr{B}_{\mathbf{0}}^D(\sqrt{l^*}),\ \boldsymbol{\theta} \mapsto \boldsymbol{\beta}^* = \frac{\boldsymbol{\theta}}{\|\boldsymbol{\theta}\|_2}((\sqrt{u^*} - \sqrt{l^*})\|\boldsymbol{\theta}\|_2 + \sqrt{l^*}) \tag{2.18}$$

In this process, we need the change of variables formula

$$\int_{l\le\boldsymbol{\beta}^{\mathsf{T}}\mathbf{A}\boldsymbol{\beta}+\mathbf{b}^{\mathsf{T}}\boldsymbol{\beta}\le u} f(\boldsymbol{\beta})d\boldsymbol{\beta} = \int_{\mathscr{S}_+^D} f(\boldsymbol{\theta}) \left|\frac{d\boldsymbol{\beta}}{d\boldsymbol{\theta}_{\mathscr{S}_c}}\right| d\boldsymbol{\theta}_{\mathscr{S}_c} \tag{2.19}$$

where $\left|\frac{d\boldsymbol{\beta}}{d\boldsymbol{\theta}_{\mathscr{S}_c}}\right| = \left|\frac{d\boldsymbol{\beta}}{d(\boldsymbol{\beta}^*)^{\mathsf{T}}}\right| \left|\frac{d\boldsymbol{\beta}^*}{d\boldsymbol{\theta}_{\mathscr{B}}^{\mathsf{T}}}\right| \left|\frac{d\boldsymbol{\theta}_{\mathscr{B}}}{d\boldsymbol{\theta}_{\mathscr{S}_c}}\right| = |\mathbf{A}|^{-\frac{1}{2}}\alpha^{D-1}(\alpha - \sqrt{l^*})|\theta_{D+1}|,\ \alpha = \sqrt{u^*} + (1/\|\boldsymbol{\theta}\|_2 - 1)\sqrt{l^*}$.

2.3.4.3 More General Constraints

We close this section with some comments on more general types of constraints. In some problems, several parameters might be unconstrained, and the type of constraints might vary across the constrained parameters. In such cases, we could group the parameters into blocks and update each block separately using the methods discussed in this section. When dealing with one-sided constraints, e.g., $\theta_i \ge l_i$, one can map the constrained domain to the whole space and sample the unconstrained parameter θ_i^*, where $\theta_i = |\theta_i^*| + l_i$. Alternatively, the one-sided constraint $\theta_i \ge l_i$ can be changed to a two-sided constraint for $\theta_i^* \in (0, 1)$ by setting $\theta_i = -\log\theta_i^* + l_i$.

2.4 Monte Carlo with Spherical Augmentation

In this section, we show how SA can be used to improve Markov Chain Monte Carlo methods applied to constrained probability distributions. In particular, we focus on two state-of-the-art sampling algorithms, namely Hamiltonian Monte Carlo [23, 46], and Lagrangian Monte Carlo [35]. Two types of Spherical HMC algorithms are designed for different types of constraints. Spherical LMC is introduced specifically for probability simplex, with application to LDA [10]. Note however that our proposed method is generic so its application goes beyond these two algorithms.

2.4.1 Common Settings

Throughout this section, we denote the original parameter vector as $\boldsymbol{\beta}$, the constrained domain as $\mathscr{D}$, the coordinate vector of sphere $\mathscr{S}^D$ as $\boldsymbol{\theta}$. All the change of variables formulae presented in the previous section can be summarized as

$$\int_{\mathscr{D}} f(\boldsymbol{\beta})d\boldsymbol{\beta}_{\mathscr{D}} = \int_{\mathscr{S}} f(\boldsymbol{\theta}) \left|\frac{d\boldsymbol{\beta}_{\mathscr{D}}}{d\boldsymbol{\theta}_{\mathscr{S}}}\right| d\boldsymbol{\theta}_{\mathscr{S}} \tag{2.20}$$

where $\left|\frac{d\boldsymbol{\beta}_{\mathscr{D}}}{d\boldsymbol{\theta}_{\mathscr{S}}}\right|$ is the Jacobian determinant of the mapping $T : \mathscr{S} \longrightarrow \mathscr{D}$ and $d\boldsymbol{\theta}_{\mathscr{S}}$ is some spherical measure.

For energy based MCMC algorithms like HMC, RHMC, and LMC, we need to investigate the change of energy under the above transformation. The original potential energy function $U(\boldsymbol{\beta}) = -\log f(\boldsymbol{\beta})$ should be transformed to the following $\phi(\boldsymbol{\theta})$

$$\phi(\boldsymbol{\theta}) = -\log f(\boldsymbol{\theta}) - \log\left|\frac{d\boldsymbol{\beta}_{\mathscr{D}}}{d\boldsymbol{\theta}_{\mathscr{S}}}\right| = U(\boldsymbol{\beta}(\boldsymbol{\theta})) - \log\left|\frac{d\boldsymbol{\beta}_{\mathscr{D}}}{d\boldsymbol{\theta}_{\mathscr{S}}}\right| \tag{2.21}$$

Consequently the total energy $H(\boldsymbol{\beta}, \mathbf{v})$ in (2.1) becomes

$$H(\boldsymbol{\theta}, \mathbf{v}) = \phi(\boldsymbol{\theta}) + \frac{1}{2}\langle \mathbf{v}, \mathbf{v}\rangle_{\mathbf{G}_{\mathscr{S}}(\boldsymbol{\theta})} \tag{2.22}$$

The gradient of potential energy U, metric and natural gradient (preconditioned gradient) under the new coordinate system $\{\boldsymbol{\theta}, \mathscr{S}^D\}$ can be calculated as follows

$$\nabla_{\boldsymbol{\theta}} U(\boldsymbol{\theta}) = \frac{d\boldsymbol{\beta}^{\mathsf{T}}}{d\boldsymbol{\theta}} \nabla_{\boldsymbol{\beta}} U(\boldsymbol{\beta}) \tag{2.23}$$

$$\mathbf{G}_{\mathscr{S}}(\boldsymbol{\theta}) = \frac{d\boldsymbol{\beta}^{\mathsf{T}}}{d\boldsymbol{\theta}} \mathbf{G}_{\mathscr{D}}(\boldsymbol{\beta}) \frac{d\boldsymbol{\beta}}{d\boldsymbol{\theta}^{\mathsf{T}}} \tag{2.24}$$

$$\mathbf{G}_{\mathscr{S}}(\boldsymbol{\theta})^{-1}\nabla_{\boldsymbol{\theta}} U(\boldsymbol{\theta}) = \left[\frac{d\boldsymbol{\beta}^{\mathsf{T}}}{d\boldsymbol{\theta}}\right]^{-1} \mathbf{G}_{\mathscr{D}}(\boldsymbol{\beta})^{-1}\nabla_{\boldsymbol{\beta}} U(\boldsymbol{\beta}) \tag{2.25}$$

In subscripts, we use notation $\mathscr{B} := \mathscr{B}_0^D(1)$; $\mathscr{S} := \mathscr{S}^D$. $\mathscr{S}_c$ indicates the sphere endowed with the canonical metric (Appendix “Canonical Metric in Cartesian Coordinates”) and $\mathscr{S}_r$ with the round metric (Appendix “Round Metric in the Spherical Coordinates”). The upper dot $\dot{}$ means taking derivative with respect to time t.

2.4.2 *Spherical Hamiltonian Monte Carlo*

We define HMC on the sphere $\mathscr{S}^D$ in two different coordinate systems: *Cartesian coordinates* and the *spherical coordinates*. The former is applied to ball type constraints or those that could be converted to ball type constraints; the later is more suited for box-type constraints. Besides the merit of implicitly handling constraints, HMC on sphere can take advantage of the splitting technique [9, 13, 56] to further improve its computational efficiency. In particular, Geodesic Monte Carlo [13] works on the cotangent bundle and relies on the embedding of the manifold into a larger Euclidean space; in contrast, our proposed methods work on the tangent bundle without such requirement.

2.4.2.1 Spherical HMC in Cartesian Coordinates

We first consider HMC for the target distribution with density $f(\boldsymbol{\theta})$ defined on the unit ball $\mathscr{B}_{\mathbf{0}}^D(1)$ endowed with the Euclidean metric $\mathbf{I}$. The potential energy is defined as $U(\boldsymbol{\theta}) := -\log f(\boldsymbol{\theta})$. Associated with the auxiliary variable $\mathbf{v}$ (i.e., velocity), we define the kinetic energy $K(\mathbf{v}) = \frac{1}{2}\mathbf{v}^{\mathsf{T}}\mathbf{I}\mathbf{v}$ for $\mathbf{v} \in T_{\boldsymbol{\theta}}\mathscr{B}_{\mathbf{0}}^D(1)$, which is a D-dimensional vector sampled from the tangent space of $\mathscr{B}_{\mathbf{0}}^D(1)$. Therefore, the Hamiltonian is defined on $\mathscr{B}_{\mathbf{0}}^D(1)$ as

$$H(\boldsymbol{\theta}, \mathbf{v}) = U(\boldsymbol{\theta}) + K(\mathbf{v}) = U(\boldsymbol{\theta}) + \frac{1}{2}\mathbf{v}^{\mathsf{T}}\mathbf{I}\mathbf{v} \tag{2.26}$$

Under the transformation $T_{\mathscr{B}\to\mathscr{S}}$ in (2.5), the above Hamiltonian (2.26) on $\mathscr{B}_{\mathbf{0}}^D(1)$ will be changed to the following Hamiltonian $H(\tilde{\boldsymbol{\theta}}, \tilde{\mathbf{v}})$ on $\mathscr{S}^D$ as in (2.22):

$$H(\tilde{\boldsymbol{\theta}}, \tilde{\mathbf{v}}) = \phi(\tilde{\boldsymbol{\theta}}) + \frac{1}{2}\mathbf{v}^{\mathsf{T}}\mathbf{G}_{\mathscr{S}_c}(\boldsymbol{\theta})\mathbf{v} = U(\tilde{\boldsymbol{\theta}}) - \log\left|\frac{d\boldsymbol{\theta}_{\mathscr{B}}}{d\boldsymbol{\theta}_{\mathscr{S}_c}}\right| + \frac{1}{2}\mathbf{v}^{\mathsf{T}}\mathbf{G}_{\mathscr{S}_c}(\boldsymbol{\theta})\mathbf{v} \tag{2.27}$$

where the potential energy $U(\tilde{\boldsymbol{\theta}}) = U(\boldsymbol{\theta})$ (i.e., the distribution is fully defined in terms of the original parameter $\boldsymbol{\theta}$, which are the first D elements of $\tilde{\boldsymbol{\theta}}$), and $\mathbf{G}_{\mathscr{S}_c}(\boldsymbol{\theta}) = \mathbf{I}_D + \boldsymbol{\theta}\boldsymbol{\theta}^{\mathsf{T}}/(1 - \|\boldsymbol{\theta}\|_2^2)$ is the *canonical spherical metric*.

Viewing $\{\boldsymbol{\theta}, \mathscr{B}_{\mathbf{0}}^D(1)\}$ as the Euclidean coordinate chart of manifold $(\mathscr{S}^D, \mathbf{G}_{\mathscr{S}_c}(\boldsymbol{\theta}))$, we have the logarithm of volume adjustment, $\log\left|\frac{d\boldsymbol{\theta}_{\mathscr{B}}}{d\boldsymbol{\theta}_{\mathscr{S}_c}}\right| = -\frac{1}{2}\log|\mathbf{G}_{\mathscr{S}_c}| = \log|\theta_{D+1}|$ (See Appendix "Canonical Metric in Cartesian Coordinates"). The last two terms in Eq. (2.27) is the minus log density of $\mathbf{v}|\boldsymbol{\theta} \sim \mathscr{N}(\mathbf{0}, \mathbf{G}_{\mathscr{S}_c}(\boldsymbol{\theta})^{-1})$ (See [28, 35] for more details). However, the derivative of log volume adjustment, θ_{D+1}^{-1}, contributes an extremely large component to the gradient of energy around the equator ($\theta_{D+1} = 0$), which in turn increases the numerical error in the discretized Hamiltonian dynamics. For the purpose of numerical stability, we instead consider the following *partial* Hamiltonian $H^*(\tilde{\boldsymbol{\theta}}, \tilde{\mathbf{v}})$ and leave the volume adjustment as weights to adjust the estimation of integration (2.20):

$$H^*(\tilde{\boldsymbol{\theta}}, \tilde{\mathbf{v}}) = U(\boldsymbol{\theta}) + \frac{1}{2}\mathbf{v}^{\mathsf{T}}\mathbf{G}_{\mathscr{S}_c}(\boldsymbol{\theta})\mathbf{v} \tag{2.28}$$

After obtaining samples $\tilde{\boldsymbol{\theta}} \sim f(\tilde{\boldsymbol{\theta}})d\boldsymbol{\theta}_{\mathscr{S}_c}$, we apply the importance weights $\left|\frac{d\boldsymbol{\theta}_{\mathscr{B}}}{d\boldsymbol{\theta}_{\mathscr{S}_c}}\right|$ to obtain samples from the correct target distribution:

$$\boldsymbol{\theta} \sim f(\tilde{\boldsymbol{\theta}})d\boldsymbol{\theta}_{\mathscr{S}_c}\left|\frac{d\boldsymbol{\theta}_{\mathscr{B}}}{d\boldsymbol{\theta}_{\mathscr{S}_c}}\right| = f(\boldsymbol{\theta})d\boldsymbol{\theta}_{\mathscr{B}} \tag{2.29}$$

If we extend the velocity as $\tilde{\mathbf{v}} = (\mathbf{v}, v_{D+1})$ with $v_{D+1} = -\boldsymbol{\theta}^{\mathsf{T}}\mathbf{v}/\theta_{D+1}$, then $\tilde{\mathbf{v}}$ falls in the tangent space of the sphere, $T_{\tilde{\boldsymbol{\theta}}}\mathscr{S}^D := \{\tilde{\mathbf{v}} \in \mathbb{R}^{D+1} | \tilde{\boldsymbol{\theta}}^{\mathsf{T}}\tilde{\mathbf{v}} = 0\}$. Therefore,

$\mathbf{v}^{\mathsf{T}}\mathbf{G}_{\mathcal{S}_c}(\boldsymbol{\theta})\mathbf{v} = \tilde{\mathbf{v}}^{\mathsf{T}}\tilde{\mathbf{v}}$. As a result, the partial Hamiltonian (2.28) can be recognized as the standard Hamiltonian (2.26) in the augmented $(D+1)$ dimensional space

$$H^*(\tilde{\boldsymbol{\theta}}, \tilde{\mathbf{v}}) = U(\tilde{\boldsymbol{\theta}}) + K(\tilde{\mathbf{v}}) = U(\tilde{\boldsymbol{\theta}}) + \frac{1}{2}\tilde{\mathbf{v}}^{\mathsf{T}}\tilde{\mathbf{v}} \tag{2.30}$$

This is due to the energy invariance presented as Proposition 2.7.1 in Appendix "Spherical Geometry". Now we can sample the velocity $\mathbf{v} \sim \mathcal{N}(\mathbf{0}, \mathbf{G}_{\mathcal{S}_c}(\boldsymbol{\theta})^{-1})$ and set $\tilde{\mathbf{v}} = \begin{bmatrix} \mathbf{I} \\ -\boldsymbol{\theta}^{\mathsf{T}}/\theta_{D+1} \end{bmatrix} \mathbf{v}$. Alternatively, since $\mathrm{Cov}[\tilde{\mathbf{v}}] = \begin{bmatrix} \mathbf{I} \\ -\boldsymbol{\theta}^{\mathsf{T}}/\theta_{D+1} \end{bmatrix} \mathbf{G}_{\mathcal{S}_c}(\boldsymbol{\theta})^{-1}$ $\begin{bmatrix} \mathbf{I} - \boldsymbol{\theta}/\theta_{D+1} \end{bmatrix} = \mathbf{I}_{D+1} - \tilde{\boldsymbol{\theta}}\tilde{\boldsymbol{\theta}}^{\mathsf{T}}$ is idempotent, we can sample $\tilde{\mathbf{v}}$ by $(\mathbf{I}_{D+1} - \tilde{\boldsymbol{\theta}}\tilde{\boldsymbol{\theta}}^{\mathsf{T}})\mathbf{z}$ with $\mathbf{z} \sim \mathcal{N}(\mathbf{0}, \mathbf{I}_{D+1})$.

The Hamiltonian function (2.28) can be used to define the Hamiltonian dynamics on the Riemannian manifold $(\mathcal{S}^D, \mathbf{G}_{\mathcal{S}_c}(\boldsymbol{\theta}))$ in terms of $(\boldsymbol{\theta}, \mathbf{p})$, or equivalently as the following Lagrangian dynamics in terms of $(\boldsymbol{\theta}, \mathbf{v})$ [35]:

$$\begin{aligned} \dot{\boldsymbol{\theta}} &= \mathbf{v} \\ \dot{\mathbf{v}} &= -\mathbf{v}^{\mathsf{T}}\boldsymbol{\Gamma}_{\mathcal{S}_c}(\boldsymbol{\theta})\mathbf{v} - \mathbf{G}_{\mathcal{S}_c}(\boldsymbol{\theta})^{-1}\nabla_{\boldsymbol{\theta}} U(\boldsymbol{\theta}) \end{aligned} \tag{2.31}$$

where $\boldsymbol{\Gamma}_{\mathcal{S}_c}(\boldsymbol{\theta})$ are the Christoffel symbols of second kind derived from $\mathbf{G}_{\mathcal{S}_c}(\boldsymbol{\theta})$. The Hamiltonian (2.28) is preserved under Lagrangian dynamics (2.31). (See [35] for more discussion).

Reference [13] split the Hamiltonian (2.28) as follows:

$$H^*(\tilde{\boldsymbol{\theta}}, \tilde{\mathbf{v}}) = U(\boldsymbol{\theta})/2 + \frac{1}{2}\mathbf{v}^{\mathsf{T}}\mathbf{G}_{\mathcal{S}_c}(\boldsymbol{\theta})\mathbf{v} + U(\boldsymbol{\theta})/2 \tag{2.32}$$

However, their approach works with $(\boldsymbol{\theta}, \mathbf{p})$ on the cotangent bundle and requires the manifold to be embedded in the Euclidean space. To avoid this assumption, instead of splitting the Hamiltonian dynamics of $(\boldsymbol{\theta}, \mathbf{p})$, we split the corresponding Lagrangian dynamics (2.31) in terms of $(\boldsymbol{\theta}, \mathbf{v})$ as follows (See Appendix "Splitting Hamiltonian (Lagrangian) Dynamics on $\mathcal{S}^D$" for more details):

$$\begin{cases} \dot{\boldsymbol{\theta}} &= \mathbf{0} \\ \dot{\mathbf{v}} &= -\frac{1}{2}\mathbf{G}_{\mathcal{S}_c}(\boldsymbol{\theta})^{-1}\nabla_{\boldsymbol{\theta}} U(\boldsymbol{\theta}) \end{cases} \tag{2.33a}$$

$$\begin{cases} \dot{\boldsymbol{\theta}} &= \mathbf{v} \\ \dot{\mathbf{v}} &= -\mathbf{v}^{\mathsf{T}}\boldsymbol{\Gamma}_{\mathcal{S}_c}(\boldsymbol{\theta})\mathbf{v} \end{cases} \tag{2.33b}$$

Note that the first dynamics (2.33a) only involves updating velocity $\tilde{\mathbf{v}}$ in the tangent space $T_{\tilde{\boldsymbol{\theta}}}\mathcal{S}^D$ and has the following solution (see Appendix "Splitting Hamiltonian (Lagrangian) Dynamics on $\mathcal{S}^D$" for more details):

$$\begin{aligned}\tilde{\boldsymbol{\theta}}(t) &= \tilde{\boldsymbol{\theta}}(0)\\ \tilde{\mathbf{v}}(t) &= \tilde{\mathbf{v}}(0) - \frac{t}{2}\left(\begin{bmatrix}\mathbf{I}_D\\ \mathbf{0}^\mathsf{T}\end{bmatrix} - \tilde{\boldsymbol{\theta}}(0)\boldsymbol{\theta}(0)^\mathsf{T}\right)\nabla_{\boldsymbol{\theta}} U(\boldsymbol{\theta}(0))\end{aligned} \tag{2.34}$$

The second dynamics (2.33b) only involves the kinetic energy and has the geodesic flow that is a *great circle* (orthodrome or Riemannian circle) on the sphere $\mathscr{S}^D$ as its analytical solution (See Appendix "Geodesic on a Sphere in Cartesian Coordinates" for more details):

$$\begin{aligned}\tilde{\boldsymbol{\theta}}(t) &= \tilde{\boldsymbol{\theta}}(0)\cos(\|\tilde{\mathbf{v}}(0)\|_2 t) + \frac{\tilde{\mathbf{v}}(0)}{\|\tilde{\mathbf{v}}(0)\|_2}\sin(\|\tilde{\mathbf{v}}(0)\|_2 t)\\ \tilde{\mathbf{v}}(t) &= -\tilde{\boldsymbol{\theta}}(0)\|\tilde{\mathbf{v}}(0)\|_2\sin(\|\tilde{\mathbf{v}}(0)\|_2 t) + \tilde{\mathbf{v}}(0)\cos(\|\tilde{\mathbf{v}}(0)\|_2 t)\end{aligned} \tag{2.35}$$

This solution defines an evolution, denoted as $g_t : (\boldsymbol{\theta}(0), \mathbf{v}(0)) \mapsto (\boldsymbol{\theta}(t), \mathbf{v}(t))$. Both (2.34) and (2.35) are symplectic. Due to the explicit formula for the geodesic flow on sphere, the second dynamics in (2.33b) is simulated exactly. Therefore, updating $\tilde{\boldsymbol{\theta}}$ does not involve discretization error so we can use large step sizes. This could lead to improved computational efficiency. Because this step is in fact a rotation on sphere, it can generate proposals that are far away from the current state. Algorithm 1 shows the steps for implementing this approach, henceforth called *Spherical HMC in Cartesian coordinates (c-SphHMC)*. It can be shown that the integrator in the algorithm has order 3 local error and order 2 global error (See the details in Appendix "Error Analysis of Spherical HMC").

Algorithm 1 Spherical HMC in Cartesian coordinates (c-SphHMC)

Initialize $\tilde{\boldsymbol{\theta}}^{(1)}$ at current $\tilde{\boldsymbol{\theta}}$ after transformation $T_{\mathscr{D}\to\mathscr{S}}$
Sample a new velocity value $\tilde{\mathbf{v}}^{(1)} \sim \mathscr{N}(\mathbf{0}, \mathbf{I}_{D+1})$
Set $\tilde{\mathbf{v}}^{(1)} \leftarrow \tilde{\mathbf{v}}^{(1)} - \tilde{\boldsymbol{\theta}}^{(1)}(\tilde{\boldsymbol{\theta}}^{(1)})^\mathsf{T}\tilde{\mathbf{v}}^{(1)}$
Calculate $H^*(\tilde{\boldsymbol{\theta}}^{(1)}, \tilde{\mathbf{v}}^{(1)}) = U(\boldsymbol{\theta}^{(1)}) + K(\tilde{\mathbf{v}}^{(1)})$
for $\ell = 1$ to L **do**

$\quad \tilde{\mathbf{v}}^{(\ell+\frac{1}{2})} = \tilde{\mathbf{v}}^{(\ell)} - \frac{\varepsilon}{2}\left(\begin{bmatrix}\mathbf{I}_D\\ \mathbf{0}^\mathsf{T}\end{bmatrix} - \tilde{\boldsymbol{\theta}}^{(\ell)}(\boldsymbol{\theta}^{(\ell)})^\mathsf{T}\right)\nabla_{\boldsymbol{\theta}} U(\boldsymbol{\theta}^{(\ell)})$

$\quad \tilde{\boldsymbol{\theta}}^{(\ell+1)} = \tilde{\boldsymbol{\theta}}^{(\ell)}\cos(\|\tilde{\mathbf{v}}^{(\ell+\frac{1}{2})}\|\varepsilon) + \frac{\tilde{\mathbf{v}}^{(\ell+\frac{1}{2})}}{\|\tilde{\mathbf{v}}^{(\ell+\frac{1}{2})}\|}\sin(\|\tilde{\mathbf{v}}^{(\ell+\frac{1}{2})}\|\varepsilon)$

$\quad \tilde{\mathbf{v}}^{(\ell+\frac{1}{2})} \leftarrow -\tilde{\boldsymbol{\theta}}^{(\ell)}\|\tilde{\mathbf{v}}^{(\ell+\frac{1}{2})}\|\sin(\|\tilde{\mathbf{v}}^{(\ell+\frac{1}{2})}\|\varepsilon) + \tilde{\mathbf{v}}^{(\ell+\frac{1}{2})}\cos(\|\tilde{\mathbf{v}}^{(\ell+\frac{1}{2})}\|\varepsilon)$

$\quad \tilde{\mathbf{v}}^{(\ell+1)} = \tilde{\mathbf{v}}^{(\ell+\frac{1}{2})} - \frac{\varepsilon}{2}\left(\begin{bmatrix}\mathbf{I}_D\\ \mathbf{0}^\mathsf{T}\end{bmatrix} - \tilde{\boldsymbol{\theta}}^{(\ell+1)}(\boldsymbol{\theta}^{(\ell+1)})^\mathsf{T}\right)\nabla_{\boldsymbol{\theta}} U(\boldsymbol{\theta}^{(\ell+1)})$

end for
Calculate $H^*(\tilde{\boldsymbol{\theta}}^{(L+1)}, \tilde{\mathbf{v}}^{(L+1)}) = U(\boldsymbol{\theta}^{(L+1)}) + K(\tilde{\mathbf{v}}^{(L+1)})$
Calculate the acceptance probability $\alpha = \min\{1, \exp[-H^*(\tilde{\boldsymbol{\theta}}^{(L+1)}, \tilde{\mathbf{v}}^{(L+1)}) + H^*(\tilde{\boldsymbol{\theta}}^{(1)}, \tilde{\mathbf{v}}^{(1)})]\}$
Accept or reject the proposal according to α for the next state $\tilde{\boldsymbol{\theta}}'$
Calculate $T_{\mathscr{S}\to\mathscr{D}}(\tilde{\boldsymbol{\theta}}')$ and the corresponding weight $|dT_{\mathscr{S}\to\mathscr{D}}|$

2.4.2.2 Spherical HMC in the Spherical Coordinates

Now we define HMC on the sphere $\mathscr{S}^D$ in the spherical coordinates $\{\boldsymbol{\theta}, \mathscr{R}_{\mathbf{0}}^D\}$. The natural metric on the sphere $\mathscr{S}^D$ induced by the coordinate mapping (2.7) is the *round spherical metric*,[1] $\mathbf{G}_{\mathscr{S}_r}(\boldsymbol{\theta}) = \operatorname{diag}[1, \sin^2(\theta_1), \ldots, \prod_{d=1}^{D-1} \sin^2(\theta_d)]$.

As in Sect. 2.4.2.1, we start with the usual Hamiltonian $H(\boldsymbol{\theta}, \mathbf{v})$ defined on $(\mathscr{R}_{\mathbf{0}}^D, \mathbf{I})$ as in (2.26) with $\mathbf{v} \in T_{\boldsymbol{\theta}}\mathscr{R}_{\mathbf{0}}^D$. Under the transformation $T_{\mathscr{R}_0 \to \mathscr{S}} : \boldsymbol{\theta} \mapsto \mathbf{x}$ in (2.7), Hamiltonian (2.26) on $\mathscr{R}_{\mathbf{0}}^D$ is changed to the following Hamiltonian $H(\mathbf{x}, \dot{\mathbf{x}})$ as in (2.22):

$$H(\mathbf{x}, \dot{\mathbf{x}}) = \phi(\mathbf{x}) + \frac{1}{2}\mathbf{v}^{\mathsf{T}}\mathbf{G}_{\mathscr{S}_c}(\boldsymbol{\theta})\mathbf{v} = U(\mathbf{x}) - \log\left|\frac{d\boldsymbol{\theta}_{\mathscr{R}_0}}{d\boldsymbol{\theta}_{\mathscr{S}_r}}\right| + \frac{1}{2}\mathbf{v}^{\mathsf{T}}\mathbf{G}_{\mathscr{S}_r}(\boldsymbol{\theta})\mathbf{v} \quad (2.36)$$

where the potential energy $U(\mathbf{x}) = U(\boldsymbol{\theta}(\mathbf{x}))$ and $\mathbf{G}_{\mathscr{S}_r}(\boldsymbol{\theta})$ is the round spherical metric.

As before, the logarithm of volume adjustment is $\log\left|\frac{d\boldsymbol{\theta}_{\mathscr{R}_0}}{d\boldsymbol{\theta}_{\mathscr{S}_r}}\right| = -\frac{1}{2}\log|\mathbf{G}_{\mathscr{S}_r}| = -\sum_{d=1}^{D-1}(D-d)\log\sin(\theta_d)$ (See Appendix "Round Metric in the Spherical Coordinates"). The last two terms in Eq. (2.36) is the minus log density of $\mathbf{v}|\boldsymbol{\theta} \sim \mathscr{N}(\mathbf{0}, \mathbf{G}_{\mathscr{S}_r}(\boldsymbol{\theta})^{-1})$. Again, for numerical stability we consider the following partial Hamiltonian $H^*(\mathbf{x}, \dot{\mathbf{x}})$ and leave the volume adjustment as weights (2.29) to adjust the estimation of integration (2.20):

$$H^*(\mathbf{x}, \dot{\mathbf{x}}) = U(\boldsymbol{\theta}) + \frac{1}{2}\mathbf{v}^{\mathsf{T}}\mathbf{G}_{\mathscr{S}_r}(\boldsymbol{\theta})\mathbf{v} \quad (2.37)$$

Taking derivative of $T_{\mathscr{R}_0 \to \mathscr{S}} : \boldsymbol{\theta} \mapsto \mathbf{x}$ in (2.7) with respect to time t we have

$$\dot{x}_d = \begin{cases} [-v_d \tan(\theta_d) + \sum_{i<d} v_i \cot(\theta_i)]x_d, & d < D+1 \\ \sum_{i<D+1} v_i \cot(\theta_i) x_{D+1}, & d = D+1 \end{cases} \quad (2.38)$$

We can show that $\mathbf{x}(\boldsymbol{\theta})^{\mathsf{T}}\dot{\mathbf{x}}(\boldsymbol{\theta}, \mathbf{v}) = 0$; that is, $\dot{\mathbf{x}} \in T_{\mathbf{x}}\mathscr{S}^D$. Taking derivative of $T_{\mathscr{S} \to \mathscr{R}_0} : \mathbf{x} \mapsto \boldsymbol{\theta}$ in (2.8) with respect to time t yields

$$v_d := \dot{\theta}_d = \begin{cases} -\frac{x_d}{\sqrt{1-\sum_{i=1}^{d} x_i^2}}\left[\frac{\dot{x}_d}{x_d} + \frac{\sum_{i=1}^{d-1} x_i \dot{x}_i}{1-\sum_{i=1}^{d-1} x_d^2}\right], & d < D \\ \frac{x_D \dot{x}_{D+1} - \dot{x}_D x_{D+1}}{x_D^2 + x_{D+1}^2}, & d = D \end{cases} \quad (2.39)$$

Further, we have $\mathbf{v}^{\mathsf{T}}\mathbf{G}_{\mathscr{S}_r}(\boldsymbol{\theta})\mathbf{v} = \dot{\mathbf{x}}^{\mathsf{T}}\dot{\mathbf{x}}$. Therefore, the partial Hamiltonian (2.37) can be recognized as the standard Hamiltonian (2.26) in the augmented $(D+1)$ dimensional space, which is again explained by the energy invariance Proposition 2.7.1 (See more details in Appendix "Spherical Geometry")

---

[1]Note, $\mathbf{v}^{\mathsf{T}}\mathbf{G}_{\mathscr{S}_r}(\boldsymbol{\theta})\mathbf{v} \le \|\mathbf{v}\|_2^2 \le \|\tilde{\mathbf{v}}\|_2^2 = \mathbf{v}^{\mathsf{T}}\mathbf{G}_{\mathscr{S}_c}(\boldsymbol{\theta})\mathbf{v}$.

$$H^*(\mathbf{x}, \dot{\mathbf{x}}) = U(\mathbf{x}) + K(\dot{\mathbf{x}}) = U(\mathbf{x}) + \frac{1}{2}\dot{\mathbf{x}}^{\mathsf{T}}\dot{\mathbf{x}} \tag{2.40}$$

Similar to the method discussed in Sect. 2.4.2.1, we split the Hamiltonian (2.37), $H^*(\tilde{\boldsymbol{\theta}}, \tilde{\mathbf{v}}) = U(\boldsymbol{\theta})/2 + \frac{1}{2}\mathbf{v}^{\mathsf{T}}\mathbf{G}_{\mathscr{S}_r}(\boldsymbol{\theta})\mathbf{v} + U(\boldsymbol{\theta})/2$, and its corresponding Lagrangian dynamics (2.31) as follows:

$$\begin{cases} \dot{\boldsymbol{\theta}} &= \mathbf{0} \\ \dot{\mathbf{v}} &= -\frac{1}{2}\mathbf{G}_{\mathscr{S}_r}(\boldsymbol{\theta})^{-1}\nabla_{\boldsymbol{\theta}}U(\boldsymbol{\theta}) \end{cases} \tag{2.41a}$$

$$\begin{cases} \dot{\boldsymbol{\theta}} &= \mathbf{v} \\ \dot{\mathbf{v}} &= -\mathbf{v}^{\mathsf{T}}\boldsymbol{\Gamma}_{\mathscr{S}_r}(\boldsymbol{\theta})\mathbf{v} \end{cases} \tag{2.41b}$$

The first dynamics (2.41a) involves updating the velocity $\mathbf{v}$ only. However, the diagonal term of $\mathbf{G}_{\mathscr{S}_r}(\boldsymbol{\theta})^{-1}$, $\prod_{i=1}^{d-1}\sin^{-2}(\theta_i)$ increases exponentially fast with respect to d. This will cause the velocity updated by (2.41a) to have extremely large components. To avoid such issue, we use small time vector $\boldsymbol{\varepsilon} = [\varepsilon, \varepsilon^2, \ldots, \varepsilon^D]$, instead of scalar ε, in updating Eq. (2.41a). The inhomogeneous discretization step sizes may not yield an accurate solution to (2.31), but they nevertheless provide a numerically stable proposal that is valid in the standard Metropolis–Hastings scheme. The second dynamics (2.41b) describes the same geodesic flow on the sphere $\mathscr{S}^D$ as (2.33b) but in the spherical coordinates $\{\boldsymbol{\theta}, \mathscr{R}_0^D\}$. Therefore it should have the same solution as (2.35) expressed in $\{\boldsymbol{\theta}, \mathscr{R}_0^D\}$. To obtain this solution, we first apply $\tilde{T}_{\mathscr{R}_0\to\mathscr{S}} : (\boldsymbol{\theta}(0), \mathbf{v}(0)) \mapsto (\mathbf{x}(0), \dot{\mathbf{x}}(0))$, which consists of (2.7), (2.38). Then, we use g_t in (2.35) to evolve $(\mathbf{x}(0), \dot{\mathbf{x}}(0))$ for some time t to find $(\mathbf{x}(t), \dot{\mathbf{x}}(t))$. Finally, we use $\tilde{T}_{\mathscr{S}\to\mathscr{R}_0} : (\mathbf{x}(t), \dot{\mathbf{x}}(t)) \mapsto (\boldsymbol{\theta}(t), \mathbf{v}(t))$, composite of (2.8), (2.39), to go back to $\mathscr{R}_0^D$.

Algorithm 2 Spherical HMC in the spherical coordinates (s-SphHMC)

Initialize $\boldsymbol{\theta}^{(1)}$ at current $\boldsymbol{\theta}$ after transformation $T_{\mathscr{D}\to\mathscr{S}}$
Sample a new velocity value $\mathbf{v}^{(1)} \sim \mathscr{N}(\mathbf{0}, \mathbf{I}_D)$
Set $v_d^{(1)} \leftarrow v_d^{(1)}\prod_{i=1}^{d-1}\sin^{-1}(\theta_i^{(1)}), d = 1, \ldots, D$
Calculate $H^*(\boldsymbol{\theta}^{(1)}, \mathbf{v}^{(1)}) = U(\boldsymbol{\theta}^{(1)}) + K(\mathbf{v}^{(1)})$
for $\ell = 1$ to L **do**
  $v_d^{(\ell+\frac{1}{2})} = v_d^{(\ell)} - \frac{\varepsilon^d}{2}\frac{\partial}{\partial\theta_d}U(\boldsymbol{\theta}^{(\ell)})\prod_{i=1}^{d-1}\sin^{-2}(\theta_i^{(\ell)}), d = 1, \ldots, D$
  $(\boldsymbol{\theta}^{(\ell+1)}, \mathbf{v}^{(\ell+\frac{1}{2})}) \leftarrow \tilde{T}_{\mathscr{S}\to\mathscr{R}_0} \circ g_\varepsilon \circ \tilde{T}_{\mathscr{R}_0\to\mathscr{S}}(\boldsymbol{\theta}^{(\ell)}, \mathbf{v}^{(\ell+\frac{1}{2})})$
  $v_d^{(\ell+1)} = v_d^{(\ell+\frac{1}{2})} - \frac{\varepsilon^d}{2}\frac{\partial}{\partial\theta_d}U(\boldsymbol{\theta}^{(\ell+1)})\prod_{i=1}^{d-1}\sin^{-2}(\theta_i^{(\ell+1)}), d = 1, \ldots, D$
end for
Calculate $H^*(\boldsymbol{\theta}^{(L+1)}, \mathbf{v}^{(L+1)}) = U(\boldsymbol{\theta}^{(L+1)}) + K(\mathbf{v}^{(L+1)})$
Calculate the acceptance probability $\alpha = \min\{1, \exp[-H^*(\boldsymbol{\theta}^{(L+1)}, \mathbf{v}^{(L+1)}) + H^*(\boldsymbol{\theta}^{(1)}, \mathbf{v}^{(1)})]\}$
Accept or reject the proposal according to α for the next state $\boldsymbol{\theta}'$
Calculate $T_{\mathscr{S}\to\mathscr{D}}(\boldsymbol{\theta}')$ and the corresponding weight $|dT_{\mathscr{S}\to\mathscr{D}}|$

Algorithm 2 summarizes the steps for this method, called *Spherical HMC in the spherical coordinates (s-SphHMC)*. In theory, the hyperrectangle $\mathscr{R}_{\mathbf{0}}^D$ can be used as a base type (as the unit ball $\mathscr{B}_{\mathbf{0}}^D(1)$ does) for general q-norm constraints for which s-SphHMC can be applied. This is because q-norm domain $\mathscr{Q}^D$ can be bijectively mapped to the hypercube $\mathscr{C}^D$, and thereafter to $\mathscr{R}_{\mathbf{0}}^D$. However the involved Jacobian matrix is rather complicated and s-SphHMC used in this way is not as efficient as c-SphHMC. Therefore, we use s-SphHMC only for box-type constraints.

2.4.3 Spherical LMC on Probability Simplex

A large class of statistical models involve defining probability distributions on the *simplex* Δ^K,

$$\Delta^K := \left\{ \boldsymbol{\pi} \in \mathbb{R}^D | \; \pi_k \geq 0, \sum_{k=1}^K \pi_d = 1 \right\} \tag{2.42}$$

As an example, we consider *latent Dirichlet allocation (LDA)* [10], which is a hierarchical Bayesian model commonly used to model document topics. This type of constraints can be viewed as a special case of the 1-norm constraint, discussed in Sect. 2.3.3.1, by identifying the first orthant (all positive components) with the others. Therefore, the underlying idea of c-SphHMC (algorithm 1) can be applied to generate samples $\{\boldsymbol{\theta}\}$ on the sphere $\mathscr{S}^{K-1}$. These samples can be transformed as $\{\boldsymbol{\theta}^2\}$ and mapped back to the simplex Δ^K.

More precisely, we should consider the following root mapping to transform Δ^K to its root space $\sqrt{\Delta}^K := \{\boldsymbol{\theta} \in \mathscr{S}^{K-1} | \theta_k \geq 0, \; \forall k = 1, \ldots, K\} \subset \mathscr{S}^{K-1}$ (i.e., the first orthant of the sphere $\mathscr{S}^{K-1}$)

$$T_{\Delta \to \sqrt{\Delta}} : \; \Delta^K \longrightarrow \sqrt{\Delta}^K, \quad \boldsymbol{\pi} \mapsto \boldsymbol{\theta} = \sqrt{\boldsymbol{\pi}} \tag{2.43}$$

Note, $\sqrt{\Delta}^K$ is introduced for the simplicity of discussion. The sampler is run on the whole sphere $\mathscr{S}^{K-1}$. Each sample on the sphere $\mathscr{S}^{K-1}$ will be mapped to a point in the simplex Δ^K via square map regardless of its corresponding orthant.

In what follows, we show that the canonical spherical metric $\mathbf{G}_{\mathscr{S}_c}(\boldsymbol{\theta})$ is the same as the Fisher metric on $\sqrt{\Delta}^K$ up to a constant. In this sense, it is more natural to define the sampling algorithms on the sphere $\mathscr{S}^{K-1}$ and it connects to Lagrangian Monte Carlo [35]. We start with the toy example discussed in [49]. Denote the observed data as $\mathbf{x} = \{x_i\}_{i=1}^N$, where each data point belongs to one of the K categories with probability $p(x_i = k|\boldsymbol{\pi}) = \pi_k$. We assume a Dirichlet prior on $\boldsymbol{\pi}$: $p(\boldsymbol{\pi}) \propto \prod_{k=1}^K \pi_k^{\alpha_k - 1}$. The posterior distribution is $p(\boldsymbol{\pi}|\mathbf{x}) \propto \prod_{k=1}^K \pi_k^{n_k + \alpha_k - 1}$, where $n_k = \sum_{i=1}^N I(x_i = k)$ counts the points x_i in category k. Denote $\mathbf{n} = [n_1, \ldots, n_K]^\mathsf{T}$ and $n := |\mathbf{n}| = \sum_{k=1}^K n_k$. For inference, we need to sample from the posterior distribution $p(\boldsymbol{\pi}|\mathbf{x})$ defined on the probability simplex.

Proposition 2.4.1 links the canonical spherical metric to the Fisher metric on $\sqrt{\Delta}^K$.

Proposition 2.4.1 *If* $\mathbf{n} \sim \text{Multinom}(n, \boldsymbol{\pi})$ *for* $\boldsymbol{\pi} \in \Delta^K$*, then the Fisher metric* $\mathbf{G}_{\sqrt{\Delta}}(\boldsymbol{\theta})$ *defined on* $\sqrt{\Delta}^K$ *has the following form:*

$$\mathbf{G}_{\sqrt{\Delta}}(\boldsymbol{\theta}) = 4n\mathbf{G}_{\mathscr{S}_c}(\boldsymbol{\theta}) \tag{2.44}$$

where $n = |\mathbf{n}|$.

Proof The Fisher metric on Δ^K is a function of $\boldsymbol{\pi}_{-K}$ (here, '$-K$' means all but the K-th components) and is calculated as follows:

$$\begin{aligned} \mathbf{G}_{\mathbf{F}}(\boldsymbol{\pi}_{-K}) &= -\mathbb{E}[\nabla^2 \log p(\mathbf{x}|\boldsymbol{\pi}_{-K})] \\ &= -\mathbb{E}[\nabla^2(\mathbf{n}_{-K}^{\mathsf{T}} \log(\boldsymbol{\pi}_{-K}) + (n - \mathbf{n}_{-K}^{\mathsf{T}}\mathbf{1}) \log(1 - \boldsymbol{\pi}_{-K}^{\mathsf{T}}\mathbf{1}))] \\ &= -\mathbb{E}[\nabla(\mathbf{n}_{-K}/\boldsymbol{\pi}_{-K} - \mathbf{1}(n - \mathbf{n}_{-K}^{\mathsf{T}}\mathbf{1})/(1 - \boldsymbol{\pi}_{-K}^{\mathsf{T}}\mathbf{1}))] \\ &= -\mathbb{E}[-\operatorname{diag}(\mathbf{n}_{-K}/\boldsymbol{\pi}_{-K}^2) - \mathbf{1}\mathbf{1}^{\mathsf{T}}(n - \mathbf{n}_{-K}^{\mathsf{T}}\mathbf{1})/(1 - \boldsymbol{\pi}_{-K}^{\mathsf{T}}\mathbf{1})^2] \\ &= n[\operatorname{diag}(1/\boldsymbol{\pi}_{-K}) + \mathbf{1}\mathbf{1}^{\mathsf{T}}/\pi_K] \end{aligned} \tag{2.45}$$

Now we use (2.43) to map the simplex to the sphere (the first orthant). Note that $\frac{d\boldsymbol{\pi}_{-K}}{d\boldsymbol{\theta}_{-K}^{\mathsf{T}}} = 2\operatorname{diag}(\boldsymbol{\theta}_{-K})$. Therefore, by coordinate transformation, Fisher metric on $\sqrt{\Delta}^K$ can be written as follows:

$$\mathbf{G}_{\sqrt{\Delta}}(\boldsymbol{\theta}) = \frac{d\boldsymbol{\pi}_{-K}^{\mathsf{T}}}{d\boldsymbol{\theta}_{-K}}\mathbf{G}_{\mathbf{F}}(\boldsymbol{\pi}_{-K})\frac{d\boldsymbol{\pi}_{-K}}{d\boldsymbol{\theta}_{-K}^{\mathsf{T}}} = 4n[\mathbf{I}_{K-1} + \boldsymbol{\theta}_{-K}\boldsymbol{\theta}_{-K}^{\mathsf{T}}/\theta_K^2] = 4n\mathbf{G}_{\mathscr{S}_c}(\boldsymbol{\theta}) \tag{2.46}$$

□

Remark 2.1 The scalar $4n$ properly scales the spherical metric in high-dimensional data intensive models. In LDA particularly, n could be the number of words counted in the selected documents. Hence, we use $\mathbf{G}_{\sqrt{\Delta}}(\boldsymbol{\theta})$ instead of $\mathbf{G}_{\mathscr{S}_c}(\boldsymbol{\theta})$.

Recall that in the development of Spherical HMC algorithms, we decided to omit the log volume adjustment term $\log\left|\frac{d\boldsymbol{\beta}_{\mathscr{D}}}{d\boldsymbol{\theta}_{\mathscr{S}}}\right|$, in the partial Hamiltonian (2.28) and (2.37), and regard it as the weight to adjust the estimate of (2.20) or resample. Although a similar numerical issue posed by the volume term still exists here, it is much alleviated by the scaling term $4n$. More importantly, afterward reweighting is not feasible if the LDA model is going to be used in an online setting. Therefore, we use $\phi(\boldsymbol{\theta})$ in (2.21), as opposed to $U(\boldsymbol{\theta})$ to avoid the reweighting step. In this case, the natural gradient in (2.34) for updating velocity becomes

$$\phi(\boldsymbol{\theta}) = U(\boldsymbol{\pi}(\boldsymbol{\theta})) - \log\left|\frac{d\boldsymbol{\pi}}{d\boldsymbol{\theta}}\right| = -2(\mathbf{n}+\alpha-0.5)^{\mathsf{T}}\log|\boldsymbol{\theta}|$$
$$\begin{bmatrix}\mathbf{I}_{K-1}\\ -\boldsymbol{\theta}_{-K}^{\mathsf{T}}/\theta_K\end{bmatrix}\mathbf{G}_{\sqrt{\Delta}}(\boldsymbol{\theta})^{-1}\nabla_{\boldsymbol{\theta}_{-K}}\phi(\boldsymbol{\theta}_{-K}) = [(\mathbf{n}+\alpha-0.5)/\boldsymbol{\theta} - \boldsymbol{\theta}*|\mathbf{n}+\alpha-0.5|](2n)^{-1} \tag{2.47}$$

We refer to the resulting method as *Spherical Lagrangian Monte Carlo (SphLMC)* which is summarized in Algorithm 3.

Algorithm 3 Spherical LMC for Simplex (SphLMC)

Initialize $\boldsymbol{\theta}^{(1)}$ at current $\boldsymbol{\theta} = \sqrt{\boldsymbol{\pi}}$
Sample a new velocity value $\mathbf{v}^{(1)} \sim \mathcal{N}(\mathbf{0}, \mathbf{I}_K)$
Set $\mathbf{v}^{(1)} \leftarrow \mathbf{v}^{(1)} - \boldsymbol{\theta}^{(1)}(\boldsymbol{\theta}^{(1)})^{\mathsf{T}}\mathbf{v}^{(1)}$
Calculate $H(\boldsymbol{\theta}^{(1)}, \mathbf{v}^{(1)}) = \phi(\boldsymbol{\theta}^{(1)}) + K(\mathbf{v}^{(1)})$
for $\ell = 1$ to L **do**

$$\mathbf{v}^{(\ell+\frac{1}{2})} = \mathbf{v}^{(\ell)} - \frac{\varepsilon}{2}\begin{bmatrix}\mathbf{I}_{K-1}\\ -(\boldsymbol{\theta})^{(\ell)\mathsf{T}}_{-K}/\theta_K^{(\ell)}\end{bmatrix}\mathbf{G}_{\sqrt{\Delta}}(\boldsymbol{\theta}^{(\ell)})^{-1}\nabla_{\boldsymbol{\theta}_{-K}}\phi(\boldsymbol{\theta}_{-K}^{(\ell)})$$
$$\boldsymbol{\theta}^{(\ell+1)} = \boldsymbol{\theta}^{(\ell)}\cos(\|\mathbf{v}^{(\ell+\frac{1}{2})}\|\varepsilon) + \frac{\mathbf{v}^{(\ell+\frac{1}{2})}}{\|\mathbf{v}^{(\ell+\frac{1}{2})}\|}\sin(\|\mathbf{v}^{(\ell+\frac{1}{2})}\|\varepsilon)$$
$$\mathbf{v}^{(\ell+\frac{1}{2})} \leftarrow -\boldsymbol{\theta}^{(\ell)}\|\mathbf{v}^{(\ell+\frac{1}{2})}\|\sin(\|\mathbf{v}^{(\ell+\frac{1}{2})}\|\varepsilon) + \mathbf{v}^{(\ell+\frac{1}{2})}\cos(\|\mathbf{v}^{(\ell+\frac{1}{2})}\|\varepsilon)$$
$$\mathbf{v}^{(\ell+1)} = \mathbf{v}^{(\ell+\frac{1}{2})} - \frac{\varepsilon}{2}\begin{bmatrix}\mathbf{I}_{K-1}\\ -(\boldsymbol{\theta})^{(\ell+1)\mathsf{T}}_{-K}/\theta_K^{(\ell+1)}\end{bmatrix}\mathbf{G}_{\sqrt{\Delta}}(\boldsymbol{\theta}^{(\ell+1)})^{-1}\nabla_{\boldsymbol{\theta}_{-K}}\phi(\boldsymbol{\theta}_{-K}^{(\ell+1)})$$

end for
Calculate $H(\boldsymbol{\theta}^{(L+1)}, \mathbf{v}^{(L+1)}) = \phi(\boldsymbol{\theta}^{(L+1)}) + K(\mathbf{v}^{(L+1)})$
Calculate the acceptance probability $\alpha = \min\{1, \exp[-H(\boldsymbol{\theta}^{(L+1)}, \mathbf{v}^{(L+1)}) + H(\boldsymbol{\theta}^{(1)}, \mathbf{v}^{(1)})]\}$
Accept or reject the proposal according to α for the next state $\boldsymbol{\theta}'$ and $\boldsymbol{\pi}' = (\boldsymbol{\theta}')^2$

To illustrate our proposed method, we consider the toy example discussed above. For this problem, [49] propose a Riemannian Langevin Dynamics (RLD) method, but use an expanded-mean parametrization to map the simplex to the whole space. As mentioned above, this approach (i.e., expanding the parameter space) might not be efficient in general. This is illustrated in Fig. 2.6. Here, we set $\alpha = 0.5$ and run RMW, WallHMC, RLD, and SphLMC for 1.1×10^5 iterations; we discard the first 10^4 samples. As we can see in Fig. 2.6, with similar acceptance rates,[2] RLD takes longer time to reach the high density region (upper left), while SphLMC reaches the region almost immediately (upper right). Compared to alternative algorithms, SphLMC method provides better probability estimates (lower left). Further, SphLMC generates samples with a substantially lower autocorrelation (lower right).

[2]Metropolis test is done for this toy example as in [49], but will be omitted in stochastic mini-batch algorithms in Sect. 2.5.5.

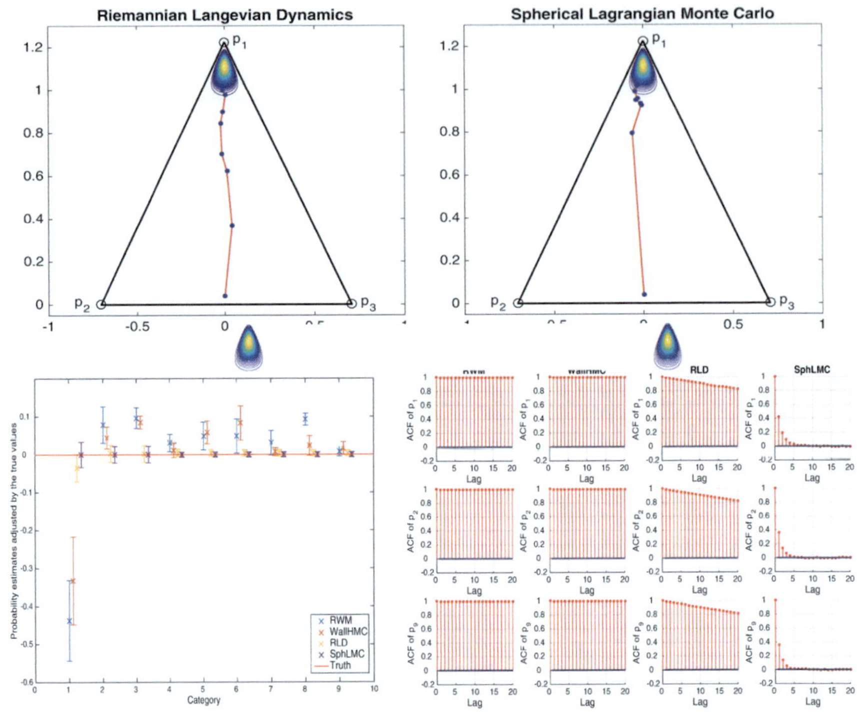

Fig. 2.6 Dirichlet-Multinomial model: first 10 steps by RLD (*upper left*), first 10 steps by SphLMC (*upper right*), probability estimates (*lower left*) and autocorrelation function for MCMC samples (*lower right*)

2.5 Experimental Results

In this section, we evaluate our proposed methods using simulated and real data. To this end, we compare their efficiency to that of RWM, Wall HMC, exact HMC [47], and the Riemannian Langevin dynamics (RLD) algorithm proposed by [49] for LDA. We define efficiency in terms of time-normalized effective sample size (ESS). Given N MCMC samples, for each parameter, we define ESS $= N[1 + 2\Sigma_{k=1}^{K} \rho(k)]^{-1}$, where $\rho(k)$ is sample autocorrelation with lag k [26]. We use the minimum ESS normalized by the CPU time, s (in seconds), as the overall measure of efficiency: min(ESS)/s. All computer codes are available online at http://www.ics.uci.edu/~slan/SphHMC/Spherical_Augmentation.html or at http://bitbucket.org/lanzithinking/sphericalaugmentation.

2.5.1 Truncated Multivariate Gaussian

For illustration purpose, we start with a truncated bivariate Gaussian distribution

$$\begin{pmatrix}\beta_1\\ \beta_2\end{pmatrix} \sim \mathcal{N}\left(\mathbf{0}, \begin{bmatrix}1 & 0.5\\ 0.5 & 1\end{bmatrix}\right), \qquad 0 \leq \beta_1 \leq 5, \quad 0 \leq \beta_2 \leq 1$$

This is box-type constraint with the lower and upper limits as $\mathbf{l} = (0, 0)$ and $\mathbf{u} = (5, 1)$ respectively. The original rectangle domain can be mapped to 2D unit disc $\mathscr{B}^2_{\mathbf{0}}(1)$ to use c-SphHMC, or mapped to 2D rectangle $\mathscr{R}^2_{\mathbf{0}}$ where s-SphHMC can be directly applied.

The upper leftmost panel of Fig. 2.7 shows the heatmap based on the exact density function, and the other panels show the corresponding heatmaps based on MCMC samples from RWM, Wall HMC, exact HMC, c-SphHMC and s-SphHMC respectively. All algorithms generate probability density estimates that visually match the true density. Table 2.1 compares the true mean and covariance of the above truncated bivariate Gaussian distribution with the point estimates using 2×10^5 (2×10^4 for each of 10 repeated experiments with different random seeds) MCMC samples in each method. Overall, all methods estimate the mean and covariance reasonably well.

To evaluate the efficiency of the above-mentioned methods, we repeat this experiment for higher dimensions, $D = 10$, and $D = 100$. As before, we set the mean to zero and set the (i, j)-th element of the covariance matrix to $\Sigma_{ij} = 1/(1 + |i - j|)$. Further, we impose the following constraints on the parameters,

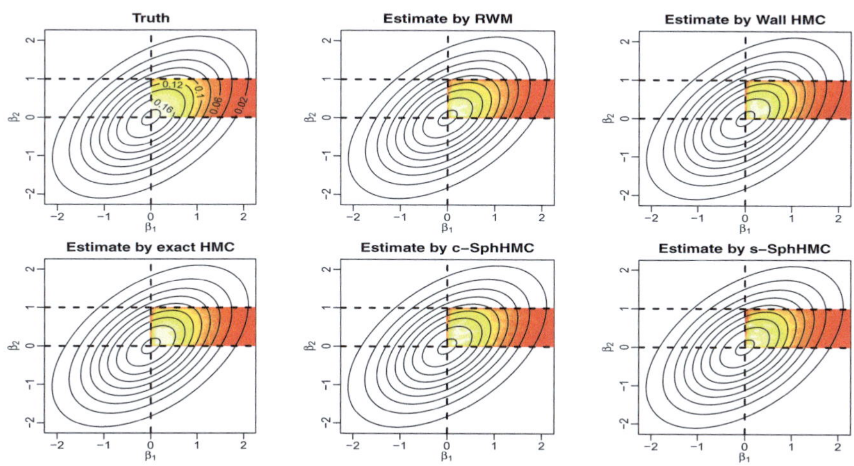

Fig. 2.7 Density plots of a truncated bivariate Gaussian using exact density function (*upper left most*) and MCMC samples from RWM, Wall HMC, exact HMC, c-SphHMC and s-SphHMC respectively. *Solid elliptical curves* always show true unconstrained probability density contours. *Dashed lines* define linear constrained domains. Colored heatmaps indicate constrained probability density based on truth or estimation from MCMC samples

Table 2.1 Comparing the point estimates for the mean and covariance of a bivariate truncated Gaussian distribution using RWM, Wall HMC, exact HMC, c-SphHMC and s-SphHMC

Method	Mean	Covariance
Truth	$\begin{bmatrix} 0.7906 \\ 0.4889 \end{bmatrix}$	$\begin{bmatrix} 0.3269 & 0.0172 \\ 0.0172 & 0.08 \end{bmatrix}$
RWM	$\begin{bmatrix} 0.7796 \pm 0.0088 \\ 0.4889 \pm 0.0034 \end{bmatrix}$	$\begin{bmatrix} 0.3214 \pm 0.009 & 0.0158 \pm 0.001 \\ 0.0158 \pm 0.001 & 0.0798 \pm 5e{-}04 \end{bmatrix}$
Wall HMC	$\begin{bmatrix} 0.7875 \pm 0.0049 \\ 0.4884 \pm 8e{-}04 \end{bmatrix}$	$\begin{bmatrix} 0.3242 \pm 0.0043 & 0.017 \pm 0.001 \\ 0.017 \pm 0.001 & 0.08 \pm 3e{-}04 \end{bmatrix}$
exact HMC	$\begin{bmatrix} 0.7909 \pm 0.0025 \\ 0.4885 \pm 0.001 \end{bmatrix}$	$\begin{bmatrix} 0.3272 \pm 0.0026 & 0.0174 \pm 7e{-}04 \\ 0.0174 \pm 7e{-}04 & 0.08 \pm 3e{-}04 \end{bmatrix}$
c-SphHMC	$\begin{bmatrix} 0.79 \pm 0.005 \\ 0.4864 \pm 0.0016 \end{bmatrix}$	$\begin{bmatrix} 0.3249 \pm 0.0045 & 0.0172 \pm 0.0012 \\ 0.0172 \pm 0.0012 & 0.0801 \pm 0.001 \end{bmatrix}$
s-SphHMC	$\begin{bmatrix} 0.7935 \pm 0.0093 \\ 0.4852 \pm 0.003 \end{bmatrix}$	$\begin{bmatrix} 0.3233 \pm 0.0062 & 0.0202 \pm 0.0018 \\ 0.0202 \pm 0.0018 & 0.0791 \pm 9e{-}04 \end{bmatrix}$

$$0 \le \beta_i \le u_i$$

where u_i (i.e., the upper bound) is set to 5 when $i = 1$; otherwise, it is set to 0.5.

For each method, we obtain 10^5 MCMC samples after discarding the initial 10^4 samples. We set the tuning parameters of algorithms such that their overall acceptance rates are within a reasonable range. As shown in Table 2.2, Spherical HMC algorithms are substantially more efficient than RWM and Wall HMC. For RWM, the proposed states are rejected about 95 % of times due to violation of the constraints. On average, Wall HMC bounces off the wall around 3.81 ($L = 2$) and 6.19 ($L = 5$) times per iteration for $D = 10$ and $D = 100$ respectively. Exact HMC is quite efficient for relatively low dimensional truncated Gaussian ($D = 10$); however it becomes very slow for higher dimensions ($D = 100$). In contrast, by augmenting the parameter space, Spherical HMC algorithms handle the constraints in a more efficient way. Since s-SphHMC is more suited for box-type constraints, it is substantially more efficient than c-SphHMC in this example.

2.5.2 *Bayesian Lasso*

In regression analysis, overly complex models tend to overfit the data. Regularized regression models control complexity by imposing a penalty on model parameters. By far, the most popular model in this group is *Lasso* (least absolute shrinkage and selection operator) proposed by [61]. In this approach, the coefficients are obtained

Table 2.2 Comparing the efficiency of RWM, Wall HMC, exact HMC, c-SphHMC and s-SphHMC in terms of sampling from truncated Gaussian distributions

Dimension	Method	AP[a]	s/iter[b]	ESS(min,med,max)[c]	Min(ESS)/s[d]	Speedup
D = 10	RWM	0.62	5.72E-05	(48, 691, 736)	7.58	1.00
	Wall HMC	0.83	1.19E-04	(31904, 86275, 87311)	2441.72	322.33
	Exact HMC	1.00	7.60E-05	(1e+05, 1e+05, 1e+05)	11960.29	1578.87
	c-SphHMC	0.82	2.53E-04	(62658, 85570, 86295)	2253.32	297.46
	s-SphHMC	0.79	2.02E-04	(76088, 1e+05, 1e+05)	3429.56	452.73
D = 100	RWM	0.81	5.45E-04	(1, 4, 54)	0.01	1.00
	Wall HMC	0.74	2.23E-03	(17777, 52909, 55713)	72.45	5130.21
	Exact HMC	1.00	4.65E-02	(97963, 1e+05, 1e+05)	19.16	1356.64
	c-SphHMC	0.73	3.45E-03	(55667, 68585, 72850)	146.75	10390.94
	s-SphHMC	0.87	2.30E-03	(74476, 99670, 1e+05)	294.31	20839.43

[a] Acceptance probability
[b] Seconds per iteration
[c] (minimum, median, maximum) effective sample size
[d] Minimal ESS per second

by minimizing the residual sum of squares (RSS) subject to a constraint on the magnitude of regression coefficients,

$$\min_{\|\beta\|_1 \le t} \mathrm{RSS}(\beta), \qquad \mathrm{RSS}(\beta) := \sum_i (y_i - \beta_0 - x_i^\mathsf{T}\beta)^2 \tag{2.48}$$

One could estimate the parameters by solving the following optimization problem:

$$\min_{\beta,\lambda} \mathrm{RSS}(\beta) + \lambda\|\beta\|_1 \tag{2.49}$$

where $\lambda \ge 0$ is the regularization parameter. [30, 48] have proposed a Bayesian alternative method, called Bayesian Lasso, where the penalty term is replaced by a prior distribution of the form $P(\beta) \propto \exp(-\lambda|\beta|)$, which can be represented as a scale mixture of normal distributions [66]. This leads to a hierarchical Bayesian model with full conditional conjugacy; therefore, the Gibbs sampler can be used for inference.

Our proposed spherical augmentation in this chapter can directly handle the constraints in Lasso models. That is, we can *conveniently* use Gaussian priors for model parameters, $\beta|\sigma^2 \sim \mathcal{N}(0, \sigma^2 I)$, and let the sampler *automatically* handle the constraint. In particular, c-SphHMC can be used to sample posterior distribution of β with the 1-norm constraint. For this problem, we modify the Wall HMC algorithm, which was originally proposed for box-type constraints [46]. See Appendix "Bounce in Diamond: Wall HMC for 1-Norm Constraint" for more details.

We evaluate our method based on the diabetes data set (N = 442, D = 10) discussed in [48]. Figure 2.8 compares coefficient estimates given by the Gibbs sampler [48], Wall HMC, and Spherical HMC, respectively, as the shrinkage factor $s := \|\hat{\beta}^{\mathrm{Lasso}}\|_1 / \|\hat{\beta}^{\mathrm{OLS}}\|_1$ changes from 0 to 1. Here, $\hat{\beta}^{\mathrm{OLS}}$ denotes the estimates obtained by ordinary least squares (OLS) regression. For the Gibbs sampler, we

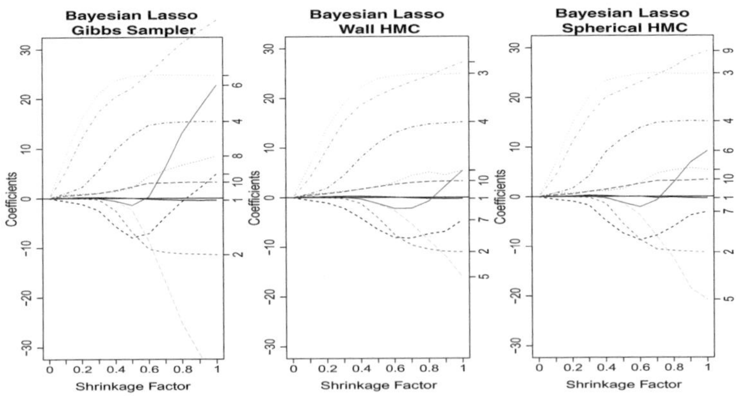

Fig. 2.8 Bayesian Lasso using three different sampling algorithms: Gibbs sampler (*left*), Wall HMC (*middle*) and Spherical HMC (*right*)

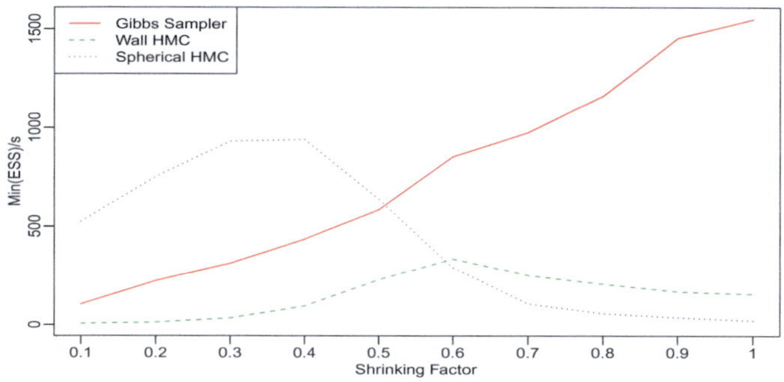

Fig. 2.9 Sampling efficiency of different algorithms for Bayesian Lasso based on the diabetes dataset

choose different λ so that the corresponding shrinkage factor s varies from 0 to 1. For Wall HMC and Spherical HMC, we fix the number of leapfrog steps to 10 and set the trajectory length such that they both have comparable acceptance rates around 70 %.

Figure 2.9 compares the sampling efficiency of these three methods. As we impose tighter constraints (i.e., lower shrinkage factors s), Spherical HMC becomes substantially more efficient than the Gibbs sampler and Wall HMC.

2.5.3 Bridge Regression

The Lasso model discussed in the previous section is in fact a member of a family of regression models called *Bridge regression* [24], where the coefficients are obtained by minimizing the residual sum of squares subject to a constraint on the magnitude of regression coefficients as follows:

$$\min_{\|\beta\|_q \leq t} \text{RSS}(\beta), \qquad \text{RSS}(\beta) := \sum_i (y_i - \beta_0 - x_i^\mathsf{T}\beta)^2 \tag{2.50}$$

For Lasso, $q = 1$, which allows the model to force some of the coefficients to become exactly zero (i.e., become excluded from the model). When $q = 2$, this model is known as *ridge regression*. Bridge regression is more flexible by allowing different q-norm constraints for different effects on shrinking the magnitude of parameters (See Fig. 2.10).

While the Gibbs sampler method of [30, 48] is limited to Lasso, our approach can be applied to all bridge regression models with different q. To handle the general q-norm constraint, one can map the constrained domain to the unit ball by (2.12) and apply c-SphHMC. Figure 2.10 compares the parameter estimates of Bayesian Lasso to the estimates obtained from two Bridge regression models with $q = 1.2$ and $q = 0.8$ for the diabetes dataset [48] using our Spherical HMC algorithm. As

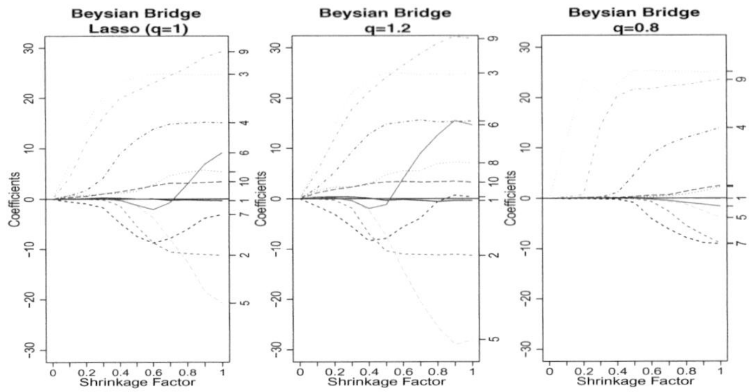

Fig. 2.10 Bayesian Bridge Regression by Spherical HMC: Lasso (q = 1, *left*), q = 1.2 (*middle*), and q = 0.8 (*right*)

expected, tighter constraints (e.g., $q = 0.8$) would lead to faster shrinkage of regression parameters as we decrease s.

2.5.4 Reconstruction of Quantized Stationary Gaussian Process

We now investigate the example of reconstructing quantized stationary Gaussian process discussed in [47]. Suppose we are given N values of a function $f(x_i), i = 1, \ldots, N$, which takes discrete values from $\{q_k\}_{k=1}^K$. We assume that this is a quantized projection of a sample $y(x_i)$ from a stationary Gaussian process with a known translation-invariant covariance kernel of the form $\Sigma_{ij} = K(|x_i - x_j|)$, and the quantization follows a known rule of the form

$$f(x_i) = q_k, \quad \text{if } z_k \leq y(x_i) < z_{k+1} \tag{2.51}$$

The objective is to sample from the posterior distribution

$$p(y(x_1), \ldots, y(x_N) | f(x_1), \ldots, f(x_N)) \sim \mathcal{N}(0, \Sigma) \quad \text{trunctated by rule (51)} \tag{2.52}$$

In this example, the function is sampled from a Gaussian process with the following kernel:

$$K(|x_i - x_j|) = \sigma^2 \exp\left\{-\frac{|x_i - x_j|^2}{2\eta^2}\right\}, \quad \sigma^2 = 0.6, \ \eta^2 = 0.2$$

We sample $N = 100$ points of $\{y(x_i)\}$ and quantize them with

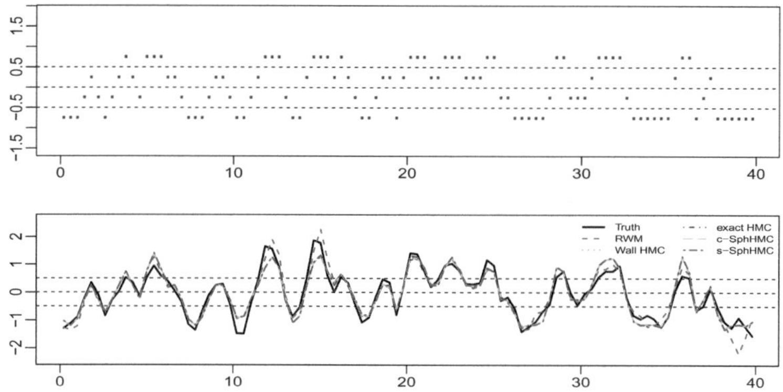

Fig. 2.11 Quantized stationary Gaussian process (*upper*) and the estimates of the process (*lower*)

$$q_1 = -0.75,\ q_2 = -0.25,\ q_3 = 0.25,\ q_4 = 0.75,$$
$$z_1 = -\infty,\ z_2 = -0.5,\ z_3 = 0,\ z_4 = 0.5,\ z_5 = +\infty$$

This example involves two types of constraints: box-type (two sided) constraints and one-sided constraints. In implementing our Spherical HMC algorithms, we transform the subspace formed by components with both finite lower and upper limits into unit ball and map the subspace formed by components with one-sided constraints to the whole space using absolute value (discussed at the end of Sect. 2.3).

Figure 2.11 shows the quantized Gaussian process (upper) and the estimates (lower) with 10^5 samples given by different MCMC algorithms. Overall, all the methods recover the truth well. Table 2.3 summarizes the efficiency of sampling 1.1×10^5 and burning the first 10^4 with RWM, Wall HMC, exact HMC, c-SphHMC and s-SphHMC. Exact HMC generates more effective samples but takes much longer time even though implemented in C. Spherical HMC algorithms outperform it in terms of time-normalized ESS. Interestingly, Wall HMC performs well in this example, even better than exact HMC and c-SphHMC.

Table 2.3 Comparing efficiency of RWM, Wall HMC, exact HMC, c-SphHMC and s-SphHMC in reconstructing a quantized stationary Gaussian process

Method	AP[a]	s/iter[b]	ESS(min,med,max)[c]	Min(ESS)/s[d]	Speedup
RWM	0.70	7.11E-05	(2,9,35)	0.22	1.00
Wall HMC	0.69	9.94E-04	(12564, 24317, 43876)	114.92	534.48
Exact HMC	1.00	1.00E-02	(72074, 1e+05, 1e+05)	65.31	303.76
c-SphHMC	0.72	1.73E-03	(13029, 26021, 56445)	68.44	318.32
s-SphHMC	0.80	1.09E-03	(14422, 31182, 81948)	120.59	560.86

[a] Acceptance probability
[b] Seconds per iteration
[c] (minimum, median, maximum) effective sample size
[d] Minimal ESS per second

2.5.5 Latent Dirichlet Allocation on Wikipedia Corpus

LDA [10] is a popular hierarchical Bayesian model for topic modeling. The model consists of K topics with probabilities $\{\pi_k\}$ drawn from a symmetric Dirichlet prior $\text{Dir}(\beta)$. A document d is modeled by a mixture of topics, with mixing proportions $\eta_d \sim \text{Dir}(\alpha)$. Document d is assumed to be generated by i.i.d. sampling of a topic assignment, z_{di}, from η_d for each word w_{di} in the document, and then drawing the word w_{di} from the assigned topic with probability $\pi_{z_{di}}$ [49]. Reference [60] integrate out η analytically to obtain the following semi-collapsed distribution:

$$p(w, z, \pi|\alpha, \beta) = \prod_{d=1}^{D} \frac{\Gamma(K\alpha)}{\Gamma(K\alpha + n_{d\cdot\cdot})} \prod_{k=1}^{K} \frac{\Gamma(\alpha + n_{dk\cdot})}{\Gamma(\alpha)} \prod_{k=1}^{K} \frac{\Gamma(W\beta)}{\Gamma(\beta)^W} \prod_{w=1}^{W} \pi_{kw}^{\beta + n_{\cdot kw} - 1} \tag{2.53}$$

where $n_{dkw} = \sum_{i=1}^{N_d} \delta(w_{di} = w, z_{di} = k)$. Here, "·" denotes the summation over the corresponding index. Given π, the documents are i.i.d so the above equation can be factorized as follows [49]:

$$\begin{aligned} p(w, z, \pi|\alpha, \beta) &= p(\pi|\beta) \prod_{d=1}^{D} p(w_d, z_d|\alpha, \pi) \\ p(w_d, z_d|\alpha, \pi) &= \prod_{k=1}^{K} \frac{\Gamma(\alpha + n_{dk\cdot})}{\Gamma(\alpha)} \prod_{w=1}^{W} \pi_{kw}^{n_{dkw}} \end{aligned} \tag{2.54}$$

To evaluate our proposed methods, we compare them with the state-of-the-art method of [49]. Their approach, called stochastic gradient Riemannian Langevin dynamics (sg-RLD) is an extension of the stochastic gradient Langevin dynamics (SGLD) proposed by [65]. Because this approach uses mini-batches of data to approximate the gradient and omits the accept/reject step of Metropolis-Hastings while decreasing the step size, we follow the same procedure to make our methods comparable. Further, because Langevin dynamics can be regarded as a single step Hamiltonian dynamics [46], we set $L = 1$. We refer the resulting algorithms as sg-SphHMC and sg-SphLMC, which are modified versions of our SphHMC and SphLMC algorithms. sg-SphLMC uses the following stochastic (natural) gradient (gradient preconditioned with metric) derived by setting $\boldsymbol{\theta} = \{\theta_{kw}\}$, $\mathbf{n} = \{n_{kw}^*\}$, and $\boldsymbol{\alpha} = \beta$ in Eq. (2.47)

$$\begin{aligned} g_{kw} &= [(n_{kw}^* + \beta - 1/2)/\theta_{kw} + \theta_{kw}(n_{k\cdot}^* + W(\beta - 1/2))]/(2 * n_{k\cdot}^*) \\ n_{kw}^* &= \frac{|D|}{|D_t|} \sum_{d \in D_t} \mathbb{E}_{z_d|w_d,\theta,\alpha}[n_{dkw}] \end{aligned} \tag{2.55}$$

where $1/2$ comes from the logarithm of volume adjustment. For sg-SphHMC, the exact Hamiltonian (2.27), as opposed to the partial Hamiltonian (2.28), is used so that the associated Hamiltonian dynamics preserves the correct target distribution. Therefore, sg-SphHMC only differs from sg-SphLMC in terms of missing the scaling term $4n$ in Eq. (2.47) and the stochastic gradient for sg-SphHMC is $4g_{kw}n^*_{k\cdot}$. (See Sect. 2.4.3) The expectation in Eq. (2.55) is calculated using Gibbs sampling on the topic assignment in each document separately, given the conditional distributions [49]

$$p(z_{di} = k|w_d, \theta, \alpha) = \frac{(\alpha + n^{\setminus i}_{dk\cdot})\pi_{kw_{di}}}{\sum_k(\alpha + n^{\setminus i}_{dk\cdot})\pi_{kw_{di}}} \tag{2.56}$$

where $\setminus i$ means a count excluding the topic assignment variable currently being updated. Step size is decreased according to $\varepsilon_t = a(1 + t/b)^{-c}$.

We use perplexity [49, 62] to compare the predictive performance of different methods in terms of the probability they assign to unseen data,

$$\begin{aligned} \mathrm{perp}(w_d|\mathscr{W}, \alpha, \beta) &= \exp\left\{-\sum_{i=1}^{n_{d\cdot\cdot}} \log p(w_{di}|\mathscr{W}, \alpha, \beta)/n_{d\cdot\cdot}\right\} \\ p(w_{di}|\mathscr{W}, \alpha, \beta) &= \mathbb{E}_{\eta_d,\pi}[\sum_k \eta_{dk}\pi_{kw_{di}}] \end{aligned} \tag{2.57}$$

where $\mathscr{W}$ is the training set and w_d is the hold-out sample. More specifically, we use the document completion approach [62], which partitions the test document w_d into two sets, w_d^{train} and w_d^{test}; we then use w_d^{train} to estimate η_d for the test document and use w_d^{test} to calculate perplexity.

We train the model online using 50000 documents randomly downloaded from Wikipedia with the vocabulary of approximately 8000 words created from Project Gutenberg texts [31]. The perplexity is evaluated on 1000 held-out documents. A mini-batch of 50 documents is used for updating the natural gradient for 4 algorithms: sg-RLD, sg-wallLMC,[3] sg-SphHMC and sg-SphLMC.

Figure 2.12 compares the above methods in terms of their perplexities. For each method, we show the best performance over different settings (Settings for best performance are listed in Table 2.4.). Both sg-wallLMC and sg-SphLMC have lower perplexity than sg-RLD at early stage, when relatively a small number of documents are used for training; as the number of training documents increases, the methods reach the same level of performance. As expected, sg-SphHMC does not perform well due to the absence of a proper scaling provided by the Fisher metric.

[3]The stochastic gradient for sg-wallLMC is $[(n^*_{kw} + \beta - 1/2) + \pi_{kw}(n^*_{k\cdot} + W(\beta - 1/2))]/n^*_{k\cdot}$ calculated with (2.45).

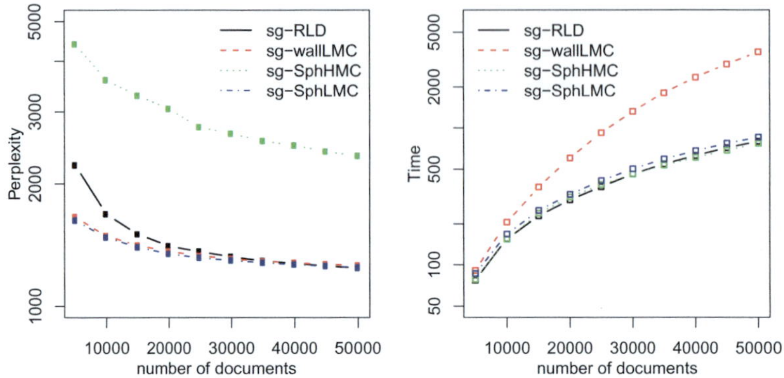

Fig. 2.12 Test-set perplexity and computation time (in log scale) based on the Wikipedia corpus

Table 2.4 Parameter settings for best performance in Wikipedia experiment

Algorithm	a	b	c	α	β	K	Gibbs samples
sg-RLD	0.01	1000	0.6	0.01	0.5000	100	100
sg-wallLMC	0.20	1000	2.0	0.01	0.5000	100	100
sg-SphHMC	0.01	1000	0.6	0.01	0.0100	100	100
sg-SphLMC	0.25	1000	1.5	0.01	0.5000	100	100

2.6 Discussion

We have introduced a new approach, *Spherical Augmentation*, for sampling from constrained probability distributions. This method maps the constrained domain to a sphere in an augmented space. Sampling algorithms can freely explore the surface of sphere to generate samples that remain within the constrained domain when mapped back to the original space. This way, our proposed method provides a mathematically natural and computationally efficient framework that can be applied to a wide range of statistical inference problems with norm constraints.

The augmentation approach proposed here is based on the change of variables theorem. We augment the original D-dimensional space with one extra dimension by either inserting slack variables (c-SphHMC) or using embedding map (s-SphHMC), The augmented Hamiltonian is the same under different representations (2.30), (2.40) due to the mathematical fact that the energy is invariant to the choice of coordinates (Proposition 2.7.1). To account for the change of geometry, a volume adjustment term needs to be used, either as a weight after obtaining all the samples (SphHMC) or as an added term to the total energy (SphLMC).

Our proposed scheme takes advantage of the splitting strategy to further improve computational efficiency. We split the Lagrangian dynamics and update velocity in the tangent space, rather than momentum in the cotangent space. Even though the embedding $\mathscr{S}^D \hookrightarrow \mathbb{R}^{D+1}$ is used to state the energy invariance (Proposition 2.7.1) and to derive spherical metrics (Appendices "Canonical Metric in Cartesian Coordinates" and "Round Metric in the Spherical Coordinates"), splitting Lagrangian dynamics alone does not require embedding as in [13] and could be used as a general method in manifold MCMC.

Although we did not explicitly explore the effect of increasing dimensionality, one would expect that the benefits of our proposed methods would diminish as dimension grows. First, standard Metropolis algorithms defined on finite-dimensions have diminishing acceptance probability for fixed step size and increasing dimension [50, 54, 55]. Second, the surface of unit D-sphere, $\mathscr{S}^D$, diminishes as $D \to +\infty$, which means an increasingly constrained space for proposals. Third, the importance weights (2.29) may become extreme around boundaries. These issues impose challenges on scaling up the proposed methods. The first issue can be resolved by recent development of dimension-independent MCMC algorithms [8, 9, 17]. The second issue could be resolved by considering a D-sphere with radius $r \propto \sqrt{D}$ (hence, the surface area $A(\mathscr{S}^D(r)) = \frac{2\pi^{(D+1)/2} r^D}{\Gamma((D+1)/2)} \nrightarrow 0$). The last one may be resolved by techniques from particle filtering [6, 16, 19].

In developing Spherical HMC, we start with the standard HMC, using the Euclidean metric $\mathbf{I}$ on unit ball $\mathbf{B}_0^D(1)$. Then, spherical geometry is introduced to handle constraints. One possible future direction could be to directly start with RHMC/LMC, which use a more informative metric (i.e., the Fisher metric $\mathbf{G_F}$), and then incorporate the spherical geometry for the constraints. For example, a possible metric for the augmented space could be $\mathbf{G_F} + \boldsymbol{\theta}\boldsymbol{\theta}^\mathsf{T}/\theta_{D+1}^2$. However, under such a metric, we might not be able to find the geodesic flow analytically, which could undermine the added benefit from using the Fisher metric. Another interesting extension could be defining HMC on D-torus $\mathscr{T}^D := \underbrace{\mathscr{S}^1 \times \cdots \mathscr{S}^1}_{D}$ for box-type constraints. Change of variables (2.7), (2.8) can be avoided but again it might be challenging to find analytical geodesic solutions on $\mathscr{T}^D$.

In future, we also intend to explore the possibility of applying the spherical augmentation to Elliptical Slice sampler [38] in order to generalize it to Spherical Slice sampler (SSS). The resulting algorithm can be applied to truncated Gaussian process models. In general, we can extend our proposed methods to infinite dimensional function spaces. This would involve the infinite dimensional manifold $\mathscr{S}^\infty := \{f \in L^2(\Omega) | \int f^2 d\mu = 1\}$. In this setting it is crucial to ensure that the acceptance probability does not drop quickly as dimension increases [9].

Acknowledgments SL is supported by EPSRC Programme Grant, Enabling Quantification of Uncertainty in Inverse Problems (EQUIP), EP/K034154/1. BS is supported by NSF grant IIS-1216045 and NIH grant R01-AI107034.

Appendix

Spherical Geometry

We first discuss the geometry of the D-dimensional sphere $\mathscr{S}^D := \{\tilde{\boldsymbol{\theta}} \in \mathbb{R}^{D+1} : \|\tilde{\boldsymbol{\theta}}\|_2 = 1\} \hookrightarrow \mathbb{R}^{D+1}$ under different coordinate systems, namely, *Cartesian coordinates* and the *spherical coordinates*. Since $\mathscr{S}^D$ can be embedded (injectively and differentiably mapped to) in $\mathbb{R}^{D+1}$, we first introduce the concept of 'induced metric'.

Definition 2.1 (*induced metric*) If $\mathscr{D}^d$ can be embedded to $\mathscr{M}^m$ $(m > d)$ by $f : U \subset \mathscr{D} \hookrightarrow \mathscr{M}$, then one can define the *induced metric*, $g_{\mathscr{D}}$, on $T\mathscr{D}$ through the metric $g_{\mathscr{M}}$ defined on $T\mathscr{M}$:

$$g_{\mathscr{D}}(\boldsymbol{\theta})(\mathbf{u}, \mathbf{v}) = g_{\mathscr{M}}(f(\boldsymbol{\theta}))(df_{\boldsymbol{\theta}}(\mathbf{u}), df_{\boldsymbol{\theta}}(\mathbf{v})), \quad \mathbf{u}, \mathbf{v} \in T_{\boldsymbol{\theta}}\mathscr{D} \tag{2.58}$$

Remark 2.2 For any $f : U \subset \mathscr{S}^D \hookrightarrow \mathbb{R}^{D+1}$, we can define the induced metric through dot product on $\mathbb{R}^{D+1}$. More specifically,

$$g_{\mathscr{S}}(\mathbf{u}, \mathbf{v}) = [(Df)\mathbf{u}]^{\mathsf{T}}(Df)\mathbf{v} = \mathbf{u}^{\mathsf{T}}[(Df)^{\mathsf{T}}(Df)]\mathbf{v} \tag{2.59}$$

where $(Df)_{(D+1)\times D}$ is the Jacobian matrix of the mapping f. A Metric induced from dot product on Euclidean space is called a "canonical metric". This observation leads to the following simple fact that lays down the foundation of Spherical HMC.

Proposition 2.7.1 (Energy invariance) *Kinetic energy* $\frac{1}{2}\langle \mathbf{v}, \mathbf{v}\rangle_{\mathbf{G}(\boldsymbol{\theta})}$ *is invariant to the choice of coordinate systems.*

Proof For any $\mathbf{v} \in T_{\boldsymbol{\theta}}\mathscr{D}$, suppose $\boldsymbol{\theta}(t)$ such that $\dot{\boldsymbol{\theta}}(0) = \mathbf{v}$. Denote the pushforward of $\mathbf{v}$ by embedding map $f : \mathscr{D} \to \mathscr{M}$ as $\tilde{\mathbf{v}} := f_*(\mathbf{v}) = \frac{d}{dt}(f \circ \boldsymbol{\theta})(0)$. Then we have

$$\frac{1}{2}\langle \mathbf{v}, \mathbf{v}\rangle_{\mathbf{G}(\boldsymbol{\theta})} = \frac{1}{2}g_{\mathscr{M}}(f(\boldsymbol{\theta}))(\tilde{\mathbf{v}}, \tilde{\mathbf{v}}) \tag{2.60}$$

That is, regardless of the form of the energy under a coordinate system, its value is the same as the one in the embedded manifold. In particular, when $\mathscr{M} = \mathbb{R}^{D+1}$, the right hand side simplifies to $\frac{1}{2}\|\tilde{\mathbf{v}}\|_2^2$. □

Canonical Metric in Cartesian Coordinates

Now consider the D-dimensional ball $\mathscr{B}_{\mathbf{0}}^D(1) := \{\boldsymbol{\theta} \in \mathbb{R}^D : \|\boldsymbol{\theta}\|_2 \leq 1\}$. Here, $\{\boldsymbol{\theta}, \mathscr{B}_{\mathbf{0}}^D(1)\}$ can be viewed as the Cartesian coordinate system for $\mathscr{S}^D$. The coordinate mapping $T_{\mathscr{B}\to\mathscr{S}_+} : \boldsymbol{\theta} \mapsto \tilde{\boldsymbol{\theta}} = (\boldsymbol{\theta}, \theta_{D+1})$ in (2.5) can be viewed as the embedding map into $\mathbb{R}^{D+1}$, and the Jacobian matrix of $T_{\mathscr{B}\to\mathscr{S}_+}$ is $dT_{\mathscr{B}\to\mathscr{S}_+} = \frac{d\tilde{\boldsymbol{\theta}}}{d\boldsymbol{\theta}^{\mathsf{T}}} = \begin{bmatrix} \mathbf{I}_D \\ -\boldsymbol{\theta}^{\mathsf{T}}/\theta_{D+1} \end{bmatrix}$. Therefore the *canonical metric* of $\mathscr{S}^D$ in Cartesian coordinates, $\mathbf{G}_{\mathscr{S}_c}(\boldsymbol{\theta})$, is

$$\mathbf{G}_{\mathscr{S}_c}(\boldsymbol{\theta}) = dT_{\mathscr{B}\to\mathscr{S}_+}^{\mathsf{T}} dT_{\mathscr{B}\to\mathscr{S}_+} = \mathbf{I}_D + \frac{\boldsymbol{\theta}\boldsymbol{\theta}^{\mathsf{T}}}{\theta_{D+1}^2} = \mathbf{I}_D + \frac{\boldsymbol{\theta}\boldsymbol{\theta}^{\mathsf{T}}}{1-\|\boldsymbol{\theta}\|_2^2} \tag{2.61}$$

Another way to obtain the metric is through the first fundamental form ds^2 (i.e., squared infinitesimal length of a curve) for $\mathscr{S}^D$, which can be expressed in terms of the differential form $d\boldsymbol{\theta}$ and the canonical metric $\mathbf{G}_{\mathscr{S}_c}(\boldsymbol{\theta})$,

$$ds^2 = \langle d\boldsymbol{\theta}, d\boldsymbol{\theta}\rangle_{\mathbf{G}_{\mathscr{S}_c}} = d\boldsymbol{\theta}^{\mathsf{T}}\mathbf{G}_{\mathscr{S}_c}(\boldsymbol{\theta})d\boldsymbol{\theta}$$

On the other hand, ds^2 can also be obtained as follows [59]:

$$ds^2 = \sum_{i=1}^{D} d\theta_i^2 + (d(\theta_{D+1}(\boldsymbol{\theta})))^2 = d\boldsymbol{\theta}^{\mathsf{T}}d\boldsymbol{\theta} + \frac{(\boldsymbol{\theta}^{\mathsf{T}}d\boldsymbol{\theta})^2}{1-\|\boldsymbol{\theta}\|_2^2} = d\boldsymbol{\theta}^{\mathsf{T}}[\mathbf{I} + \boldsymbol{\theta}\boldsymbol{\theta}^{\mathsf{T}}/\theta_{D+1}^2]d\boldsymbol{\theta}$$

Equating the above two quantities yields the form of the canonical metric $\mathbf{G}_{\mathscr{S}_c}(\boldsymbol{\theta})$ as in Eq. (2.61). This viewpoint provides a natural way to explain the length of tangent vector. For any vector $\tilde{\mathbf{v}} = (\mathbf{v}, v_{D+1}) \in T_{\tilde{\boldsymbol{\theta}}}\mathscr{S}^D = \{\tilde{\mathbf{v}} \in \mathbb{R}^{D+1} : \tilde{\boldsymbol{\theta}}^{\mathsf{T}}\tilde{\mathbf{v}} = 0\}$, one could think of $\mathbf{G}_{\mathscr{S}_c}(\boldsymbol{\theta})$ as a mean to express the length of $\tilde{\mathbf{v}}$ in terms of $\mathbf{v}$,

$$\mathbf{v}^{\mathsf{T}}\mathbf{G}_{\mathscr{S}_c}(\boldsymbol{\theta})\mathbf{v} = \|\mathbf{v}\|_2^2 + \frac{\mathbf{v}^{\mathsf{T}}\boldsymbol{\theta}\boldsymbol{\theta}^{\mathsf{T}}\mathbf{v}}{\theta_{D+1}^2} = \|\mathbf{v}\|_2^2 + \frac{(-\theta_{D+1}v_{D+1})^2}{\theta_{D+1}^2} = \|\mathbf{v}\|_2^2 + v_{D+1}^2 = \|\tilde{\mathbf{v}}\|_2^2 \tag{2.62}$$

This indeed verifies the energy invariance Proposition 2.7.1.

The following proposition provides the analytic forms of the determinant and the inverse of $\mathbf{G}_{\mathscr{S}_c}(\boldsymbol{\theta})$.

Proposition 2.7.2 *The determinant and the inverse of the* canonical metric *are as follows:*

$$|\mathbf{G}_{\mathscr{S}_c}(\boldsymbol{\theta})| = \theta_{D+1}^{-2}, \qquad \mathbf{G}_{\mathscr{S}_c}(\boldsymbol{\theta})^{-1} = \mathbf{I}_D - \boldsymbol{\theta}\boldsymbol{\theta}^{\mathsf{T}} \tag{2.63}$$

Proof The determinant of the canonical metric $\mathbf{G}_{\mathscr{S}_c}(\boldsymbol{\theta})$ is given by the matrix determinant lemma

$$|\mathbf{G}_{\mathscr{S}_c}(\boldsymbol{\theta})| = \det\left[\mathbf{I}_D + \frac{\boldsymbol{\theta}\boldsymbol{\theta}^{\mathsf{T}}}{\theta_{D+1}^2}\right] = 1 + \frac{\boldsymbol{\theta}^{\mathsf{T}}\boldsymbol{\theta}}{\theta_{D+1}^2} = \frac{1}{\theta_{D+1}^2}$$

The inverse of $\mathbf{G}_{\mathscr{S}_c}(\boldsymbol{\theta})$ is obtained by the Sherman-Morrison-Woodbury formula [29]

$$\mathbf{G}_{\mathscr{S}_c}(\boldsymbol{\theta})^{-1} = \left[\mathbf{I}_D + \frac{\boldsymbol{\theta}\boldsymbol{\theta}^{\mathsf{T}}}{\theta_{D+1}^2}\right]^{-1} = \mathbf{I}_D - \frac{\boldsymbol{\theta}\boldsymbol{\theta}^{\mathsf{T}}/\theta_{D+1}^2}{1+\boldsymbol{\theta}^{\mathsf{T}}\boldsymbol{\theta}/\theta_{D+1}^2} = \mathbf{I}_D - \boldsymbol{\theta}\boldsymbol{\theta}^{\mathsf{T}}$$

□

Corollary 2.1 *The volume adjustment of changing measure in* (2.6) *is*

$$\left|\frac{d\boldsymbol{\theta}_{\mathscr{B}}}{d\boldsymbol{\theta}_{\mathscr{S}_c}}\right| = |\mathbf{G}_{\mathscr{S}_c}(\boldsymbol{\theta})|^{-\frac{1}{2}} = |\theta_{D+1}| \tag{2.64}$$

Proof Canonical measure can be defined through the Riesz representation theorem by using a positive linear functional on the space $C_0(\mathscr{S}^D)$ of compactly supported continuous functions on $\mathscr{S}^D$ [21, 59]. More precisely, there is a unique positive Borel measure μ_c such that for (any) coordinate chart $(\mathscr{B}_{\mathbf{0}}^D(1), T_{\mathscr{B}\to\mathscr{S}_+})$,

$$\int_{\mathscr{S}_+^D} f(\tilde{\boldsymbol{\theta}})d\boldsymbol{\theta}_{\mathscr{S}_c} = \int_{\mathscr{B}_{\mathbf{0}}^D(1)} f(\boldsymbol{\theta})\sqrt{|\mathbf{G}_{\mathscr{S}_c}(\boldsymbol{\theta})|}d\boldsymbol{\theta}_{\mathscr{B}}$$

where $\mu_c = d\boldsymbol{\theta}_{\mathscr{S}_c}$, and $d\boldsymbol{\theta}_{\mathscr{B}}$ is the Euclidean measure. Therefore we have

$$\left|\frac{d\boldsymbol{\theta}_{\mathscr{S}_c}}{d\boldsymbol{\theta}_{\mathscr{B}}}\right| = |\mathbf{G}_{\mathscr{S}_c}(\boldsymbol{\theta})|^{\frac{1}{2}} = |\theta_{D+1}|^{-1}$$

Alternatively, $\left|\frac{d\boldsymbol{\theta}_{\mathscr{B}}}{d\boldsymbol{\theta}_{\mathscr{S}_c}}\right| = |\theta_{D+1}|$. □

Geodesic on a Sphere in Cartesian Coordinates

To find the geodesic on a sphere, we need to solve the following equations:

$$\dot{\boldsymbol{\theta}} = \mathbf{v} \tag{2.65}$$

$$\dot{\mathbf{v}} = -\mathbf{v}^\mathsf{T}\boldsymbol{\Gamma}_{\mathscr{S}_c}(\boldsymbol{\theta})\mathbf{v} \tag{2.66}$$

for which we need to calculate the Christoffel symbols, $\boldsymbol{\Gamma}_{\mathscr{S}_c}(\boldsymbol{\theta})$, first. Note that the (i, j)-th element of $\mathbf{G}_{\mathscr{S}_c}$ is $g_{ij} = \delta_{ij} + \theta_i\theta_j/\theta_{D+1}^2$, and the (i, j, k)-th element of $d\mathbf{G}_{\mathscr{S}_c}$ is $g_{ij,k} = (\delta_{ik}\theta_j + \theta_i\delta_{jk})/\theta_{D+1}^2 + 2\theta_i\theta_j\theta_k/\theta_{D+1}^4$. Therefore

$$\begin{aligned}\Gamma_{ij}^k &= \frac{1}{2}g^{kl}[g_{lj,i} + g_{il,j} - g_{ij,l}]\\ &= (\delta^{kl} - \theta^k\theta^l)\theta_l/\theta_{D+1}^2[\delta_{ij} + \theta_i\theta_j/\theta_{D+1}^2]\\ &= \theta_k[\delta_{ij} + \theta_i\theta_j/\theta_{D+1}^2] = [\mathbf{G}_{\mathscr{S}_c}(\boldsymbol{\theta}) \otimes \boldsymbol{\theta}]_{ijk}\end{aligned}$$

Using these results, we can write Eq. (2.66) as $\dot{\mathbf{v}} = -\mathbf{v}^\mathsf{T}\mathbf{G}_{\mathscr{S}_c}(\boldsymbol{\theta})\mathbf{v}\boldsymbol{\theta} = -\|\tilde{\mathbf{v}}\|_2^2\boldsymbol{\theta}$. Further, we have

$$\begin{aligned}\dot{\theta}_{D+1} &= \frac{d}{dt}\sqrt{1 - \|\boldsymbol{\theta}\|_2^2} &&= -\frac{\boldsymbol{\theta}^\mathsf{T}}{\theta_{D+1}}\dot{\boldsymbol{\theta}} &&= v_{D+1}\\ \dot{v}_{D+1} &= -\frac{d}{dt}\frac{\boldsymbol{\theta}^\mathsf{T}\mathbf{v}}{\theta_{D+1}} &&= -\frac{\dot{\boldsymbol{\theta}}^\mathsf{T}\mathbf{v} + \boldsymbol{\theta}^\mathsf{T}\dot{\mathbf{v}}}{\theta_{D+1}} + \frac{\boldsymbol{\theta}^\mathsf{T}\mathbf{v}}{\theta_{D+1}^2}\dot{\theta}_{D+1} &&= -\|\tilde{\mathbf{v}}\|_2^2\theta_{D+1}\end{aligned}$$

Therefore, we can rewrite the geodesic equations (2.65), (2.66) with augmented components as

$$\dot{\tilde{\boldsymbol{\theta}}} = \tilde{\mathbf{v}} \tag{2.67}$$

$$\dot{\tilde{\mathbf{v}}} = -\|\tilde{\mathbf{v}}\|_2^2 \tilde{\boldsymbol{\theta}} \tag{2.68}$$

Multiplying both sides of Eq. (2.68) by $\tilde{\mathbf{v}}^\mathsf{T}$ to obtain $\frac{d}{dt}\|\tilde{\mathbf{v}}\|_2^2 = 0$, we can solve the above system of differential equations as follows:

$$\tilde{\boldsymbol{\theta}}(t) = \tilde{\boldsymbol{\theta}}(0)\cos(\|\tilde{\mathbf{v}}(0)\|_2 t) + \frac{\tilde{\mathbf{v}}(0)}{\|\tilde{\mathbf{v}}(0)\|_2}\sin(\|\tilde{\mathbf{v}}(0)\|_2 t)$$
$$\tilde{\mathbf{v}}(t) = -\tilde{\boldsymbol{\theta}}(0)\|\tilde{\mathbf{v}}(0)\|_2 \sin(\|\tilde{\mathbf{v}}(0)\|_2 t) + \tilde{\mathbf{v}}(0)\cos(\|\tilde{\mathbf{v}}(0)\|_2 t)$$

Round Metric in the Spherical Coordinates

Consider the D-dimensional hyperrectangle $\mathscr{R}_\mathbf{0}^D := [0, \pi]^{D-1} \times [0, 2\pi)$ and the corresponding spherical coordinate system, $\{\boldsymbol{\theta}, \mathscr{R}_\mathbf{0}^D\}$, for $\mathscr{S}^D$. The coordinate mapping $T_{\mathscr{R}_\mathbf{0}\to\mathscr{S}} : \boldsymbol{\theta} \mapsto \mathbf{x},\ x_d = \cos(\theta_d)\prod_{i=1}^{d-1}\sin(\theta_i),\ d = 1, \ldots, D+1,\ (\theta_{D+1} = 0)$ can be viewed as the embedding map into $\mathbb{R}^{D+1}$, and the Jacobian matrix of $T_{\mathscr{R}_\mathbf{0}\to\mathscr{S}}$ is $\frac{d\mathbf{x}}{d\boldsymbol{\theta}^\mathsf{T}}$ with the (d, j)-th element $[-\tan(\theta_d)\delta_{dj} + \cot(\theta_j)I(j < d)]x_d$. The induced metric of $\mathscr{S}^D$ in the spherical coordinates is called *round metric*, denoted as $\mathbf{G}_{\mathscr{S}_r}(\boldsymbol{\theta})$, whose (i, j)-th element is as follows

$$\begin{aligned}
&\mathbf{G}_{\mathscr{S}_r}(\boldsymbol{\theta})_{ij} \\
&= \sum_{d=1}^{D+1}[-\tan(\theta_d)\delta_{di} + \cot(\theta_j)I(i < d)][-\tan(\theta_d)\delta_{dj} + \cot(\theta_j)I(j < d)]x_d^2 \\
&= \begin{cases} -\tan(\theta_j)\cot(\theta_i)x_j^2 + \cot(\theta_i)\cot(\theta_j)\sum_{d>j} x_d^2 = 0, & i < j \\ \tan^2(\theta_i)x_i^2 + \cot^2(\theta_i)\sum_{d>i} x_d^2 = (\tan^2(\theta_i) + 1)x_i^2 = \prod_{i=1}^{d-1}\sin^2(\theta_i), & i = j \end{cases} \\
&= \prod_{i=1}^{d-1}\sin^2(\theta_i)\delta_{ij}
\end{aligned} \tag{2.69}$$

Therefore, $\mathbf{G}_{\mathscr{S}_r}(\boldsymbol{\theta}) = \mathrm{diag}[1, \sin^2(\theta_1), \ldots, \prod_{d=1}^{D-1}\sin^2(\theta_d)]$. Another way to obtain $\mathbf{G}_{\mathscr{S}_r}(\boldsymbol{\theta})$ is through the coordinate change

$$\mathbf{G}_{\mathscr{S}_r}(\boldsymbol{\theta}) = \frac{d\boldsymbol{\theta}_{\mathscr{S}_c}^\mathsf{T}}{d\boldsymbol{\theta}_{\mathscr{S}_r}}\mathbf{G}_{\mathscr{S}_c}(\boldsymbol{\theta})\frac{d\boldsymbol{\theta}_{\mathscr{S}_c}}{d\boldsymbol{\theta}_{\mathscr{S}_r}^\mathsf{T}} \tag{2.70}$$

Similar to Corollary (2.1), we have

Proposition 2.7.3 *The volume adjustment of changing measure in* (2.9) *is*

$$\left|\frac{d\boldsymbol{\theta}_{\mathscr{R}_0}}{d\boldsymbol{\theta}_{\mathscr{S}_r}}\right| = |\mathbf{G}_{\mathscr{S}_r}(\boldsymbol{\theta})|^{-\frac{1}{2}} = \prod_{d=1}^{D-1} \sin^{-(D-d)}(\theta_d) \tag{2.71}$$

Jacobian of the Transformation Between q-Norm Domains

The following proposition gives the weights needed for the transformation from $\mathscr{Q}^D$ to $\mathscr{B}_{\mathbf{0}}^D(1)$.

Proposition 2.7.4 *The Jacobian determinant (weight) of $T_{\mathscr{B}\to\mathscr{Q}}$ is as follows:*

$$|dT_{\mathscr{S}\to\mathscr{Q}}| = \left(\frac{2}{q}\right)^D \left(\prod_{i=1}^{D} |\theta_i|\right)^{2/q-1} \tag{2.72}$$

Proof Note

$$T_{\mathscr{B}\to\mathscr{Q}} : \boldsymbol{\theta} \mapsto \boldsymbol{\beta} = \operatorname{sgn}(\boldsymbol{\theta})|\boldsymbol{\theta}|^{2/q}$$

The Jacobian matrix for $T_{\mathscr{B}\to\mathscr{Q}}$ is

$$\frac{d\boldsymbol{\beta}}{d\boldsymbol{\theta}^{\mathsf{T}}} = \frac{2}{q}\operatorname{diag}(|\boldsymbol{\theta}|^{2/q-1})$$

Therefore the Jacobian determinant of $T_{\mathscr{B}\to\mathscr{Q}}$ is

$$|dT_{\mathscr{B}\to\mathscr{Q}}| = \left|\frac{d\boldsymbol{\beta}}{d\boldsymbol{\theta}^{\mathsf{T}}}\right| = \left(\frac{2}{q}\right)^D \left(\prod_{i=1}^{D} |\theta_i|\right)^{2/q-1}$$

□

The following proposition gives the weights needed for the change of domains from $\mathscr{R}^D$ to $\mathscr{B}_{\mathbf{0}}^D(1)$.

Proposition 2.7.5 *The Jacobian determinant (weight) of $T_{\mathscr{B}\to\mathscr{R}}$ is as follows:*

$$|dT_{\mathscr{B}\to\mathscr{R}}| = \frac{\|\boldsymbol{\theta}\|_2^D}{\|\boldsymbol{\theta}\|_\infty^D} \prod_{i=1}^{D} \frac{u_i - l_i}{2} \tag{2.73}$$

Proof First, we note

$$T_{\mathscr{B}\to\mathscr{R}} = T_{\mathscr{C}\to\mathscr{R}} \circ T_{\mathscr{B}\to\mathscr{C}} : \boldsymbol{\theta} \mapsto \boldsymbol{\beta}' = \boldsymbol{\theta}\frac{\|\boldsymbol{\theta}\|_2}{\|\boldsymbol{\theta}\|_\infty} \mapsto \boldsymbol{\beta} = \frac{\mathbf{u}-\mathbf{l}}{2}\boldsymbol{\beta}' + \frac{\mathbf{u}+\mathbf{l}}{2}$$

The corresponding Jacobian matrices are

$$T_{\mathscr{B}\to\mathscr{C}}: \quad \frac{d\boldsymbol{\beta}'}{d\boldsymbol{\theta}^{\mathsf{T}}} = \frac{\|\boldsymbol{\theta}\|_2}{\|\boldsymbol{\theta}\|_\infty}\left[\mathbf{I}+\boldsymbol{\theta}\left(\frac{\boldsymbol{\theta}^{\mathsf{T}}}{\|\boldsymbol{\theta}\|_2^2} - \frac{\mathbf{e}^{\mathsf{T}}_{\arg\max|\boldsymbol{\theta}|}}{\boldsymbol{\theta}_{\arg\max|\boldsymbol{\theta}|}}\right)\right]$$

$$T_{\mathscr{C}\to\mathscr{R}}: \quad \frac{d\boldsymbol{\beta}}{d(\boldsymbol{\beta}')^{\mathsf{T}}} = \operatorname{diag}\left(\frac{\mathbf{u}-\mathbf{l}}{2}\right)$$

where $\mathbf{e}_{\arg\max|\boldsymbol{\theta}|}$ is a vector with ($\arg\max|\boldsymbol{\theta}|$)-th element 1 and all others 0. Therefore,

$$|dT_{\mathscr{B}\to\mathscr{R}}| = |dT_{\mathscr{C}\to\mathscr{R}}|\,|dT_{\mathscr{B}\to\mathscr{C}}| = \left|\frac{d\boldsymbol{\beta}}{d(\boldsymbol{\beta}')^{\mathsf{T}}}\right|\left|\frac{d\boldsymbol{\beta}'}{d\boldsymbol{\theta}^{\mathsf{T}}}\right| = \frac{\|\boldsymbol{\theta}\|_2^D}{\|\boldsymbol{\theta}\|_\infty^D}\prod_{i=1}^{D}\frac{u_i-l_i}{2}$$

□

Splitting Hamiltonian (Lagrangian) Dynamics on $\mathscr{S}^D$

Splitting the Hamiltonian dynamics and its usefulness in improving HMC is a well-studied topic of research [13, 36, 56]. Splitting the Lagrangian dynamics (used in our approach), on the other hand, has not been discussed in the literature, to the best of our knowledge. Therefore, we prove the validity of our splitting method by starting with the well-understood method of splitting Hamiltonian [13],

$$H^*(\boldsymbol{\theta},\mathbf{p}) = \frac{1}{2}U(\boldsymbol{\theta}) + \frac{1}{2}\mathbf{p}^{\mathsf{T}}\mathbf{G}_{\mathscr{S}_c}(\boldsymbol{\theta})^{-1}\mathbf{p} + \frac{1}{2}U(\boldsymbol{\theta})$$

The corresponding systems of differential equations,

$$\begin{cases} \dot{\boldsymbol{\theta}} &= \mathbf{0} \\ \dot{\mathbf{p}} &= -\frac{1}{2}\nabla_{\boldsymbol{\theta}}U(\boldsymbol{\theta}) \end{cases}$$

$$\begin{cases} \dot{\boldsymbol{\theta}} &= \mathbf{G}_{\mathscr{S}_c}(\boldsymbol{\theta})^{-1}\mathbf{p} \\ \dot{\mathbf{p}} &= -\frac{1}{2}\mathbf{p}^{\mathsf{T}}\mathbf{G}_{\mathscr{S}_c}(\boldsymbol{\theta})^{-1}d\mathbf{G}_{\mathscr{S}_c}(\boldsymbol{\theta})\mathbf{G}_{\mathscr{S}_c}(\boldsymbol{\theta})^{-1}\mathbf{p} \end{cases}$$

can be written in terms of Lagrangian dynamics in $(\boldsymbol{\theta},\mathbf{v})$ as follows:

$$\begin{cases} \dot{\boldsymbol{\theta}} &= \mathbf{0} \\ \dot{\mathbf{v}} &= -\frac{1}{2}\mathbf{G}_{\mathscr{S}_c}(\boldsymbol{\theta})^{-1}\nabla_{\boldsymbol{\theta}}U(\boldsymbol{\theta}) \end{cases}$$

$$\begin{cases} \dot{\boldsymbol{\theta}} & = \mathbf{v} \\ \dot{\mathbf{v}} & = -\mathbf{v}^{\mathsf{T}}\boldsymbol{\Gamma}_{\mathscr{S}_c}(\boldsymbol{\theta})\mathbf{v} \end{cases}$$

We have solved the second dynamics (on the right) in Appendix "Geodesic on a Sphere in Cartesian Coordinates". To solve the first dynamics, we note that

$$\begin{aligned} \dot{\theta}_{D+1} &= \frac{d}{dt}\sqrt{1-\|\boldsymbol{\theta}\|_2^2} &&= -\frac{\boldsymbol{\theta}^{\mathsf{T}}}{\theta_{D+1}}\dot{\boldsymbol{\theta}} &&= 0 \\ \dot{v}_{D+1} &= -\frac{d}{dt}\frac{\boldsymbol{\theta}^{\mathsf{T}}\mathbf{v}}{\theta_{D+1}} &&= -\frac{\dot{\boldsymbol{\theta}}^{\mathsf{T}}\mathbf{v}+\boldsymbol{\theta}^{\mathsf{T}}\dot{\mathbf{v}}}{\theta_{D+1}} + \frac{\boldsymbol{\theta}^{\mathsf{T}}\mathbf{v}}{\theta_{D+1}^2}\dot{\theta}_{D+1} &&= \frac{1}{2}\frac{\boldsymbol{\theta}^{\mathsf{T}}}{\theta_{D+1}}\mathbf{G}_{\mathscr{S}_c}(\boldsymbol{\theta})^{-1}\nabla_{\boldsymbol{\theta}}U(\boldsymbol{\theta}) \end{aligned}$$

Therefore, we have

$$\begin{aligned} \tilde{\boldsymbol{\theta}}(t) &= \tilde{\boldsymbol{\theta}}(0) \\ \tilde{\mathbf{v}}(t) &= \tilde{\mathbf{v}}(0) - \frac{t}{2}\begin{bmatrix} \mathbf{I} \\ -\frac{\boldsymbol{\theta}(0)^{\mathsf{T}}}{\theta_{D+1}(0)} \end{bmatrix}[\mathbf{I}-\boldsymbol{\theta}(0)\boldsymbol{\theta}(0)^{\mathsf{T}}]\nabla_{\boldsymbol{\theta}}U(\boldsymbol{\theta}) \end{aligned}$$

where $\begin{bmatrix} \mathbf{I} \\ -\frac{\boldsymbol{\theta}(0)^{\mathsf{T}}}{\theta_{D+1}(0)} \end{bmatrix}[\mathbf{I}-\boldsymbol{\theta}(0)\boldsymbol{\theta}(0)^{\mathsf{T}}] = \begin{bmatrix} \mathbf{I}-\boldsymbol{\theta}(0)\boldsymbol{\theta}(0)^{\mathsf{T}} \\ -\theta_{D+1}(0)\boldsymbol{\theta}(0)^{\mathsf{T}} \end{bmatrix} = \begin{bmatrix} \mathbf{I} \\ \mathbf{0}^{\mathsf{T}} \end{bmatrix} - \tilde{\boldsymbol{\theta}}(0)\boldsymbol{\theta}(0)^{\mathsf{T}}$.

Finally, we note that $\|\tilde{\boldsymbol{\theta}}(t)\|_2 = 1$ if $\|\tilde{\boldsymbol{\theta}}(0)\|_2 = 1$ and $\tilde{\mathbf{v}}(t) \in T_{\tilde{\boldsymbol{\theta}}(t)}\mathscr{S}_c^D$ if $\tilde{\mathbf{v}}(0) \in T_{\tilde{\boldsymbol{\theta}}(0)}\mathscr{S}_c^D$.

Error Analysis of Spherical HMC

Following [36], we now show that the discretization error $e_n = \|\mathbf{z}(t_n) - \mathbf{z}^{(n)}\| = \|(\boldsymbol{\theta}(t_n), \mathbf{v}(t_n)) - (\boldsymbol{\theta}^{(n)}, \mathbf{v}^{(n)})\|$ (i.e., the difference between the true solution and the numerical solution) is $\mathscr{O}(\varepsilon^3)$ locally and $\mathscr{O}(\varepsilon^2)$ globally, where ε is the discretization step size. Here, we assume that $\mathbf{f}(\boldsymbol{\theta}, \mathbf{v}) := \mathbf{v}^{\mathsf{T}}\boldsymbol{\Gamma}(\boldsymbol{\theta})\mathbf{v} + \mathbf{G}(\boldsymbol{\theta})^{-1}\nabla_{\boldsymbol{\theta}}U(\boldsymbol{\theta})$ is smooth; hence, $\mathbf{f}$ and its derivatives are uniformly bounded as $\mathbf{z} = (\boldsymbol{\theta}, \mathbf{v})$ evolves within finite time duration T. We expand the true solution $\mathbf{z}(t_{n+1})$ at t_n:

$$\begin{aligned} \mathbf{z}(t_{n+1}) &= \mathbf{z}(t_n) + \dot{\mathbf{z}}(t_n)\varepsilon + \frac{1}{2}\ddot{\mathbf{z}}(t_n)\varepsilon^2 + \mathscr{O}(\varepsilon^3) \\ &= \begin{bmatrix} \boldsymbol{\theta}(t_n) \\ \mathbf{v}(t_n) \end{bmatrix} + \begin{bmatrix} \mathbf{v}(t_n) \\ -\mathbf{f}(\boldsymbol{\theta}(t_n), \mathbf{v}(t_n)) \end{bmatrix}\varepsilon + \frac{1}{2}\begin{bmatrix} -\mathbf{f}(\boldsymbol{\theta}(t_n), \mathbf{v}(t_n)) \\ -\dot{\mathbf{f}}(\boldsymbol{\theta}(t_n), \mathbf{v}(t_n)) \end{bmatrix}\varepsilon^2 + \mathscr{O}(\varepsilon^3) \end{aligned} \tag{2.76}$$

We first consider Spherical HMC in Cartesian coordinates, where $\mathbf{f}(\boldsymbol{\theta}, \mathbf{v}) = \|\tilde{\mathbf{v}}\|^2\boldsymbol{\theta} + [\mathbf{I} - \boldsymbol{\theta}\boldsymbol{\theta}^{\mathsf{T}}]\nabla_{\boldsymbol{\theta}}U(\boldsymbol{\theta})$. From Eq. (2.34) we have

$$\begin{aligned} \mathbf{v}^{(n+1/2)} &= \mathbf{v}^{(n)} - \frac{\varepsilon}{2}(\mathbf{I} - \boldsymbol{\theta}^{(n)}(\boldsymbol{\theta}^{(n)})^{\mathsf{T}})\nabla_{\boldsymbol{\theta}}U(\boldsymbol{\theta}^{(n)}) \\ \|\tilde{\mathbf{v}}^{(n+1/2)}\|^2 &= \|\tilde{\mathbf{v}}^{(n)}\|^2 - \varepsilon(\mathbf{v}^{(n)})^{\mathsf{T}}\nabla_{\boldsymbol{\theta}}U(\boldsymbol{\theta}^{(n)}) + \mathscr{O}(\varepsilon^2) \end{aligned} \tag{2.77}$$

Now we expand Eq. (2.35) using Taylor series as follows:

$$\begin{aligned}\boldsymbol{\theta}^{(n+1)} &= \boldsymbol{\theta}^{(n)}[1 - \|\tilde{\mathbf{v}}^{(n+1/2)}\|^2\varepsilon^2/2 + \mathscr{O}(\varepsilon^4)] \\ &\quad + \mathbf{v}^{(n+1/2)}\varepsilon[1 - \|\tilde{\mathbf{v}}^{(n+1/2)}\|^2\varepsilon^2/3! + \mathscr{O}(\varepsilon^4)] \\ \mathbf{v}^{(n+3/4)} &= -\boldsymbol{\theta}^{(n)}\|\tilde{\mathbf{v}}^{(n+1/2)}\|^2\varepsilon[1 - \|\tilde{\mathbf{v}}^{(n+1/2)}\|^2\varepsilon^2/3! + \mathscr{O}(\varepsilon^4)] \\ &\quad + \mathbf{v}^{(n+1/2)}[1 - \|\tilde{\mathbf{v}}^{(n+1/2)}\|^2\varepsilon^2/2 + \mathscr{O}(\varepsilon^4)]\end{aligned}$$

Substituting (2.77) in the above equations yields

$$\begin{aligned}\boldsymbol{\theta}^{(n+1)} &= \boldsymbol{\theta}^{(n)} + \mathbf{v}^{(n+1/2)}\varepsilon - \boldsymbol{\theta}^{(n)}\|\tilde{\mathbf{v}}^{(n+1/2)}\|^2\varepsilon^2/2 + \mathscr{O}(\varepsilon^3) \\ &= \boldsymbol{\theta}^{(n)} + \mathbf{v}^{(n)}\varepsilon - \frac{1}{2}\mathbf{f}(\boldsymbol{\theta}^{(n)}, \mathbf{v}^{(n)})\varepsilon^2 + \mathscr{O}(\varepsilon^3) \\ \mathbf{v}^{(n+3/4)} &= \mathbf{v}^{(n+1/2)} - \boldsymbol{\theta}^{(n)}\|\tilde{\mathbf{v}}^{(n+1/2)}\|^2\varepsilon - \mathbf{v}^{(n+1/2)}\|\tilde{\mathbf{v}}^{(n+1/2)}\|^2\varepsilon^2/2 + \mathscr{O}(\varepsilon^3) \\ &= \mathbf{v}^{(n)} - [(\mathbf{I} - \boldsymbol{\theta}^{(n)}(\boldsymbol{\theta}^{(n)})^{\mathsf{T}})\nabla_{\boldsymbol{\theta}}U(\boldsymbol{\theta}^{(n)})/2 + \boldsymbol{\theta}^{(n)}\|\tilde{\mathbf{v}}^{(n)}\|^2]\varepsilon \\ &\quad + [\boldsymbol{\theta}^{(n)}(\mathbf{v}^{(n)})^{\mathsf{T}}\nabla_{\boldsymbol{\theta}}U(\boldsymbol{\theta}^{(n)}) - \mathbf{v}^{(n)}\|\tilde{\mathbf{v}}^{(n)}\|^2/2]\varepsilon^2 + \mathscr{O}(\varepsilon^3)\end{aligned}$$

With the above results, we have

$$\begin{aligned}\mathbf{v}^{(n+1)} &= \mathbf{v}^{(n+3/4)} - \frac{\varepsilon}{2}(\mathbf{I} - \boldsymbol{\theta}^{(n+1)}(\boldsymbol{\theta}^{(n+1)})^{\mathsf{T}})\nabla_{\boldsymbol{\theta}}U(\boldsymbol{\theta}^{(n+1)}) \\ &= \mathbf{v}^{(n)} - \mathbf{f}(\boldsymbol{\theta}^{(n)}, \mathbf{v}^{(n)})\varepsilon + [\boldsymbol{\theta}^{(n)}(\mathbf{v}^{(n)})^{\mathsf{T}}\nabla_{\boldsymbol{\theta}}U(\boldsymbol{\theta}^{(n)}) - \mathbf{v}^{(n)}\|\tilde{\mathbf{v}}^{(n)}\|^2/2]\varepsilon^2 + \mathscr{O}(\varepsilon^3) \\ &\quad - \frac{1}{2}[(\mathbf{I} - \boldsymbol{\theta}^{(n)}(\boldsymbol{\theta}^{(n)})^{\mathsf{T}})\nabla_{\boldsymbol{\theta}}^2U(\boldsymbol{\theta}^{(n)})\mathbf{v}^{(n)} - (\boldsymbol{\theta}^{(n)}(\mathbf{v}^{(n)})^{\mathsf{T}} + \mathbf{v}^{(n)}(\boldsymbol{\theta}^{(n)})^{\mathsf{T}})\nabla_{\boldsymbol{\theta}}U(\boldsymbol{\theta}^{(n)})]\varepsilon^2 \\ &= \mathbf{v}^{(n)} - \mathbf{f}(\boldsymbol{\theta}^{(n)}, \mathbf{v}^{(n)})\varepsilon - \frac{1}{2}\dot{\mathbf{f}}(\boldsymbol{\theta}^{(n)}, \mathbf{v}^{(n)})\varepsilon^2 + \mathscr{O}(\varepsilon^3)\end{aligned}$$

where for the last equality we need to show $(\mathbf{v}^{(n)})^{\mathsf{T}}\nabla_{\boldsymbol{\theta}}U(\boldsymbol{\theta}^{(n)}) = -2\frac{d}{dt}\|\tilde{\mathbf{v}}^{(n)}\|^2$. This can be proved as follows:

$$\begin{aligned}\frac{d}{dt}\|\tilde{\mathbf{v}}\|^2 &= \frac{d}{dt}[\|\tilde{\mathbf{v}}\|^2 + v_{D+1}^2] = 2[-\mathbf{v}^{\mathsf{T}}\mathbf{f} + v_{D+1}\dot{v}_{D+1}] \\ &= 2\left[-\mathbf{v}^{\mathsf{T}}\mathbf{f} + \left(-\frac{\dot{\boldsymbol{\theta}}^{\mathsf{T}}\mathbf{v} + \boldsymbol{\theta}^{\mathsf{T}}\dot{\mathbf{v}}}{\theta_{D+1}} + \frac{\boldsymbol{\theta}^{\mathsf{T}}\mathbf{v}}{\theta_{D+1}^2}\dot{\theta}_{D+1}\right)v_{D+1}\right] \\ &= -2\left[\left(\mathbf{v} - \frac{v_{D+1}}{\theta_{D+1}}\boldsymbol{\theta}\right)^{\mathsf{T}}\mathbf{f} + \frac{v_{D+1}}{\theta_{D+1}}\|\tilde{\mathbf{v}}\|^2\right] \\ &= -2\left[\left(\mathbf{v}^{\mathsf{T}}\boldsymbol{\theta} - \frac{v_{D+1}}{\theta_{D+1}}(\|\boldsymbol{\theta}\|^2 - 1)\right)\|\tilde{\mathbf{v}}\|^2 + \left(\mathbf{v} - \frac{v_{D+1}}{\theta_{D+1}}\boldsymbol{\theta}\right)^{\mathsf{T}}[\mathbf{I} - \boldsymbol{\theta}\boldsymbol{\theta}^{\mathsf{T}}]\nabla_{\boldsymbol{\theta}}U(\boldsymbol{\theta})\right] \\ &= -2\left[\mathbf{v}^{\mathsf{T}}\nabla_{\boldsymbol{\theta}}U(\boldsymbol{\theta}) + \left(-\mathbf{v}^{\mathsf{T}}\boldsymbol{\theta} - \frac{v_{D+1}}{\theta_{D+1}}(1 - \|\boldsymbol{\theta}\|^2)\right)\boldsymbol{\theta}^{\mathsf{T}}\nabla_{\boldsymbol{\theta}}U(\boldsymbol{\theta})\right] \\ &= -2\mathbf{v}^{\mathsf{T}}\nabla_{\boldsymbol{\theta}}U(\boldsymbol{\theta})\end{aligned}$$

Therefore we have

$$\mathbf{z}^{(n+1)} := \begin{bmatrix} \boldsymbol{\theta}^{(n+1)} \\ \mathbf{v}^{(n+1)} \end{bmatrix} = \begin{bmatrix} \boldsymbol{\theta}^{(n)} \\ \mathbf{v}^{(n)} \end{bmatrix} + \begin{bmatrix} \mathbf{v}^{(n)} \\ -\mathbf{f}(\boldsymbol{\theta}^{(n)}, \mathbf{v}^{(n)}) \end{bmatrix} \varepsilon + \frac{1}{2} \begin{bmatrix} -\mathbf{f}(\boldsymbol{\theta}^{(n)}, \mathbf{v}^{(n)}) \\ -\dot{\mathbf{f}}(\boldsymbol{\theta}^{(n)}, \mathbf{v}^{(n)}) \end{bmatrix} \varepsilon^2 + \mathscr{O}(\varepsilon^3) \tag{2.78}$$

The local error is

$$\begin{aligned} e_{n+1} &= \|\mathbf{z}(t_{n+1}) - \mathbf{z}^{(n+1)}\| \\ &= \left\| \begin{bmatrix} \boldsymbol{\theta}(t_n) - \boldsymbol{\theta}^{(n)} \\ \mathbf{v}(t_n) - \mathbf{v}^{(n)} \end{bmatrix} + \begin{bmatrix} \mathbf{v}(t_n) - \mathbf{v}^{(n)} \\ -[\mathbf{f}(t_n) - \mathbf{f}^{(n)}] \end{bmatrix} \varepsilon + \frac{1}{2} \begin{bmatrix} -[\mathbf{f}(t_n) - \mathbf{f}^{(n)}] \\ -[\dot{\mathbf{f}}(t_n) - \dot{\mathbf{f}}^{(n)}] \end{bmatrix} \varepsilon^2 + \mathscr{O}(\varepsilon^3) \right\| \\ &\leq (1 + M_1\varepsilon + M_2\varepsilon^2) e_n + \mathscr{O}(\varepsilon^3) \end{aligned} \tag{2.79}$$

where $M_k = c_k \sup_{t\in[0,T]} \|\nabla^k \mathbf{f}(\boldsymbol{\theta}(t), \mathbf{v}(t))\|$, $k = 1, 2$ for some constants $c_k > 0$. Accumulating the local errors by iterating the above inequality for $L = T/\varepsilon$ steps provides the following global error:

$$\begin{aligned} e_{L+1} &\leq (1 + M_1\varepsilon + M_2\varepsilon^2) e_L + \mathscr{O}(\varepsilon^3) \leq (1 + M_1\varepsilon + M_2\varepsilon^2)^2 e_{L-1} + 3\mathscr{O}(\varepsilon^3) \leq \cdots \\ &\leq (1 + M_1\varepsilon + M_2\varepsilon^2)^L e_1 + L\mathscr{O}(\varepsilon^3) \leq (e^{M_1 T} + T)\varepsilon^2 \to 0, \quad as\ \varepsilon \to 0 \end{aligned} \tag{2.80}$$

For Spherical HMC in the spherical coordinates, we conjecture that the integrator of Algorithm 2 still has order 3 local error and order 2 global error. One can follow the same argument as above to verify this.

Bounce in Diamond: Wall HMC for 1-Norm Constraint

Reference [46] discusses the *Wall* HMC method for ∞-norm constraint only. We can however derive a similar approach for 1-norm constraint. As shown in the left panel of Fig. 2.13, given the current state $\boldsymbol{\theta}_0$, HMC makes a proposal $\boldsymbol{\theta}$. It will hit the boundary to move from $\boldsymbol{\theta}_0$ towards $\boldsymbol{\theta}$. To determine the hit point 'X', we are required to solve for $t \in (0, 1)$ such that

$$\|\boldsymbol{\theta}_0 + (\boldsymbol{\theta} - \boldsymbol{\theta}_0)t\|_1 = \sum_{d=1}^{D} |\theta_0^d + (\theta^d - \theta_0^d)t| = 1 \tag{2.81}$$

One can find the hitting time using the bisection method. However, a more efficient method is to find the orthant in which the sampler hits the boundary, i.e., find the normal direction $\mathbf{n}$ with elements being ± 1. Then, we can find t,

$$\|\boldsymbol{\theta}_0 + (\boldsymbol{\theta} - \boldsymbol{\theta}_0)t\|_1 = \mathbf{n}^\mathsf{T}[\boldsymbol{\theta}_0 + (\boldsymbol{\theta} - \boldsymbol{\theta}_0)t] = 1 \implies t^* = \frac{1 - \mathbf{n}^\mathsf{T}\boldsymbol{\theta}_0}{\mathbf{n}^\mathsf{T}(\boldsymbol{\theta} - \boldsymbol{\theta}_0)} \tag{2.82}$$

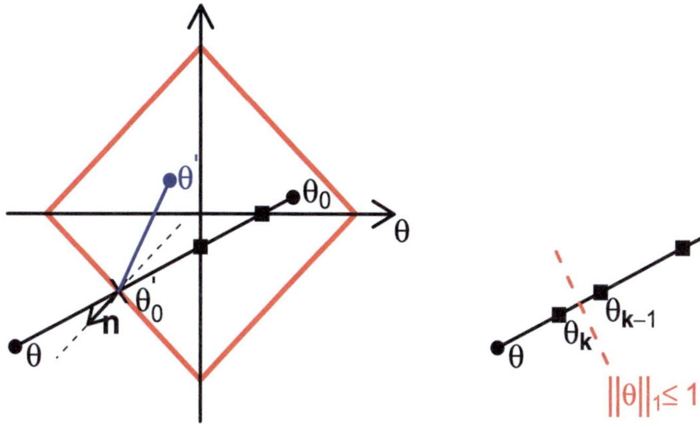

Fig. 2.13 Wall HMC bounces in the 1-norm constraint domain. *Left* given the current state $\boldsymbol{\theta}_0$, Wall HMC proposes $\boldsymbol{\theta}$, but bounces of the boundary and reaches $\boldsymbol{\theta}'$ instead. *Right* determining the hitting time by monitoring the first intersection point with coordinate planes that violates the constraint

Therefore the hit point is $\boldsymbol{\theta}_0' = \boldsymbol{\theta}_0 + (\boldsymbol{\theta} - \boldsymbol{\theta}_0)t^*$ and consequently the reflection point is

$$\boldsymbol{\theta}' = \boldsymbol{\theta} - 2\mathbf{n}^*\langle \mathbf{n}^*, \boldsymbol{\theta} - \boldsymbol{\theta}_0' \rangle = \boldsymbol{\theta} - 2\mathbf{n}(\mathbf{n}^\mathsf{T}\boldsymbol{\theta} - 1)/D \tag{2.83}$$

where $\mathbf{n}^* := \mathbf{n}/\|\mathbf{n}\|_2$ and $\mathbf{n}^\mathsf{T}\boldsymbol{\theta}_0' = 1$ because $\boldsymbol{\theta}_0'$ is on the boundary with the normal direction $\mathbf{n}^*$.

It is in general difficult to directly determine the intersection of $\boldsymbol{\theta} - \boldsymbol{\theta}_0$ with boundary. Instead, we can find its intersections with coordinate planes $\{\boldsymbol{\pi}_d\}_{d=1}^D$, where $\boldsymbol{\pi}_d := \{\boldsymbol{\theta} \in \mathbb{R}^D | \theta^d = 0\}$. The intersection times are defined as $\mathbf{T} = \{\theta_0^d/(\theta_0^d - \theta^d) | \theta_0^d \neq \theta^d\}$. We keep those between 0 and 1 and sort them in ascending order (Fig. 2.13, right panel). Then, we find the intersection points $\{\boldsymbol{\theta}_k := \boldsymbol{\theta}_0 + (\boldsymbol{\theta} - \boldsymbol{\theta}_0)T_k\}$ that violate the constraint $\|\boldsymbol{\theta}\| \leq 1$. Denote the first intersection point outside the constrained domain as $\boldsymbol{\theta}_k$. The signs of $\boldsymbol{\theta}_k$ and $\boldsymbol{\theta}_{k-1}$ determine the orthant of the hitting point $\boldsymbol{\theta}_0'$.

Note, for each $d \in \{1, \ldots D\}$, $(\text{sign}(\boldsymbol{\theta}_k^d), \text{sign}(\boldsymbol{\theta}_{k-1}^d))$ cannot be $(+, -)$ or $(-, +)$, otherwise there exists an intersection point $\boldsymbol{\theta}^* := \boldsymbol{\theta}_0 + (\boldsymbol{\theta} - \boldsymbol{\theta}_0)T^*$ with some coordinate plane $\boldsymbol{\pi}_{d^*}$ between $\boldsymbol{\theta}_k$ and $\boldsymbol{\theta}_{k-1}$. Then $T_{k-1} < T^* < T_k$ contradicts the order of $\mathbf{T}$.[4] Therefore any point (including $\boldsymbol{\theta}_0'$) between $\boldsymbol{\theta}_k$ and $\boldsymbol{\theta}_{k-1}$ must have the same sign as $\text{sign}(\text{sign}(\boldsymbol{\theta}_k) + \text{sign}(\boldsymbol{\theta}_{k-1}))$; that is

$$\mathbf{n} = \text{sign}(\text{sign}(\boldsymbol{\theta}_k) + \text{sign}(\boldsymbol{\theta}_{k-1})) \tag{2.84}$$

After moving from $\boldsymbol{\theta}$ to $\boldsymbol{\theta}'$, we examine whether $\boldsymbol{\theta}'$ satisfies the constraint. If it does not satisfy the constraint, we repeat above procedure with $\boldsymbol{\theta}_0 \leftarrow \boldsymbol{\theta}_0'$ and $\boldsymbol{\theta} \leftarrow \boldsymbol{\theta}'$

[4]The same argument applies when $T_k = 1$, i.e. $\boldsymbol{\theta}$ is the first point outside the domain among $\{\boldsymbol{\theta}_k\}$.

until the final state is inside the constrained domain. Then we adjust the velocity direction by

$$\mathbf{v} \leftarrow (\boldsymbol{\theta}' - \boldsymbol{\theta}'_0)\frac{\|\mathbf{v}\|}{\|\boldsymbol{\theta}' - \boldsymbol{\theta}'_0\|} \tag{2.85}$$

Algorithm 4 summarizes the above steps.

Algorithm 4 Wall HMC for 1-norm constraint (Wall HMC)

Initialize $\boldsymbol{\theta}^{(1)}$ at the current state $\boldsymbol{\theta}$ after transformation
Sample a new velocity value $\mathbf{v}^{(1)} \sim \mathcal{N}(\mathbf{0}, \mathbf{I}_D)$
Calculate $H(\boldsymbol{\theta}^{(1)}, \mathbf{v}^{(1)}) = U(\boldsymbol{\theta}^{(1)}) + K(\mathbf{v}^{(1)})$
for $\ell = 1$ to L **do**
  $\mathbf{v}^{(\ell+\frac{1}{2})} = \mathbf{v}^{(\ell)} - \frac{\varepsilon}{2}\nabla_{\boldsymbol{\theta}} U(\boldsymbol{\theta}^{(\ell)})$
  $\boldsymbol{\theta}^{(\ell+1)} = \boldsymbol{\theta}^{(\ell)} + \varepsilon \mathbf{v}^{(\ell+\frac{1}{2})}$
  set hit $\leftarrow$ false
  while $\|\boldsymbol{\theta}^{(\ell)}\| > 1$ **do**
    find times intersecting with coordinate planes: $\mathbf{T} = \{T_d := \theta_d^{(\ell)}/(\theta_d^{(\ell)} - \theta_d^{(\ell+1)}) | \theta_d^{(\ell)} \neq \theta_d^{(\ell+1)}\}$
    sort those between 0 and 1 in ascending order: $\mathbf{T} = \{0 \leq T_k \uparrow \leq 1\}$
    find the first point in $\{\boldsymbol{\theta}_k := \boldsymbol{\theta}^{(\ell)} + (\boldsymbol{\theta}^{(\ell+1)} - \boldsymbol{\theta}^{(\ell)})T_k\}$ that violates $\|\boldsymbol{\theta}\| \leq 1$ and denote it as $\boldsymbol{\theta}_k$
    set normal direction as $\mathbf{n} = \mathrm{sign}(\mathrm{sign}(\boldsymbol{\theta}_k) + \mathrm{sign}(\boldsymbol{\theta}_{k-1}))$
    find the wall hitting time $t^* = (1 - \mathbf{n}^{\mathsf{T}}\boldsymbol{\theta}^{(\ell)})/(\mathbf{n}^{\mathsf{T}}(\boldsymbol{\theta}^{(\ell+1)} - \boldsymbol{\theta}^{(\ell)}))$
    $\boldsymbol{\theta}^{(\ell)} \leftarrow \boldsymbol{\theta}^{(\ell)} + (\boldsymbol{\theta}^{(\ell+1)} - \boldsymbol{\theta}^{(\ell)})t^*$ and $\boldsymbol{\theta}^{(\ell+1)} \leftarrow \boldsymbol{\theta}^{(\ell+1)} - 2\mathbf{n}\langle \mathbf{n}, \boldsymbol{\theta}^{(\ell+1)} - \boldsymbol{\theta}^{(\ell)}\rangle/\|\mathbf{n}\|_2^2$
    set hit $\leftarrow$ true
  end while
  if hit **then**
    $\mathbf{v}^{(\ell+\frac{1}{2})} \leftarrow (\boldsymbol{\theta}^{(\ell+1)} - \boldsymbol{\theta}^{(\ell)})\|\mathbf{v}^{(\ell+\frac{1}{2})}\|/\|\boldsymbol{\theta}^{(\ell+1)} - \boldsymbol{\theta}^{(\ell)}\|$
  end if
  $\mathbf{v}^{(\ell+1)} = \mathbf{v}^{(\ell+\frac{1}{2})} - \frac{\varepsilon}{2}\nabla_{\boldsymbol{\theta}} U(\boldsymbol{\theta}^{(\ell+1)})$
end for
Calculate $H(\boldsymbol{\theta}^{(L+1)}, \mathbf{v}^{(L+1)}) = U(\boldsymbol{\theta}^{(L+1)}) + K(\mathbf{v}^{(L+1)})$
Calculate the acceptance probability $\alpha = \min\{1, \exp[-H(\boldsymbol{\theta}^{(L+1)}, \mathbf{v}^{(L+1)}) + H(\boldsymbol{\theta}^{(1)}, \mathbf{v}^{(1)})]\}$
Accept or reject the proposal according to α for the next state $\boldsymbol{\theta}'$

References

1. Y. Ahmadian, J.W. Pillow, L. Paninski, Efficient Markov chain Monte Carlo methods for decoding neural spike trains. Neural Comput. **23**(1), 46–96 (2011)
2. S. Ahn, Y. Chen, M. Welling, Distributed and adaptive darting Monte Carlo through regenerations, in *Proceedings of the 16th International Conference on Artificial Intelligence and Statistics (AI Stat)* (2013)
3. S. Ahn, B. Shahbaba, M. Welling, Distributed stochastic gradient MCMC, in *International Conference on Machine Learning* (2014)
4. S. Amari, H. Nagaoka, *Methods of Information Geometry, in Translations of Mathematical Monographs*, vol. 191 (Oxford University Press, Oxford, 2000)

5. C. Andrieu, E. Moulines, On the ergodicity properties of some adaptive MCMC algorithms. Ann. Appl. Probab. **16**(3), 1462–1505 (2006)
6. C. Andrieu, A. Doucet, R. Holenstein, Particle Markov chain Monte Carlo methods. J. R. Stat. Soc.: Ser. B (Stat. Methodol.) **72**(3), 269–342 (2010)
7. M.J. Beal, Variational algorithms for approximate Bayesian inference. Ph.D. thesis, University College London, London (2003)
8. A. Beskos, G. Roberts, A. Stuart, J. Voss, MCMC methods for diffusion bridges. Stoch. Dyn. **8**(03), 319–350 (2008)
9. A. Beskos, F.J. Pinski, J.M. Sanz-Serna, A.M. Stuart, Hybrid Monte Carlo on Hilbert spaces. Stoch. Process. Appl. **121**(10), 2201–2230 (2011)
10. D.M. Blei, A.Y. Ng, M.I. Jordan, Latent Dirichlet allocation. J. Mach. Learn. Res. **3**, 993–1022 (2003)
11. A.E. Brockwell, Parallel markov chain monte carlo simulation by Pre-Fetching. J. Comput. Gr. Stat. **15**, 246–261 (2006)
12. M.A. Brubaker, M. Salzmann, R. Urtasun, A family of MCMC methods on implicitly defined manifolds, in *Proceedings of the Fifteenth International Conference on Artificial Intelligence and Statistics (AISTATS-12)*, vol. 22, ed. by N.D. Lawrence, M.A. Girolami (2012), pp. 161–172
13. S. Byrne, M. Girolami, Geodesic Monte Carlo on embedded manifolds. Scand. J. Stat. **40**(4), 825–845 (2013)
14. B. Calderhead, M. Sustik, Sparse approximate manifolds for differential geometric MCMC, in *Advances in Neural Information Processing Systems*, vol. 25, ed. by P. Bartlett, F. Pereira, C. Burges, L. Bottou, K. Weinberger (2012), pp. 2888–2896
15. O. Cappé, R. Douc, A. Guillin, J.M. Marin, C.P. Robert, Adaptive importance sampling in general mixture classes. Stat. Comput. **18**(4), 447–459 (2008)
16. N. Chopin, A sequential particle filter method for static models. Biometrika **89**(3), 539–552 (2002)
17. S.L. Cotter, G.O. Roberts, A. Stuart, D. White et al., MCMC methods for functions: modifying old algorithms to make them faster. Stat. Sci. **28**(3), 424–446 (2013)
18. R.V. Craiu, J. Rosenthal, C. Yang, Learn from thy neighbor: parallel-chain and regional adaptive MCMC. J. Am. Stat. Assoc. **104**(488), 1454–1466 (2009)
19. P. Del Moral, A. Doucet, A. Jasra, Sequential Monte Carlo samplers. J. R. Stat. Soc.: Ser. B (Stat. Methodol.) **68**(3), 411–436 (2006)
20. N. de Freitas, P. Højen-Sørensen, M. Jordan, R. Stuart, Variational MCMC, in *Proceedings of the 17th Conference in Uncertainty in Artificial Intelligence, UAI '01* (Morgan Kaufmann Publishers Inc., San Francisco, 2001), pp. 120–127
21. M.P. do Carmo, *Riemannian Geometry*, 1st edn (Birkhäuser, Boston, 1992)
22. R. Douc, C.P. Robert, A vanilla rao-blackwellization of metropolis-hastings algorithms. Ann. Stat. **39**(1), 261–277 (2011)
23. S. Duane, A.D. Kennedy, B.J. Pendleton, D. Roweth, Hybrid Monte Carlo. Phys. Lett. B **195**(2), 216–222 (1987)
24. I.E. Frank, J.H. Friedman, A statistical view of some chemometrics regression tools. Technometrics **35**(2), 109–135 (1993)
25. A. Gelfand, L. van der Maaten, Y. Chen, M. Welling, On herding and the cycling perceptron theorem. Adv. Neural Inf. Process. Syst. **23**, 694–702 (2010)
26. C.J. Geyer, Practical Markov Chain Monte Carlo. Stat. Sci. **7**(4), 473–483 (1992)
27. W.R. Gilks, G.O. Roberts, S.K. Sahu, Adaptive Markov chain Monte Carlo through regeneration. J. Am. Stat. Assoc. **93**(443), 1045–1054 (1998)
28. M. Girolami, B. Calderhead, Riemann manifold Langevin and Hamiltonian Monte Carlo methods. J. R. Stat. Soc. Ser. B (with discussion) **73**(2), 123–214 (2011)
29. G.H. Golub, C.F. Van Loan, *Matrix Computations*, 3rd edn. (Johns Hopkins University Press, Baltimore, 1996)
30. C. Hans, Bayesian lasso regression. Biometrika **96**(4), 835–845 (2009)

31. M. Hoffman, F.R. Bach, D.M. Blei, Online learning for latent dirichlet allocation, in *Advances in Neural Information Processing Systems* (2010), pp. 856–864
32. M.D. Hoffman, A. Gelman, The No-U-Turn sampler: adaptively setting path lengths in Hamiltonian Monte Carlo. J. Mach. Learn. Res. **15**(1), 1593–1623 (2014)
33. K. Kurihara, M. Welling, N. Vlassis, Accelerated variational Dirichlet process mixtures, in *Advances of Neural Information Processing Systems – NIPS*, vol. 19 (2006)
34. S. Lan, B. Shahbaba, Spherical hamiltonian Monte Carlo for constrained target distributions, in *The 31st International Conference on Machine Learning, Beijing* (2014), pp. 629–637
35. S. Lan, V. Stathopoulos, B. Shahbaba, M. Girolami, Markov chain Monte Carlo from Lagrangian dynamics. J. Comput. Gr. Stat. **24**(2), 357–378 (2015)
36. B. Leimkuhler, S. Reich, *Simulating Hamiltonian Dynamics* (Cambridge University Press, Cambridge, 2004)
37. J. Møller, A. Pettitt, K. Berthelsen, R. Reeves, An efficient Markov chain Monte Carlo method for distributions with intractable normalisation constants. Biometrica **93**, 451–458 (2006) (To appear)
38. I. Murray, R.P. Adams, D.J. MacKay, Elliptical slice sampling. JMLR:W&CP **9**, 541–548 (2010)
39. P. Mykland, L. Tierney, B. Yu, Regeneration in Markov chain samplers. J. Am. Stat. Assoc. **90**(429), 233–241 (1995)
40. P. Neal, G.O. Roberts, Optimal scaling for random walk metropolis on spherically constrained target densities. Methodol. Comput. Appl. Probab. **10**(2), 277–297 (2008)
41. P. Neal, G.O. Roberts, W.K. Yuen, Optimal scaling of random walk metropolis algorithms with discontinuous target densities. Ann. Appl. Probab. **22**(5), 1880–1927 (2012)
42. R.M. Neal, Probabilistic inference using Markov Chain Monte Carlo methods. Technical Report CRG-TR-93-1, Department of Computer Science, University of Toronto (1993)
43. R.M. Neal, *Bayesian Learning for Neural Networks* (Springer, Secaucus, 1996)
44. R.M. Neal, Slice sampling. Ann. Stat. **31**(3), 705–767 (2003)
45. R.M. Neal, The short-cut metropolis method. Technical Report 0506, Department of Statistics, University of Toronto (2005)
46. R.M. Neal, MCMC using Hamiltonian dynamics, in: *Handbook of Markov Chain Monte Carlo*, ed. by S. Brooks, A. Gelman, G. Jones, X.L. Meng. Chapman and Hall/CRC (2011), pp. 113–162
47. A. Pakman, L. Paninski, Exact Hamiltonian Monte Carlo for truncated multivariate Gaussians. J. Comput. Gr. Stat. **23**(2), 518–542 (2014)
48. T. Park, G. Casella, The Bayesian lasso. J. Am. Stat. Assoc. **103**(482), 681–686 (2008)
49. S. Patterson, Y.W. Teh, Stochastic gradient riemannian langevin dynamics on the probability simplex, in *Advances in Neural Information Processing Systems* (2013), pp. 3102–3110
50. N.S. Pillai, A.M. Stuart, A.H. Thiéry et al., Optimal scaling and diffusion limits for the Langevin algorithm in high dimensions. Ann. Appl. Probab. **22**(6), 2320–2356 (2012)
51. J.G. Propp, D.B. Wilson, Exact sampling with coupled Markov chains and applications to statistical mechanics (1996), pp. 223–252
52. D. Randal, G. Arnaud, M. Jean-Michel, R.P. Christian, Minimum variance importance sampling via population Monte Carlo. ESAIM: Probab. Stat. **11**, 427–447 (2007)
53. G.O. Roberts, S.K. Sahu, Updating schemes, correlation structure, blocking and parameterisation for the Gibbs sampler. J. R. Stat. Soc. Ser. B **59**, 291–317 (1997)
54. G.O. Roberts, A. Gelman, W.R. Gilks, Weak convergence and optimal scaling of random walk metropolis algorithms. Ann. Appl. Probab. **7**(1), 110–120 (1997)
55. G.O. Roberts, J.S. Rosenthal et al., Optimal scaling for various metropolis-hastings algorithms. Stat. Sci. **16**(4), 351–367 (2001)
56. B. Shahbaba, S. Lan, W.O. Johnson, R.M. Neal, Split Hamiltonian Monte Carlo. Stat. Comput. **24**(3), 339–349 (2014)
57. C. Sherlock, G.O. Roberts, Optimal scaling of the random walk metropolis on elliptically symmetric unimodal targets. Bernoulli **15**(3), 774–798 (2009)
58. M. Spivak, *Calculus on Manifolds*, vol. 1 (WA Benjamin, New York, 1965)

59. M. Spivak, *A Comprehensive Introduction to Differential Geometry*, vol. 1, 2nd edn (Publish or Perish, Inc., Houston, 1979)
60. Y.W. Teh, D. Newman, M. Welling, A collapsed variational bayesian inference algorithm for latent dirichlet allocation, in *Advances in Neural Information Processing Systems* (2006), pp. 1353–1360
61. R. Tibshirani, Regression shrinkage and selection via the lasso. J. R. Stat. Soc. Ser. B **58**(1), 267–288 (1996)
62. H.M. Wallach, I. Murray, R. Salakhutdinov, D. Mimno, Evaluation methods for topic models, in *Proceedings of the 26th Annual International Conference on Machine Learning* (ACM, 2009), pp. 1105–1112
63. G.R. Warnes, The normal kernel coupler: an adaptive Markov Chain Monte Carlo method for efficiently sampling from multi-modal distributions. Technical Report No. 395, University of Washington (2001)
64. M. Welling, Herding dynamic weights to learn, in *Proceeding of International Conference on Machine Learning* (2009)
65. M. Welling, Y.W. Teh, Bayesian learning via stochastic gradient Langevin dynamics, in *Proceedings of the 28th International Conference on Machine Learning (ICML)* (2011), pp. 681–688
66. M. West, On scale mixtures of normal distributions. Biometrika **74**(3), 646–648 (1987)
67. Y. Zhang, C. Sutton, Quasi-Newton methods for Markov Chain Monte Carlo, in *Advances in Neural Information Processing Systems*, ed. by J. Shawe-Taylor, R.S. Zemel, P. Bartlett, F.C.N. Pereira, K.Q. Weinberger, vol. 24 (2011), pp. 2393–2401

Chapter 3
Geometric Optimization in Machine Learning

Suvrit Sra and Reshad Hosseini

Abstract Machine learning models often rely on sparsity, low-rank, orthogonality, correlation, or graphical structure. The structure of interest in this chapter is geometric, specifically the manifold of positive definite (PD) matrices. Though these matrices recur throughout the applied sciences, our focus is on more recent developments in machine learning and optimization. In particular, we study (i) models that might be nonconvex in the Euclidean sense but are convex along the PD manifold; and (ii) ones that are neither Euclidean nor geodesic convex but are nevertheless amenable to global optimization. We cover basic theory for (i) and (ii); subsequently, we present a scalable Riemannian limited-memory BFGS algorithm (that also applies to other manifolds). We highlight some applications from statistics and machine learning that benefit from the geometric structure studies.

3.1 Introduction

Fitting mathematical models to data invariably requires numerical optimization. The field is thus replete with tricks and techniques for better modeling, analysis, and implementation of optimization algorithms. Among other aspects, the notion of "structure," is of perennial importance: its knowledge often helps us obtain faster algorithms, attain scalability, gain insights, or capture a host of other attributes.

Structure has multifarious meanings, of which perhaps the best known is sparsity [5, 54]. But our focus is different: we study geometric structure, in particular where model parameters lie on a Riemannian manifold.

Geometric structure has witnessed increasing interest, for instance in optimization over matrix manifolds (including orthogonal, low-rank, positive definite matrices,

---

S. Sra (✉)
Laboratory for Information & Decision Systems (LIDS),
Massachusetts Institute of Technology, Cambridge, MA, USA
e-mail: suvrit@mit.edu

R. Hosseini
School of ECE, College of Engineering, University of Tehran, Tehran, Iran
e-mail: reshad.hosseini@ut.ac.ir

H.Q. Minh and V. Murino (eds.), *Algorithmic Advances in Riemannian Geometry and Applications*, Advances in Computer Vision and Pattern Recognition, DOI 10.1007/978-3-319-45026-1_3

among others) [1, 12, 51, 57]. However, in distinction to general manifold optimization, which extends most Euclidean schemes to Riemannian manifolds [1, 55], we focus on the specific case of "geometric optimization" problems that exploit the special structure of the manifold of positive definite (PD) matrices. Two geometric aspects play a crucial role here: (i) non-positive curvature of the manifold, which allows defining a global curved notion of convexity along geodesics on the manifold; and (ii) convex (Euclidean) conic structure (e.g., as used in Perron–Frobenius theory, which includes the famous PageRank algorithm as a special case).

One of our key motivations for studying geometric optimization is that for many problems it may help uncover hidden (geodesic) convexity, and thus provably place global optimization of certain nonconvex problems within reach [51]. Moreover, exploiting geometric convexity can have remarkable empirical consequences for problems involving PD matrices [48]; which persist even without overall (geodesic) convexity, as will be seen in Sect. 3.5.1.

Finally, since PD matrices are ubiquitous in not only machine learning and statistics, but throughout the applied sciences, the new modeling and algorithmic tools offered by geometric optimization should prove to be valuable in many other settings. To stoke the reader's imagination beyond the material described in this chapter, we close with a short list of further applications in Sect. 3.5.3. We also refer the reader to our recent work [51], that develops the theoretical material related to geometric optimization in greater detail.

With this background, we are now ready to recall the geometric concepts at the heart of our presentation, before moving on to detailed applications of our ideas.

3.2 Manifolds and Geodesic Convexity

A smooth manifold is a space that locally resembles Euclidean space [30]. We focus on Riemannian manifolds (smooth manifolds equipped with a smoothly varying inner product on the tangent space) as their geometry permits a natural extension of many nonlinear optimization algorithms [1, 55].

In particular, we focus on the Riemannian manifold of real symmetric positive definite (PD) matrices. Most of the ideas that we describe apply more broadly to *Hadamard manifolds* (i.e., Riemannian manifolds with non-positive curvature), but we limit attention to the PD manifold for concreteness and due to its vast importance in machine learning and beyond [6, 17, 46].

A key concept on manifolds is that of *geodesics*, which are curves that join points along shortest paths. Geodesics help one extend the notion of convexity to *geodesic convexity*. Formally, suppose $\mathcal{M}$ is a Riemmanian manifold, and x, y are two points on $\mathcal{M}$. Say γ is a unit speed geodesic joining x to y, such that

$$\gamma_{xy} : [0, 1] \to \mathcal{M}, \quad \text{s.t.} \quad \gamma_{xy}(0) = x, \ \gamma_{xy}(1) = y.$$

Table 3.1 Summary of key Riemannian objects for the PD matrix manifold

Definition	Expression for PD matrices
Tangent space	Space of symmetric matrices
Metric between two tangent vectors ξ, η at Σ	$g_\Sigma(\xi, \eta) = \mathrm{tr}(\Sigma^{-1}\xi\Sigma^{-1}\eta)$
Gradient at Σ if Euclidean gradient is $\nabla f(\Sigma)$	$\mathrm{grad} f(\Sigma) = \frac{1}{2}\Sigma(\nabla f(X) + \nabla f(X)^T)\Sigma$
Exponential map at point Σ in direction ξ	$R_\Sigma(\xi) = \Sigma \exp(\Sigma^{-1}\xi)$
Parallel transport of tangent vector ξ from Σ_1 to Σ_2	$\mathscr{T}_{\Sigma_1,\Sigma_2}(\xi) = E\xi E^T, \quad E = (\Sigma_2\Sigma_1^{-1})^{1/2}$

Then, we call a set $\mathscr{A} \subseteq \mathscr{M}$ *geodesically convex*, henceforth g-convex, if geodesics between arbitrary pairs of points in $\mathscr{A}$ lie completely in $\mathscr{A}$. We note that the PD manifold has unique geodesics globally. We say $f : \mathscr{A} \to \mathbb{R}$ is g-convex if for all $x, y \in \mathscr{A}$, the composition $f \circ \gamma_{xy} : [0, 1] \to \mathbb{R}$ is convex in the usual sense. On the manifold $\mathbb{P}_d$ under its usual trace Riemannian metric (see Table 3.1) the geodesic γ_{XY} between $X, Y \in \mathbb{P}_d$ has the beautiful closed-form [6, Chap. 6]:

$$\gamma_{XY}(t) := X^{1/2}(X^{-1/2}YX^{-1/2})^t X^{1/2}, \quad 0 \le t \le 1. \tag{3.1}$$

It is common to write $X\sharp_t Y \equiv \gamma_{XY}(t)$, and we also use this notation for brevity. Therewith, a function $f : \mathbb{P}_d \to \mathbb{R}$ is *g-convex* if on a g-convex set $\mathscr{A}$ it satisfies

$$f(X\sharp_t Y) \le (1-t)f(X) + tf(Y), \qquad t \in [0, 1], \; X, Y \in \mathscr{A}. \tag{3.2}$$

G-convex functions are remarkable in that they can be nonconvex in the Euclidean sense, but can still be globally optimized. Such functions on PD matrices have already proved important in several recent applications [17, 18, 23, 45, 49, 50, 59, 60, 64]. We provide several examples below drawn from [51]; the reader should consult [51] for a detailed and more systematic development of g-convexity on PD matrices.

Example 1 The following functions are g-convex on $\mathbb{P}_d$: (i) $\mathrm{tr}(e^A)$; (ii) $\mathrm{tr}(A^\alpha)$ for $\alpha \ge 1$; (iii) $\lambda_{\max}(e^A)$; (iv) $\lambda_{\max}(A^\alpha)$ for $\alpha \ge 1$.

Example 2 Let $X \in \mathbb{C}^{d\times k}$ be an arbitrary rank-k matrix ($k \le d$), then $A \mapsto \mathrm{tr}\, X^*AX$ is log-g-convex, that is,

$$\mathrm{tr}\, X^*(A\sharp_t B)X \le [\mathrm{tr}\, X^*AX]^{1-t}[\mathrm{tr}\, X^*BX]^t, \qquad t \in [0, 1]. \tag{3.3}$$

Inequality (3.3) depends on a nontrivial property of $\sharp_t$ proved e.g., in [51, Theorem 2.8].

Example 3 If $h : \mathbb{R}_+ \to \mathbb{R}_+$ is nondecreasing and log-convex, then the map $A \mapsto \sum_{i=1}^k \log h(\lambda_i(A))$ is g-convex. For instance, if $h(x) = e^x$, we obtain the special case that $A \mapsto \log \mathrm{tr}(e^A)$ is g-convex.

Example 4 Let $A_i \in \mathbb{C}^{d\times k}$ with $k \le d$ be such that that map $X \mapsto \sum_{i=1}^m A_i^*XA_i$ is strictly positive; let $B \succeq 0$. Then $\phi(X) := \log\det(B + \sum_i A_i^*XA_i)$ is g-convex on $\mathbb{P}_d$.

Example 5 The *Riemannian distance* $\delta_R(A, X) := \|\log(X^{-1/2}AX^{-1/2})\|_F$ between $A, X \in \mathbb{P}_d$ [6, Chap. 6] is well-known to be jointly g-convex, see e.g., [6, Corollary 6.1.11]. To obtain an infinite family of such g-convex distances see [51, Corollary 2.19].

Consequently, the Fréchet (Karcher) mean and median of PD matrices are g-convex optimization problems. Formally, these problems seek to solve

$$\min_{X \succ 0} \sum\nolimits_{i=1}^{m} w_i \delta_R(X, A_i), \qquad \text{(Geometric Median)},$$

$$\min_{X \succ 0} \sum\nolimits_{i=1}^{m} w_i \delta_R^2(X, A_i), \qquad \text{(Geometric Mean)},$$

where $\sum_i w_i = 1, w_i \geq 0$, and $A_i \succ 0$ for $1 \leq i \leq m$. The latter problem has received extensive interest in the literature [6–8, 26, 40, 42, 49]. Its optimal solution is unique owing to the strict g-convexity of its objective.

3.3 Beyond g-Convexity: Thompson Nonexpansivity

We highlight now a special class of nonconvex functions that is amenable to global optimization without requiring g-convexity. Specifically, we consider functions that admit "sup norm" contractions, namely contractions under the *Thompson metric*:

$$\delta_T(X, Y) := \|\log(Y^{-1/2}XY^{-1/2})\|, \tag{3.4}$$

where $\|\cdot\|$ is the usual operator norm (hence the 'sup'). This metric is an object of great interest in nonlinear Perron–Frobenius theory [29, 31].

We consider maps nonexpansive under the Thompson metric (3.4). Since the metric space $(\mathbb{P}_d, \delta_T)$ is complete [53], nonexpansive maps under this metric provide fertile grounds for designing convergent fixed-point algorithms *without* necessarily relying on g-convexity. We say $\Phi : \mathbb{P}_d \to \mathbb{P}_d$ is *Thompson nonexpansive* if

$$\delta_T(\Phi(X), \Phi(Y)) \leq q\delta_T(X, Y), \qquad 0 \leq q \leq 1. \tag{3.5}$$

If $q < 1$, then Φ is called *q-contractive*. Since the Thompson metric is generated by the operator norm, it turns out to satisfy a larger body of properties (than δ_R) that are useful for analyzing fixed-point iterations. We recall some of these properties below—for details please see [29, 31, 32, 51].

Proposition 1 *Unless noted otherwise, all matrices are assumed to be PD.*

$$\delta_T(X^{-1}, Y^{-1}) = \delta_T(X, Y) \tag{3.6a}$$

$$\delta_T(B^*XB, B^*YB) = \delta_T(X, Y), \qquad B \in GL_n(\mathbb{C}) \tag{3.6b}$$

$$\delta_T(X^t, Y^t) \leq |t|\delta_T(X, Y), \qquad \textit{for } t \in [-1, 1] \tag{3.6c}$$

$$\delta_T\Big(\sum\nolimits_i w_iX_i, \sum\nolimits_i w_iY_i\Big) \leq \max_{1\leq i\leq m} \delta_T(X_i, Y_i), \qquad w_i \geq 0, w \neq 0 \tag{3.6d}$$

$$\delta_T(X + A, Y + A) \leq \frac{\alpha}{\alpha+\beta}\delta_T(X, Y), \qquad A \succeq 0, \tag{3.6e}$$

where $\alpha = \max\{\|X\|, \|Y\|\}$ *and* $\beta = \lambda_{\min}(A)$. *Moreover, for* $X \in \mathbb{C}^{d\times k}$ *(*$k \leq d$*) with full column rank we have the* compression inequality *[51, Theorem 4.3]:*

$$\delta_T(X^*AX, X^*BX) \leq \delta_T(A, B). \tag{3.6f}$$

3.3.1 Why Thompson Nonexpansivity?

Below we review a setting where Thompson nonexpansivity is useful. Consider the optimization problem $\min_{S\succ 0} \Phi(S)$, where Φ is continuously differentiable on $\mathbb{P}_d$. Since the constraint set is open, a necessary condition of optimality of a point S^* is that its gradient vanishes, that is,

$$\nabla\Phi(S^*) = 0. \tag{3.7}$$

Various approaches could be used for solving the nonlinear (matrix) equation (3.7). And among these, fixed-point iterations may be particularly attractive. Here, one designs a map $\mathscr{G} : \mathbb{P}_d \to \mathbb{P}_d$, using which we can rewrite (3.7) in the form

$$S^* = \mathscr{G}(S^*), \tag{3.8}$$

that is, S^* is a fixed-point of $\mathscr{G}$, and by construction a stationary point of Φ.

Typically, finding fixed-points is difficult. However, if the map $\mathscr{G}$ can be chosen such that it is Thompson-contractive, then simply running the Picard iteration

$$S_{k+1} \leftarrow \mathscr{G}(S_k), \quad k = 0, 1, \ldots, \tag{3.9}$$

will yield a unique solution to (3.7)—both existence and uniqueness follow from the Banach contraction theorem. The reason we insist on Thompson contractivity is because many of our examples fail to be Euclidean contractions (or even Riemannian contractions) but end up being Thompson contractions. Thus, studying Thompson nonexpansivity is valuable. We highlight below a concrete example that arises in some applications [15–18, 63], and is not a Euclidean but a Thompson contraction.

3.3.1.1 Application: Geometric Mean of PD Matrices

Let $A_1, \ldots, A_n \in \mathbb{P}_d$ be input matrices and $w_i \geq 0$ be nonnegative weights such that $\sum_{i=1}^n w_i = 1$. A particular geometric mean of the $\{A_i\}_{i=1}^n$, called the *S-mean*, is obtained by computing [15, 18]

$$\min_{X \succ 0} \; h(X) := \sum\nolimits_{i=1}^n w_i \delta_S^2(X, A_i), \tag{3.10}$$

where δ_S^2 is the squared *Stein-distance*

$$\delta_S^2(X, Y) := \log\det\left(\tfrac{X+Y}{2}\right) - \tfrac{1}{2}\log\det(XY), \qquad X, Y \succ 0. \tag{3.11}$$

It can be shown that δ_S^2 is strictly g-convex (in both arguments) [49]. Thus, Problem (3.10) is a g-convex optimization problem. It is easily seen to possess a solution, whence the strict g-convexity of $h(X)$ immediately implies that this solution must be unique. What remains is to obtain an algorithm to compute this solution.

Noting (3.11) while differentiating $h(X)$, we obtain the nonlinear matrix equation

$$0 = \nabla h(X) \quad \Longleftrightarrow \quad X^{-1} = \sum\nolimits_i w_i \left(\tfrac{X+A_i}{2}\right)^{-1},$$

from which we naively obtain the Picard iteration (see e.g., [15, 49] for details):

$$X_{k+1} \leftarrow \left[\sum\nolimits_i w_i \left(\tfrac{X_k+A_i}{2}\right)^{-1}\right]^{-1}. \tag{3.12}$$

Applying (3.6a), (3.6d), and (3.6e) in sequence one sees that (3.12) is a Thompson (but not Riemannian) contraction, which immediately shows its validity as a Picard iteration and its linear rate of convergence to the global optimum of (3.10).

3.4 Manifold Optimization

Creating fixed-point iterations is somewhat of an art, and it is not always clear how to obtain one for a given problem. Therefore, developing general purpose iterative optimization algorithms is of great practical importance.

For Euclidean optimization a common recipe is to iteratively perform the following:(a) find a descent direction; and (b) obtain sufficient decrease via line-search which also helps ensure convergence. We follow a similar recipe for Riemannian manifolds by replacing Euclidean concepts by their Riemannian counterparts. For example, we now compute descent directions in the tangent space. At a point X, the tangent space T_X is the approximating vector space (see Fig. 3.1). Given a descent direction, $\xi_X \in T_X$, we perform line-search along a smooth curve on the manifold (red curve in Fig. 3.1). The derivative of this curve at X provides the descent direction ξ_X. We refer the reader to [1, 55] for an in depth introduction to manifold optimization.

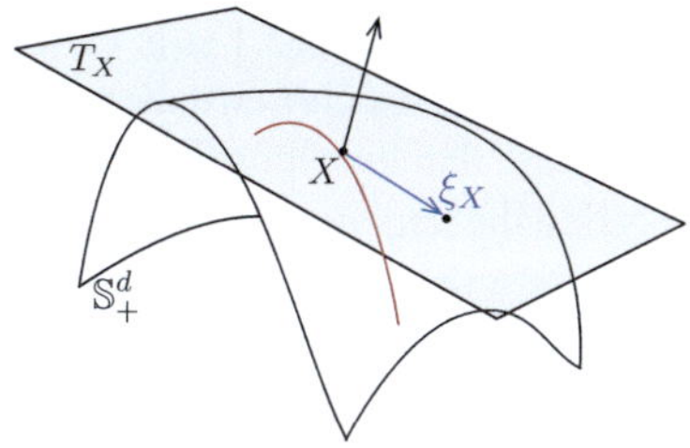

Fig. 3.1 Line-search on a manifold: X is a point on the manifold, T_X is the tangent space at the point X, ξ_X is a descent direction at X; the *red curve* is the curve along which line-search is performed

Euclidean methods such as conjugate-gradient and LBFGS combine gradients at the current point with gradients and descent directions at previous points to obtain a new descent direction. To adapt such algorithms to manifolds we need to define how to transport vectors in a tangent space at one point to vectors in a tangent space at another point.

On Riemannian manifolds, the gradient is a direction in the tangent space, where the inner product of the gradient with another direction in the tangent space gives the directional derivative of the function. Formally, if g_X defines the inner product in the tangent space T_X, then

$$Df(X)\xi = g_X(\mathrm{grad}f(X), \xi), \quad \text{for } \xi \in T_X.$$

Given a descent direction the curve along which we perform line-search can be a geodesic. A map that combines the direction and a step-size to obtain a corresponding point on the geodesic is called an *exponential map*. Riemannian manifolds also come equipped with a natural way to transport vectors on geodesics that is called parallel transport. Intuitively, a parallel transport is a differential map with zero derivative along the geodesics.

Using these ideas, and in particular deciding where to perform the parallel transport we can obtain different variants of Riemannian LBFGS. We recall one specific LBFGS variant from [51] (presented as Algorithm 1), which yields the best performance in our applications, once we combine it with a suitable line-search algorithm. The LBFGS method explained in [25] is the same as the method in [51]. We also implemented LBFGS variant of BFGS method explained in [45] and observed that it has worse running time.

In particular, to ensure Riemannian LBFGS always produces a descent direction, we must ensure that the line-search algorithm satisfies the *Wolfe conditions* [45]:

$$f(R_{X_k}(\alpha\xi_k)) \le f(X_k) + c_1\alpha Df(X_k)\xi_k, \tag{3.13}$$

$$Df(R_{X_k}(\alpha\xi_k))\mathscr{T}_{X_k,R_{X_k}(\alpha\xi_k)}(\xi_k) \ge c_2 Df(X_k)\xi_k, \tag{3.14}$$

where $0 < c_1 < c_2 < 1$.The first condition (3.13) is called the sufficient-decrease condition and the coefficient c_1 is typically chosen to be a small number. The sufficient decrease condition does not ensure that the algorithm makes sufficient progress. The second condition (3.14) is called the curvature condition and together with the other condition, they ensure convergence to a stationary point. We tried different values of c_1 and c_2 and observed that like in Euclidean case choosing $c_1 = 10^{-4}$ and $c_2 = 0.9$ works reasonable well in practice. Note that $\alpha Df(X_k)\xi_k = g_{X_k}(\text{grad}f(X_k), \alpha\xi_k)$, i.e., the derivative of $f(X_k)$ in the direction $\alpha\xi_k$ is the inner product of descent direction and gradient of the function. Practical line-search algorithms implement a stronger (Wolfe) version of (3.14) that enforces

$$|Df(R_{X_k}(\alpha\xi_k))\mathscr{T}_{X_k,R_{X_k}(\alpha\xi_k)}(\xi_k)| \leq c_2 Df(X_k)\xi_k. \tag{3.15}$$

We also implemented a line-search algorithm satisfying strong Wolfe conditions. Key details of a practical way to implement this line-search may be found in [23].

Algorithm 1 Pseudocode for Riemannian LBFGS

Given: Riemannian manifold $\mathscr{M}$ with Riemannian metric g; parallel transport $\mathscr{T}$ on $\mathscr{M}$; exp-map R; initial value X_0; a smooth function f

Set initial $H_{\text{diag}} = 1/\sqrt{g_{X_0}(\text{grad}f(X_0), \text{grad}f(X_0))}$

for $k = 0, 1, \ldots$ **do**

  Obtain descent direction ξ_k by unrolling the RBFGS method

  Compute $\xi_k \leftarrow$ HESSMUL($-\text{grad}f(X_k), k$)

  Use line-search to find α such that it satisfies Wolfe conditions

  Calculate $X_{k+1} = R_{X_k}(\alpha\xi_k)$

  Define $S_k = \mathscr{T}_{X_k,X_{k+1}}(\alpha\xi_k)$

  Define $Y_k = \text{grad}f(X_{k+1}) - \mathscr{T}_{X_k,X_{k+1}}(\text{grad}f(X_k))$

  Update $H_{\text{diag}} = g_{X_{k+1}}(S_k, Y_k)/g_{X_{k+1}}(Y_k, Y_k)$

  Store Y_k; S_k; $g_{X_{k+1}}(S_k, Y_k)$; $g_{X_{k+1}}(S_k, S_k)/g_{X_{k+1}}(S_k, Y_k)$; H_{diag}

end for

return X_k

function HESSMUL(P, k)

if $k > 0$ **then**

  $\tilde{P} = P - \frac{g_{X_{k+1}}(S_k,P)}{g_{X_{k+1}}(Y_k,S_k)} Y_k$

  $\hat{P} = \mathscr{T}_{X_k,X_{k+1}}$ HESSMUL$(\mathscr{T}^{-1}_{X_k,X_{k+1}}\tilde{P}, k-1)$ **return** $\hat{P} - \frac{g_{X_{k+1}}(Y_k,\hat{P})}{g_{X_{k+1}}(Y_k,S_k)} S_k + \frac{g_{X_{k+1}}(S_k,S_k)}{g_{X_{k+1}}(Y_k,S_k)} P$

else

  return $H_{\text{diag}} P$

end if

end function

3.5 Applications

We are ready to present two applications of geometric optimization. Section 3.5.1 summarizes recent progress in fitting Gaussian Mixture Models (GMMs), for which g-convexity proves remarkably useful and ultimately helps Algorithm 1 to greatly outperform the famous Expectation Maximization (EM) algorithm—this is remarkable as previously many believed it impossible to outdo EM via general nonlinear optimization techniques. Next, in Sect. 3.5.2 we present an application to maximum likelihood parameter estimation for non-Gaussian elliptically contoured distributions. These problems are Euclidean nonconvex but often either g-convex or Thompson nonexpansive, and thus amenable to geometric optimization.

3.5.1 Gaussian Mixture Models

The material of this section is based on the authors' recent work [23]; the interested reader is encouraged to consult that work for additional details.

Gaussian mixture models (GMMs) have a long history in machine learning and signal processing and continue to enjoy widespread use [10, 21, 37, 41]. For GMM parameter estimation, expectation maximization (EM) [20] still seems to be the de facto choice—although other approaches have also been considered [44], typical nonlinear programming methods such as conjugate gradients, quasi-Newton, Newton, are usually viewed as inferior to EM [61].

One advantage that EM enjoys is that its M-Step satisfies the PD constraint on covariances by construction. Other methods often struggle when dealing with this constraint. An approach is to make the problem unconstrained by performing a change-of-variables using Cholesky decompositions (as also exploited in semidefinite programming [14]). Another possibility is to formulate the PD constraint via a set of smooth convex inequalities [56] or to use log-barriers and to invoke interior-point methods. But such methods tend to be much slower than EM-like iterations, especially in higher dimensions [50].

Driven by these concerns the authors view GMM fitting as a manifold optimization problem in [23]. But surprisingly, an out-of-the-box invocation of manifold optimization completely fails to compete with and to outdo EM, further work is required: *g-convexity supplies the missing link.*

3.5.1.1 Problem Setup

Let $\mathcal{N}$ denote the Gaussian density with mean $\mu \in \mathbb{R}^d$ and covariance $\Sigma \in \mathbb{P}_d$, i.e.,

$$\mathcal{N}(x; \mu, \Sigma) := \det(\Sigma)^{-1/2}(2\pi)^{-d/2}\exp\big(-\tfrac{1}{2}(x-\mu)^T\Sigma^{-1}(x-\mu)\big).$$

A Gaussian mixture model has the probability density

$$p(x) := \sum\nolimits_{j=1}^{K} \alpha_j \mathcal{N}(x; \mu_j, \Sigma_j), \qquad x \in \mathbb{R}^d,$$

where $\alpha \in \Delta_K$, the K-dimensional probability simplex, and $\{\mu_j \in \mathbb{R}^d, \Sigma_j \succ 0\}_{j=1}^K$. Given i.i.d. samples $\{x_1, \ldots, x_n\}$, we wish to estimate these parameters by maximum likelihood. This leads to the *GMM optimization* problem

$$\max_{\alpha \in \Delta_K, \{\mu_j, \Sigma_j \succ 0\}_{j=1}^K} \quad \sum\nolimits_{i=1}^{n} \log\Big(\sum\nolimits_{j=1}^{K} \alpha_j \mathcal{N}(x_i; \mu_j, \Sigma_j)\Big). \tag{3.16}$$

Problem (3.16) is well-known to be a difficult nonconvex problem. So like EM, we also seek only efficient computation of local solutions. As alluded to above, before we can successfully apply manifold optimization (in particular, our LBFGS algorithm) to solve (3.16), we need to expose its g-convexity.

To that end, we begin with maximum likelihood estimation for a single Gaussian

$$\max_{\mu, \Sigma \succ 0} \; \mathcal{L}(\mu, \Sigma) := \sum\nolimits_{i=1}^{n} \log \mathcal{N}(x_i; \mu, \Sigma). \tag{3.17}$$

Although (3.17) is Euclidean convex, it is *not* g-convex. In [23] a simple reformulation[1] is used that makes (3.17) g-convex and ends up having far-reaching impact on the overall GMM problem. More precisely, we augment the sample vectors x_i to instead consider $y_i^T = [x_i^T \; 1]$. Therewith, problem (3.17) turns into

$$\max_{S \succ 0} \; \widehat{\mathcal{L}}(S) := \sum\nolimits_{i=1}^{n} \log \widehat{\mathcal{N}}(y_i; S), \tag{3.18}$$

where $\widehat{\mathcal{N}}(y_i; S) := \sqrt{2\pi} \exp(\tfrac{1}{2}) \mathcal{N}(y_i; 0, S)$. Theorem 1 shows that (3.18) is g-convex and its solution yields the solution to the original problem (3.17).

Theorem 1 ([23]) *The map $-\widehat{\mathcal{L}}(S)$ is g-convex. Moreover, if μ^*, Σ^* maximize (3.17), and S^* maximizes (3.18), then $\widehat{\mathcal{L}}(S^*) = \mathcal{L}(\mu^*, \Sigma^*)$ for* $S^* = \begin{pmatrix} \Sigma^* + \mu^* \mu^{*T} & \mu^* \\ \mu^{*T} & 1 \end{pmatrix}$.

Theorem 2 states a local version of this result for GMMs.

Theorem 2 ([23]) *A local maximum of the reparameterized GMM log-likelihood*

$$\widehat{\mathcal{L}}(\{S_j\}_{j=1}^K) := \sum\nolimits_{i=1}^{n} \log\Big(\sum\nolimits_{j=1}^{K} \alpha_j \widehat{\mathcal{N}}(y_i; S_j)\Big) \tag{3.19}$$

is a local maximum of the original log-likelihood (3.16).

[1] This reformulation essentially uses the "natural parameters."

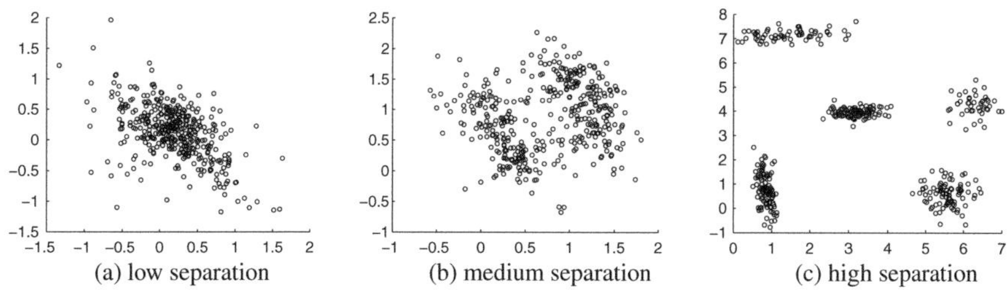

Fig. 3.2 Data clouds for three levels of separation: $c = 0.2$ (low); $c = 1$ (medium); $c = 5$ (high)

3.5.1.2 Numerical Results

We solve the reparameterized problem (3.19)[2] using Algorithm 1. We illustrate the performance through experiments on real and simulated data. All compared methods are initialized using k-means++ [3], and all share the same stopping criterion. The methods stop when the difference of *average log-likelihood* (i.e., log-likelihood/n) between iterations falls below 10^{-6}, or when the iteration count exceeds 1500.

Since EM's performance depends on the degree of separation of the mixture components [33, 61], we also assess the impact of separation on our methods. We generate data as proposed in [19, 58]. The distributions are chosen so that their means satisfy the following separation inequality:

$$\forall_{i \neq j} : \|\mu_i - \mu_j\| \geq c \max_{i,j}\{\mathrm{tr}(\Sigma_i), \mathrm{tr}(\Sigma_j)\}.$$

The parameter c shows level of separation; we use e to denote *eccentricity*, i.e., the ratio of the largest eigenvalue of the covariance matrix to its smallest eigenvalue. A typical 2D data with $K = 5$ created for different separations is shown in Fig. 3.2.

We tested both high eccentricity ($e = 10$) and spherical ($e = 1$) Gaussians. Table 3.2 reports the results, which are obtained after running 20 different random initializations. Without our reformulation Riemannian optimization is not competitive (we omit the results), while with the reformulation our Riemannian LBFGS matches or exceeds EM. We note in passing that a Cholesky decomposition based formulation ends up being vastly inferior to both EM and our Riemannian methods. Numerical results supporting this claim may be found in [23].

Next, we present an evaluation on a natural image dataset, for which GMMs have been reported to be effective [66]. GMMs is used to fit the density of natural images patches. We extracted 200K image patches of size 6×6 from random locations in the images and subtracted the DC component. Therefore, each training datum is a 35-dimensional vector corresponding to a DC-subtracted image patch. GMM fitting results obtained by different algorithms are reported in Table 3.3. As can be seen, manifold LBFGS performs better than EM and manifold CG. Moreover, our

[2] Actually, we solve a slightly different *unconstrained* problem that also reparameterizes α_j.

Table 3.2 Speed and average log-likelihood (ALL) comparisons for $d = 20$, $e = 10$, and $e = 1$. The numbers are averaged values for 20 runs over different sampled datasets, therefore the ALL values are not comparable to each other. The standard-deviation are also reported in the table

		EM ($e = 10$)		LBFGS ($e = 10$)		EM ($e = 1$)		LBFGS ($e = 1$)	
		Time (s)	ALL	Time (s)	ALL	Time (s)	ALL	Time (s)	ALL
$c = 0.2$	$K = 2$	1.1 ± 0.4	-10.7	5.6 ± 2.7	-10.7	65.7 ± 33.1	17.6	39.4 ± 19.3	17.6
	$K = 5$	30.0 ± 45.5	-12.7	49.2 ± 35.0	-12.7	365.6 ± 138.8	17.5	160.9 ± 65.9	17.5
$c = 1$	$K = 2$	0.5 ± 0.2	-10.4	3.1 ± 0.8	-10.4	6.0 ± 7.1	17.0	12.9 ± 13.0	17.0
	$K = 5$	104.1 ± 113.8	-13.4	79.9 ± 62.8	-13.3	40.5 ± 61.1	16.2	51.6 ± 39.5	16.2
$c = 5$	$K = 2$	0.2 ± 0.2	-11.0	3.4 ± 1.4	-11.0	0.2 ± 0.1	17.1	3.0 ± 0.5	17.1
	$K = 5$	38.8 ± 65.8	-12.8	41.0 ± 45.7	-12.8	17.5 ± 45.6	16.1	20.6 ± 22.5	16.1

reformulation proves crucial: without it manifold optimization is substantially slower. The Cholesky-based model without our reformulation has the worst performance (not reported), and even with reformulation it is inferior to the other approaches.

Figure 3.3 visualizes evolution of the objective function with the number of iterations (i.e., the number of log-likelihood and gradient evaluations, or the number of E- and M-steps). The datasets used in Fig. 3.3 are the "magic telescope" and "year prediction" datasets,[3] as well as natural image data used in Table 3.2. It can be seen that although manifold optimization methods spend time doing line-search they catch up with the EM algorithm in few iterations.

3.5.2 *MLE for Elliptically Contoured Distributions*

Our next application is maximum likelihood parameter estimation for Kotz-type distributions. Here given i.i.d. samples $(x_1, \ldots, x_n)$ from an Elliptically Contoured Distribution $\mathscr{E}_\varphi(S)$, up to constants the log-likelihood is of the form

$$\mathscr{L}(x_1, \ldots, x_n; S) = -\tfrac{1}{2} n \log\det S + \sum\nolimits_{i=1}^{n} \log \varphi(x_i^T S^{-1} x_i), \tag{3.20}$$

where φ is a so-called *density generating function* (dgf). We write $\Phi \equiv -\mathscr{L}$, so that computing the MLE amounts to minimizing Φ. But this is in general difficult: Φ can be nonconvex and may have multiple local minima. However, under suitable assumptions on φ, we can still maximize (3.20) to global optimality. Some examples are already known [27, 43, 65], and geometric optimization yields results that are more general than previously known examples. We refer the reader to [51] for the precise details, and provide a quick summary of the main ideas below.

The "suitable assumptions" alluded to above cover two main classes of dgfs:

[3]Available at UCI machine learning dataset repository via https://archive.ics.uci.edu/ml/datasets.

Table 3.3 Speed and ALL comparisons for natural image data $d = 35$

	EM Algorithm		LBFGS reformulated		CG reformulated		CG original		CG Cholesky reformulated	
	Time (s)	ALL	Time (s)	ALL	Time (s)	ALL	Time (s)	ALL	Time (s)	ALL
$K = 2$	16.61	29.28	14.23	29.28	17.52	29.28	947.35	29.28	476.77	29.28
$K = 4$	165.77	31.65	106.53	31.65	153.94	31.65	6380.01	31.64	2673.21	31.65
$K = 8$	596.01	32.81	332.85	32.81	536.94	32.81	14282.80	32.58	9306.33	32.81
$K = 10$	2159.47	33.05	658.34	33.06	1048.00	33.06	17711.87	33.03	7463.72	33.05

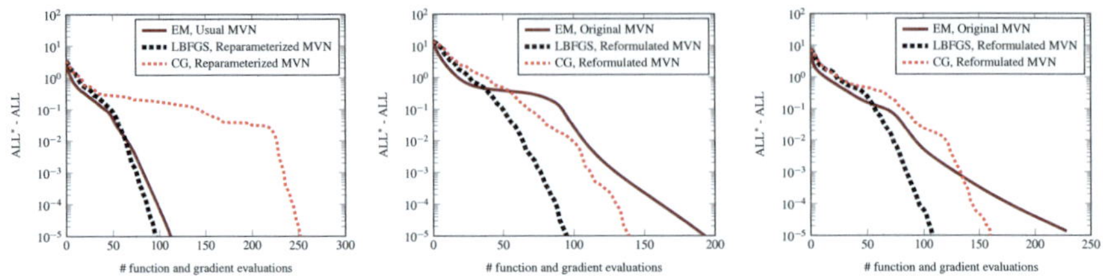

Fig. 3.3 The objective function with respect to the number of function and gradient evaluations. The objective function is the Best ALL minus current ALL values. *Left* "magic telescope" ($K = 5, d = 10$). *Middle* "year predict" ($K = 6, d = 90$). *Right* natural images ($K = 8, d = 35$)

(i) Geodesically convex (g-convex): This class contains functions for which the negative log-likelihood $\Phi(S)$ is g-convex. Some members of this class have been previously studied (possibly without exploiting g-convexity);

(ii) Log-Nonexpansive (LN): This class was introduced in [51]. It exploits the "non-positive curvature" property of the PD manifold and it covers several ECDs outside the scope of previous methods [27, 59, 65]. This class is essentially the same as what we call Thompson nonexpansive in this chapter.

In [51], the authors also discuss the class Log-Convex (LC), for which the dgf φ is log-convex, whereby Φ is nonconvex. But since Φ is now a difference of convex functions it is amenable to majorization-minimization.

Several examples of strictly g-convex ECDs are (i) Multivariate Gaussian; (ii) Kotz with $\alpha \leq \frac{d}{2}$ (its special cases include Gaussian, multivariate power exponential, multivariate W-distribution with shape parameter smaller than one, elliptical gamma with shape parameter $\nu \leq \frac{d}{2}$); (iii) Multivariate-t; (iv) Multivariate Pearson type II with positive shape parameter; (v) Elliptical multivariate logistic distribution.

For the class LN, we can circumvent the machinery of manifold optimization and obtain simple fixed-point algorithms as alluded to in Sect. 3.3.

As an illustrative example, consider the problem of finding the minimum of negative log-likelihood solution of *Kotz-type distribution* (which is a particular ECD):

$$\Phi(S) = \tfrac{n}{2}\log\det(S) + (\tfrac{d}{2} - \alpha)\sum\nolimits_{i=1}^{n} \log(x_i^T S^{-1} x_i) + \sum\nolimits_{i=1}^{n} \left(\tfrac{x_i^T S^{-1} x_i}{b}\right)^{\beta}, \quad (3.21)$$

where α, β, and b are (known) fixed parameters. To minimize Φ, following Sect. 3.3, we seek to solve $\nabla\Phi(S) = 0$. This amounts to the nonlinear matrix equation

$$S = \tfrac{2}{n}\sum\nolimits_{i=1}^{n} x_i h(x_i^T S^{-1} x_i) x_i^T, \quad (3.22)$$

where $h(\cdot) = (\frac{d}{2} - \alpha)(\cdot)^{-1} + \frac{\beta}{b^\beta}(\cdot)^{\beta-1}$. If (3.22) has positive definite solution, then it is a candidate MLE. If it is unique, then it is the desired minimum of (3.21).

The question now is whether upon setting $\mathscr{G} := \frac{2}{n}\sum_{i=1}^{n} x_i h(x_i^T S^{-1} x_i) x_i^T$ and simply iterating $S_{k+1} \leftarrow \mathscr{G}(S_k)$, we can obtain a solution to (3.22). This is where the theory developed in Sect. 3.3 comes into play. We mention below a slightly stronger result.

Let $\tau = 1 - \beta$ and $c = \frac{b^{\beta}(d/2-\alpha)}{\beta}$. Knowing that $h(\cdot) = g((\frac{d}{2} - \alpha)(\cdot)^{-1})$ has the same contraction factor as $g(\cdot)$, it can be shown that h in the iteration (3.22) for Kotz-type distributions for which $0 < \beta < 2$ and $\alpha < \frac{d}{2}$ is Thompson-contractive. Therewith, one can show the following convergence result.

Theorem 3 ([51]) *For Kotz-type distributions with* $0 < \beta < 2$ *and* $\alpha < \frac{d}{2}$, *Iteration* (3.22) *converges to a unique fixed point.*

3.5.2.1 Numerical Results

We compare now the convergence speed of fixed-point (FP) MLE iterations for different sets of parameters α and β. For our experiments, we sample 10,000 points from a Kotz-type distribution with a random scatter matrix and prescribed values of α and β. We compare the fixed-point approach with four different manifold optimization methods: (i) steepest descent (SD); (ii) conjugate gradient (CG); (iii) limited-memory RBFGS (LBFGS); (iv) trust-region (TR). All methods are initialized with the same random covariance matrix.

The first experiment (Fig. 3.4) fixes α, β, and shows the effect of dimension on convergence. Next, in Fig. 3.5, we fix dimension and consider the effect of varying α and β. As it is evident from the figures, FP and steepest descent method could have very slow convergence in some cases. FP2 denotes a re-scaled version of the basic fixed-point iteration FP (see [51] for details); the scaling improves conditioning and accelerates the method, leading to an overall best performance.

3.5.3 Other Applications

To conclude we briefly mention below additional applications that rely on geometric optimization. Our listing is by no means complete, and is biased toward work more

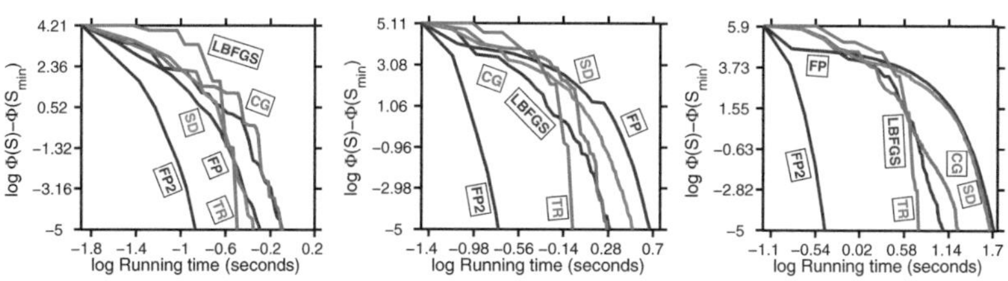

Fig. 3.4 Running times comparison between normal fixed-point iteration (FP), fixed-point iteration with scaling factor (FP2) and four different manifold optimization methods. The objective function is Kotz-type negative log-likelihood with parameters $\beta = 0.5$ and $\alpha = 1$. The plots show (from *left* to *right*), running times for estimating $S \in \mathbb{P}_d$, for $d \in \{4, 16, 64\}$

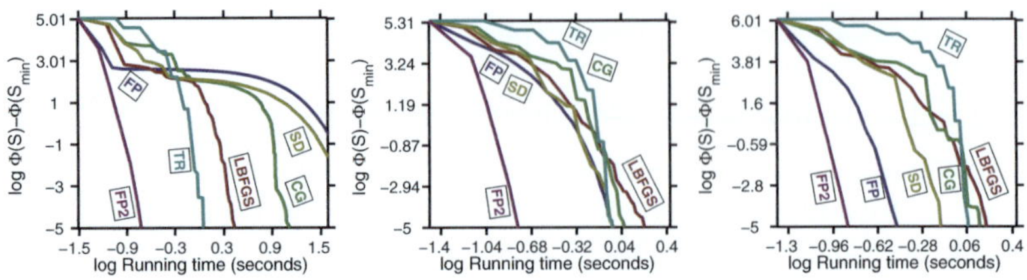

Fig. 3.5 Running time variance for Kotz-type distributions with $d = 16$ and $\alpha = 2\beta$ for different values of $\beta \in \{0.1, 1, 1.7\}$

closely related to machine learning. However, it should provide a starting point for the interested reader in exploring other applications and aspects of the rapidly evolving area of geometric optimization.

Computer vision. Chapter 4 (see references therein) describes applications to image retrieval, dictionary learning, and other problems in computer vision that involve PD matrix data, and therefore directly or indirectly rely on geometric optimization.

Signal processing. Diffusion Tensor Imaging (DTI) [28]; Radar and signal processing [2, 42]; Brain Computer Interfaces (BCI) [62].

ML and Statistics. Social networks [47]; Deep learning [34, 36]; Determinantal point processes [24, 35]; Fitting elliptical gamma distributions [52]; Fitting mixture models [22, 38]; see also [11].

Others. Structured PD matrices [9]; Manifold optimization with rank constraints [57] and symmetries [39]. We also mention here two key theoretical references: general g-convex optimization [4], and wider mathematical background [13].

Acknowledgments SS acknowledges partial support from NSF grant IIS-1409802.

References

1. P.A. Absil, R. Mahony, R. Sepulchre, *Optimization Algorithms on Matrix Manifolds* (Princeton University Press, Princeton, 2009)
2. M. Arnaudon, F. Barbaresco, L. Yang, Riemannian medians and means with applications to radar signal processing. IEEE J. Sel. Top. Signal Process. **7**(4), 595–604 (2013)
3. D. Arthur, S. Vassilvitskii, K-means++: the advantages of careful seeding, in *Proceedings of the Eighteenth Annual ACM-SIAM Symposium on Discrete Algorithms (SODA)* (2007), pp. 1027–1035
4. M. Bacák, *Convex Analysis and Optimization in Hadamard Spaces*, vol. 22 (Walter de Gruyter GmbH & Co KG, Berlin, 2014)
5. F. Bach, R. Jenatton, J. Mairal, G. Obozinski, Optimization with sparsity-inducing penalties. Foundations and Trends® in Machine Learning **4**(1), 1–106 (2012)
6. R. Bhatia, *Positive Definite Matrices* (Princeton University Press, Princeton, 2007)
7. R. Bhatia, R.L. Karandikar, The matrix geometric mean. Technical report, isid/ms/2-11/02, Indian Statistical Institute (2011)

8. D.A. Bini, B. Iannazzo, Computing the Karcher mean of symmetric positive definite matrices. Linear Algebra Appl. **438**(4), 1700–1710 (2013)
9. D.A. Bini, B. Iannazzo, B. Jeuris, R. Vandebril, Geometric means of structured matrices. BIT Numer. Math. **54**(1), 55–83 (2014)
10. C.M. Bishop, *Pattern Recognition and Machine Learning* (Springer, New York, 2007)
11. N. Boumal, Optimization and estimation on manifolds. Ph.D. thesis, Université catholique de Louvain (2014)
12. N. Boumal, B. Mishra, P.A. Absil, R. Sepulchre, Manopt, a matlab toolbox for optimization on manifolds. J. Mach. Learn. Res. **15**(1), 1455–1459 (2014)
13. M.R. Bridson, A. Haefliger, *Metric Spaces of Non-positive Curvature*, vol. 319 (Springer Science & Business Media, Berlin, 1999)
14. S. Burer, R.D. Monteiro, Y. Zhang, Solving semidefinite programs via nonlinear programming. part i: transformations and derivatives. Technical report, TR99-17, Rice University, Houston TX (1999)
15. Z. Chebbi, M. Moahker, Means of Hermitian positive-definite matrices based on the log-determinant α-divergence function. Linear Algebra Appl. **436**, 1872–1889 (2012)
16. A. Cherian, S. Sra, Riemannian dictionary learning and sparse coding for positive definite matrices. IEEE Trans. Neural Netw. Learn. Syst. (2015) (Submitted)
17. A. Cherian, S. Sra, Positive definite matrices: data representation and applications to computer vision, *Riemannian Geometry in Machine Learning, Statistics, Optimization, and Computer Vision*, Advances in Computer Vision and Pattern Recognition (Springer, New York, 2016) (this book)
18. A. Cherian, S. Sra, A. Banerjee, N. Papanikolopoulos, Jensen-Bregman logdet divergence for efficient similarity computations on positive definite tensors. IEEE Trans. Pattern Anal. Mach. Intell. (2012)
19. S. Dasgupta, Learning mixtures of Gaussians, in *40th Annual Symposium on Foundations of Computer Science* (IEEE, 1999), pp. 634–644
20. A.P. Dempster, N.M. Laird, D.B. Rubin, Maximum likelihood from incomplete data via the EM algorithm. J. R. Stat. Soc. Ser. B **39**, 1–38 (1977)
21. R.O. Duda, P.E. Hart, D.G. Stork, *Pattern Classification*, 2nd edn. (Wiley, New York, 2000)
22. R. Hosseini, M. Mash'al, Mixest: an estimation toolbox for mixture models (2015). arXiv:1507.06065
23. R. Hosseini, S. Sra, Matrix manifold optimization for Gaussian mixtures, in *Advances in Neural Information Processing Systems (NIPS)* (2015)
24. J.B. Hough, M. Krishnapur, Y. Peres, B. Virág et al., Determinantal processes and independence. Probab. Surv. **3**, 206–229 (2006)
25. W. Huang, K.A. Gallivan, P.A. Absil, A Broyden class of quasi-Newton methods for Riemannian optimization. SIAM J. Optim. **25**(3), 1660–1685 (2015)
26. B. Jeuris, R. Vandebril, B. Vandereycken, A survey and comparison of contemporary algorithms for computing the matrix geometric mean. Electron. Trans. Numer. Anal. **39**, 379–402 (2012)
27. J.T. Kent, D.E. Tyler, Redescending M-estimates of multivariate location and scatter. Ann. Stat. **19**(4), 2102–2119 (1991)
28. D. Le Bihan, J.F. Mangin, C. Poupon, C.A. Clark, S. Pappata, N. Molko, H. Chabriat, Diffusion tensor imaging: concepts and applications. J. Magn. Reson. Imaging **13**(4), 534–546 (2001)
29. H. Lee, Y. Lim, Invariant metrics, contractions and nonlinear matrix equations. Nonlinearity **21**, 857–878 (2008)
30. J.M. Lee, *Introduction to Smooth Manifolds*, vol. 218, GTM (Springer, New York, 2012)
31. B. Lemmens, R. Nussbaum, *Nonlinear Perron-Frobenius Theory* (Cambridge University Press, Cambridge, 2012)
32. Y. Lim, M. Pálfia, Matrix power means and the Karcher mean. J. Funct. Anal. **262**, 1498–1514 (2012)
33. J. Ma, L. Xu, M.I. Jordan, Asymptotic convergence rate of the EM algorithm for Gaussian mixtures. Neural Comput. **12**(12), 2881–2907 (2000)
34. Z. Mariet, S. Sra, Diversity networks (2015). arXiv:1511.05077

35. Z. Mariet, S. Sra, Fixed-point algorithms for learning determinantal point processes, in *International Conference on Machine Learning (ICML)* (2015)
36. J. Masci, D. Boscaini, M.M. Bronstein, P. Vandergheynst, ShapeNet: convolutional neural networks on non-Euclidean manifolds (2015). arXiv:1501.06297
37. G.J. McLachlan, D. Peel, *Finite Mixture Models* (Wiley, New Jersey, 2000)
38. A. Mehrjou, R. Hosseini, B.N. Araabi, Mixture of ICAs model for natural images solved by manifold optimization method, in *7th International Conference on Information and Knowledge Technology* (2015)
39. B. Mishra, A Riemannian approach to large-scale constrained least-squares with symmetries. Ph.D. thesis, Université de Namur (2014)
40. M. Moakher, A differential geometric approach to the geometric mean of symmetric positive-definite matrices. SIAM J. Matrix Anal. Appl. (SIMAX) **26**, 735–747 (2005)
41. K.P. Murphy, *Machine Learning: A Probabilistic Perspective* (MIT Press, Cambridge, 2012)
42. F. Nielsen, R. Bhatia (eds.), *Matrix Information Geometry* (Springer, New York, 2013)
43. E. Ollila, D. Tyler, V. Koivunen, H.V. Poor, Complex elliptically symmetric distributions: survey, new results and applications. IEEE Trans. Signal Process. **60**(11), 5597–5625 (2011)
44. R.A. Redner, H.F. Walker, Mixture densities, maximum likelihood, and the EM algorithm. Siam Rev. **26**, 195–239 (1984)
45. W. Ring, B. Wirth, Optimization methods on Riemannian manifolds and their application to shape space. SIAM J. Optim. **22**(2), 596–627 (2012)
46. B. Schölkopf, A.J. Smola, *Learning with Kernels* (MIT Press, Cambridge, 2002)
47. A. Shrivastava, P. Li, A new space for comparing graphs, in *IEEE/ACM International Conference on Advances in Social Networks Analysis and Mining (ASONAM)* (IEEE, 2014), pp. 62–71
48. S. Sra, On the matrix square root and geometric optimization (2015). arXiv:1507.08366
49. S. Sra, Positive definite matrices and the S-divergence, in *Proceedings of the American Mathematical Society* (2015). arXiv:1110.1773v4
50. S. Sra, R. Hosseini, Geometric optimisation on positive definite matrices for elliptically contoured distributions, in *Advances in Neural Information Processing Systems* (2013), pp. 2562–2570
51. S. Sra, R. Hosseini, Conic geometric optimisation on the manifold of positive definite matrices. SIAM J. Optim. **25**(1), 713–739 (2015)
52. S. Sra, R. Hosseini, L. Theis, M. Bethge, Data modeling with the elliptical gamma distribution, in *Artificial Intelligence and Statistics (AISTATS)*, vol. 18 (2015)
53. A.C. Thompson, On certain contraction mappings in partially ordered vector space. Proc. AMS **14**, 438–443 (1963)
54. R. Tibshirani, Regression shrinkage and selection via the Lasso. J. R. Stat. Soc. Ser. B (Methodol.) **58**, 267–288 (1996)
55. C. Udrişte, *Convex Functions and Optimization Methods on Riemannian Manifolds* (Kluwer, Dordrecht, 1994)
56. R.J. Vanderbei, H.Y. Benson, On formulating semidefinite programming problems as smooth convex nonlinear optimization problems. Technical report, Princeton (2000)
57. B. Vandereycken, Riemannian and multilevel optimization for rank-constrained matrix problems. Ph.D. thesis, Department of Computer Science, KU Leuven (2010)
58. J.J. Verbeek, N. Vlassis, B. Kröse, Efficient greedy learning of Gaussian mixture models. Neural Comput. **15**(2), 469–485 (2003)
59. A. Wiesel, Geodesic convexity and covariance estimation. IEEE Trans. Signal Process. **60**(12), 6182–6189 (2012)
60. A. Wiesel, Unified framework to regularized covariance estimation in scaled Gaussian models. IEEE Trans. Signal Process. **60**(1), 29–38 (2012)
61. L. Xu, M.I. Jordan, On convergence properties of the EM algorithm for Gaussian mixtures. Neural Comput. **8**, 129–151 (1996)
62. F. Yger, A review of kernels on covariance matrices for BCI applications, in *IEEE International Workshop on Machine Learning for Signal Processing (MLSP)* (IEEE, 2013), pp. 1–6

63. J. Zhang, L. Wang, L. Zhou, W. Li, Learning discriminative Stein Kernel for SPD matrices and its applications (2014). arXiv:1407.1974
64. T. Zhang, Robust subspace recovery by geodesically convex optimization (2012). arXiv:1206.1386
65. T. Zhang, A. Wiesel, S. Greco, Multivariate generalized Gaussian distribution: convexity and graphical models. IEEE Trans. Signal Process. **60**(11), 5597–5625 (2013)
66. D. Zoran, Y. Weiss, Natural images, Gaussian mixtures and dead leaves, in *Advances in Neural Information Processing Systems* (2012), pp. 1736–1744

Chapter 4
Positive Definite Matrices: Data Representation and Applications to Computer Vision

Anoop Cherian and Suvrit Sra

Abstract Numerous applications in computer vision and machine learning rely on representations of data that are compact, discriminative, and robust while satisfying several desirable invariances. One such recently successful representation is offered by *symmetric positive definite* (SPD) matrices. However, the modeling power of SPD matrices comes at a price: rather than a flat Euclidean view, SPD matrices are more naturally viewed through curved geometry (Riemannian or otherwise) which often complicates matters. We focus on models and algorithms that rely on the geometry of SPD matrices, and make our discussion concrete by casting it in terms of covariance descriptors for images. We summarize various commonly used distance metrics on SPD matrices, before highlighting formulations and algorithms for solving sparse coding and dictionary learning problems involving SPD data. Through empirical results, we showcase the benefits of mathematical models that exploit the curved geometry of SPD data across a diverse set of computer vision applications.

4.1 Introduction

Efficient representations that compactly capture salient properties of data form the basis of every algorithm in computer vision and machine learning. Consider for instance the task of tracking a person across video frames given (i) a video sequence, and (ii) a bounding box around the object to be tracked (see Fig. 4.1). Methods for this task typically first generate a representation (descriptor) of the image patch inside the bounding box and then proceed further. Success of the tracking application relies on several desirable properties, for instance (i) ability of the representation to uniquely

A. Cherian (✉)
ARC Centre of Excellence for Robotic Vision, Australian National University, Canberra, Australia
e-mail: anoop.cherian@anu.edu.au
URL: http://www.roboticvision.org

S. Sra
Massachusetts Institute of Technology, Cambridge, MA, USA
e-mail: suvrit@mit.edu

H.Q. Minh and V. Murino (eds.), *Algorithmic Advances in Riemannian Geometry and Applications*, Advances in Computer Vision and Pattern Recognition, DOI 10.1007/978-3-319-45026-1_4

Fig. 4.1 An illustrative application: people tracking in a video sequence. We are given a video sequence and a bounding box of the person to be tracked (*dark box*), and the problem is to track the person along the sequence, i.e., to generate the lighter (*yellow*) boxes in subsequent frames

characterize the contents of the tracked patch against any other patch in a video frame; (ii) robustness of the descriptor to camera sensor noise; (iii) stability despite changes in illuminations; (iv) tolerance for occlusions; and (v) robustness to affine deformations of the object.

A simple descriptor for tracking could be the normalized color histogram of an image patch. This choice partially satisfies properties such as (i), (ii), and (iii). We can improve this naive color descriptor by adding additional statistics of the image patch, such as the shape and pose of the object, texture of the patch, edge orientations, etc. However, each of these additional data features requires high-dimensional descriptors for its representation, whereby the final augmented data descriptor can be extremely large. This in turn raises storage concerns, while also complicating learning, recognition, and tracking due to the curse of dimensionality.

A simple but effective alternative representation that fuses such multi-modal cues was introduced in [53], dubbed *region covariance descriptors*. The key idea is to model an image patch using correlations between different low-level features. To this end, we extract raw image features such as color, intensity gradients, etc., from every pixel within the desired patch. Then we stack these raw features and compute their covariance matrix, resulting in the *covariance descriptor* (for the patch).

Although it seems surprising that such a simple approach can lead to a powerful descriptor, we must note that the diagonal of the covariance matrix captures the statistical variance of each individual feature and the off-diagonals capture the correlation between different features. Thus, this descriptor captures second-order co-occurrences of multimodal features. Since this matrix has size quadratic only in the number of features, it is independent of the number of pixels in the patch and is thus very compact. Also, since the features means are subtracted when computing the covariance, this descriptor implicitly applies mean-filtering to the data, providing noise robustness. Due to these unique properties, the use of covariance descriptors has proliferated into several applications in computer vision and beyond.

We note that SPD matrices also play an important role as data descriptors in several other non-vision applications, such as, diffusion tensor imaging [3], brain-computer interfaces [17], sound compression [45], polarimetric image modeling [24], virus classification [21], and as quantum density matrices [20].

While constructing covariance descriptors is simple (see Sect. 4.1.1), using them can be demanding. The key difficulty arises due to the non-Euclidean nature of covariances. Although it is tempting to simply view SPD matrices using the geometry of their embedding (Euclidean) space [1], the corresponding manifold is not complete (i.e., under this geometry, not all Cauchy sequences will converge within the SPD cone) which can lead to irrelevant solutions for certain applications [4]. Therefore, several alternative geometries for SPD matrices have been considered—Sect. 4.1.2 outlines some of the more widely used choices.

Subsequently, in Sect. 4.2 we illustrate use of these geometries on the concrete tasks of sparse coding and dictionary learning. These problems have been widely studied (for vectors) and are important to a variety of applications [36]. However, their extension to SPD matrix data is non-trivial and less studied; we describe two different frameworks that achieve these extensions.

4.1.1 Covariance Descriptors and Example Applications

Before delving into theoretical details, we recall below details on construction of covariance descriptors and summarize two illustrative applications.

Definition 1 (*Covariance Descriptor*) Suppose we have a data instance $\mathbf{I} \in \mathbb{R}^n$ (image patch, video snippet, etc.). Let $f_1^j, f_2^j, \ldots, f_d^j$, (each $f_i^j \in \mathbb{R}, j = 1, 2, \ldots, n$) represent d features computed for the j-th component of $\mathbf{I}$ (such as image intensity gradients, filter outputs, etc.). Let $\boldsymbol{f}^j = \left[f_1^j, f_2^j, \ldots, f_d^j\right]^T$; then the *Covariance Descriptor* S for $\mathbf{I}$ is the $d \times d$ covariance matrix given by:

$$S = \frac{1}{n}\sum_{j=1}^{n}(\boldsymbol{f}^j - \boldsymbol{\mu})(\boldsymbol{f}^j - \boldsymbol{\mu})^T, \tag{4.1}$$

where $\boldsymbol{\mu} = \frac{1}{n}\sum_{j=1}^{n}\boldsymbol{f}^j$ is the mean feature vector. We will assume that the features are linearly independent, in which case $S \in \mathbb{S}_+^d$, the cone of $d \times d$ SPD matrices.

4.1.1.1 Illustrative Application 1: People Tracking

We continue with our example of people tracking described in Fig. 4.1. One important choice that we must make is what features to use for constructing the descriptor. For people tracking, we will first consider a texture based approach as described in [53] that uses raw image intensity features from each pixel. Given a patch $\mathbf{I}$, let $I_r(x, y), I_g(x, y), I_b(x, y)$ represent the red, green, and blue color intensities respectively of pixel at spatial coordinates (x, y) in the patch. Further, let $I_x(x, y), I_y(x, y)$ represent the gray scale image intensity gradients in the horizontal and vertical directions respectively. Then, we define

$$f(x, y) = \left[I_r(x, y), I_g(x, y), I_b(x, y), I_x(x, y), I_y(x, y)\right]^T \tag{4.2}$$

as the feature vector. A covariance descriptor capturing color and shape of the objects in $\mathbf{I}$ is then given by:

$$S_{tracking} = \frac{1}{|\mathbf{I}|}\sum_{x,y}(f(x, y) - \boldsymbol{\mu})(f(x, y) - \boldsymbol{\mu})^T, \tag{4.3}$$

where $\boldsymbol{\mu} = \frac{1}{|\mathbf{I}|}\sum_{x,y} f(x, y)$ is the mean feature, and $|\mathbf{I}|$ is the patch size. Note that the order of the pixels in the patch are ignored when computing the covariance matrix, thus providing invariance to changes in the pose of the object. However, we may associate additional spatial correlations between the pixel features by including the (x, y) coordinates as well into (4.2). In case, rotational invariance is desired, instead of including (x, y) coordinates, we could include $\sqrt{x^2 + y^2}$ as an additional feature for a pixel at location (x, y). Curvature information could be easily included by computing the second-order gradients for every pixel and augmenting (4.2). This illustration shows the flexibility of the covariance descriptor to blend any feature of choice into the framework easily. The next example shows a richer set of features.

4.1.1.2 Illustrative Application 2: Face Recognition

While using raw features such as gradients could be sufficient to capture textures, recognizing faces require much more expressive and discriminative feature sets. In [39], the authors propose to use covariance descriptors for this task where the features are generated from Gabor wavelets. These filters are believed to reflect the human visual system [37] and measures the energy distribution in the image patch at various scales and orientations. These filters have been previously shown to be useful to extract discriminative and subtle features suitable for recognition [33]; using them within a covariance descriptor setup is shown to lead to significantly better recognition performance in [39], via capturing their second-order statistics. To this end, 40 Gabor wavelet filters are first designed consisting of 8 orientations and 5 different scales. Next, small patches centered at each pixel in the face patch are convolved with these filters, thus forming a 40 dimensional feature vector for each pixel, which precedes applying the steps described above to generate 40×40 covariance descriptors.

Covariance descriptors have been found to be useful in several other vision applications, such as visual surveillance [35, 51, 52], object recognition [23, 29], action recognition [25, 47, 50], image set classification [54], and emotion classification [55], to name a few. Almost all these works use a similar setup for computing the covariances, except that they use various features suitable for the application.

4.1.2 Geometry of SPD Matrices

To effectively use covariance descriptors we must next define schemes to compute similarity/distance between two descriptors. Intuitively, the distance measure enforces some geometry on the space of these descriptors, and which particular geometry is preferable depends on the application. As alluded to previously, a variety of SPD geometries have been explored, the simplest of which is the usual Euclidean geometry where the distance between SPD matrices $\boldsymbol{X}$ and $\boldsymbol{Y}$ is simply the Frobenius distance given by

$$\mathrm{d}_{\mathrm{F}}(\boldsymbol{X}, \boldsymbol{Y}) := \|\boldsymbol{X} - \boldsymbol{Y}\|_F . \tag{4.4}$$

However, d_{F} is neither affine invariant nor does it lead to a complete metric space (due to singular boundary). Affine invariance is important in applications such as diffusion MRI [40] while completeness is crucial when defining convergent sequences on the SPD manifold.

Two basic alternative distances popular in computer vision are (i) the affine invariant Riemannian metric (AIRM) [40], and (ii) the log-Euclidean Riemannian metric (LERM) [4]. Both measures induce a Riemannian geometry; the former induces a curved geometry while the latter "flattens" the manifold by mapping into the tangent space at the identity matrix (which is Euclidean). The corresponding distance functions are

$$\mathrm{d}_{\mathrm{R}}(\boldsymbol{X}, \boldsymbol{Y}) := \left\|\mathrm{Log}\boldsymbol{X}^{-1/2}\boldsymbol{Y}\boldsymbol{X}^{-1/2}\right\|_F , \tag{4.5}$$

$$\mathrm{d}_{\mathrm{LE}}(\boldsymbol{X}, \boldsymbol{Y}) := \|\mathrm{Log}\boldsymbol{X} - \mathrm{Log}\boldsymbol{Y}\|_F , \tag{4.6}$$

where $\boldsymbol{X}^{-1/2}$ is the matrix square-root of SPD matrix $\boldsymbol{X}^{-1}$ and Log is the principal matrix logarithm. Distance (4.5) is affine invariant, and enjoys a host of other remarkable properties [7, Chap. 6]. LERM, however, is not affine invariant though it is rotation and scale invariant separately. Both AIRM and LERM are computationally demanding: for $d \times d$ SPD matrices, assuming approximately $4d^3/3$ flops for eigendecomposition and d^3 for matrix multiplication, both these distances require approximately $14d^3/3$ flops.

To reduce the computational cost, while preserving affine invariance and other geometric properties, the Stein distance was introduced in [15, 48]; it is defined as

$$\mathrm{d}_{\mathrm{S}}(\boldsymbol{X}, \boldsymbol{Y}) := \left[\log\det\left(\tfrac{\boldsymbol{X}+\boldsymbol{Y}}{2}\right) - \tfrac{1}{2}\log\det(\boldsymbol{X}\boldsymbol{Y})\right]^{1/2}. \tag{4.7}$$

Computing d_{S} requires only three Cholesky decompositions, at a total cost of d^3 flops. Moreover, d_{S} is analytically simpler to work with, as its gradients are also simpler to obtain than either AIRM or LERM. Consequently, it has found application is a number of recent works, some of which we will refer to in the sequel.

4.2 Application to Sparse Coding and Dictionary Learning

After outlining a few basic SPD geometries, we are now ready to describe concrete applications where specific properties of SPD matrices play a role. In particular, we discuss sparse coding and dictionary learning for SPD valued data. The first model we cover is *generalized dictionary learning* (GDL), a direct extension of usual (vector) dictionary learning.

Our second model uses the natural Riemannian geometry of SPD matrices, and measures sparse reconstruction loss using the AIRM distance. There are other formulations for dictionary learning and sparse coding on SPD matrices, for instance [25–28, 46]; these differ from our model primarily in the formulation of the sparse reconstruction loss.

4.2.1 *Dictionary Learning with SPD Atoms*

Traditional dictionary learning takes vectors $\boldsymbol{x}_i \in \mathbb{R}^p$ $(1 \leq i \leq m)$ and constructs a matrix $\boldsymbol{B} \in \mathbb{R}^{p\times n}$ and code vectors $\boldsymbol{\alpha}_i \in \mathbb{R}^n$ (usually $n \gg p$), so that

$$\boldsymbol{x}_i \approx \boldsymbol{B}\boldsymbol{\alpha}_i, \quad \text{and} \quad \boldsymbol{\alpha}_i \text{ is sparse}, \quad \text{for } 1 \leq i \leq m.$$

The sparsity requirement on $\boldsymbol{\alpha}_i$ is commonly enforced using ℓ_0- or ℓ_1-norm penalties or constraints. Since both $\boldsymbol{B}$ and $\boldsymbol{\alpha}_i$ are unknown dictionary learning usually results in a difficult nonconvex optimization task. Nevertheless, it has found remarkable success toward sparse coding and other applications [18].

We depart from the above setup in that we consider input matrices $\boldsymbol{X}_i \in \mathbb{R}^{p\times q}$, $1 \leq i \leq m$. Then, instead of a dictionary matrix $\boldsymbol{B}$ we learn a tensor B, which we identify with a linear operator $\mathsf{B} : \mathbb{R}^{n\times r} \to \mathbb{R}^{p\times q}$ so that

$$\boldsymbol{X}_i \approx \mathsf{B}(\boldsymbol{A}_i), \quad \text{and} \quad \boldsymbol{A}_i \text{ is sparse}, \quad \text{for } 1 \leq i \leq m. \tag{4.8}$$

Using (4.8), one model of GDL is the following [49]:

$$\min_{\boldsymbol{A}_1,\ldots,\boldsymbol{A}_m,\mathsf{B}} \quad \tfrac{1}{2}\sum\nolimits_{i=1}^{m} \|\boldsymbol{X}_i - \mathsf{B}(\boldsymbol{A}_i)\|_{\mathrm{F}}^2 + \sum\nolimits_{i=1}^{m} \beta_i \mathrm{sp}(\boldsymbol{A}_i), \tag{4.9}$$

where $\beta_i > 0$ are scalar hyperparameters while $\mathrm{sp}(\boldsymbol{A})$ enforces some notion of sparsity. For instance, $\mathrm{sp}(\boldsymbol{A})$ could be the cardinality function $\|\boldsymbol{A}\|_0$ (which computes the number of non-zero entries in A), its convex relaxation $\|\boldsymbol{A}\|_1$, the matrix-rank $\mathrm{rank}(\boldsymbol{A})$ or its convex relaxation, the trace-norm $\|\boldsymbol{A}\|_{\mathrm{tr}}$.

Formulation (4.9) requires one more modification for SPD valued inputs. Specifically, we need to ensure that the approximation $\mathsf{B}(\boldsymbol{A}_i) \succeq 0$. In this chapter, we explore two variants of this approximation: (i) based on conic combinations of rank-one positive semi-definite matrices (as detailed below) and (ii) based on conic combinations of full-rank SPD matrices (as explored in Sect. 4.2.2). For the first option, let us define B via

$$\mathsf{B}(\boldsymbol{A}) := \boldsymbol{B}\boldsymbol{A}\boldsymbol{B}^T, \quad \text{for some matrix } \boldsymbol{B}, \tag{4.10}$$

and additionally *restricting* to $\boldsymbol{A} \succeq 0$. Observe that (4.10) can be written as

$$\mathrm{vec}(\boldsymbol{B}\boldsymbol{A}\boldsymbol{B}^T) = (\boldsymbol{B} \otimes \boldsymbol{B})\,\mathrm{vec}(\boldsymbol{A}), \tag{4.11}$$

where vec stacks columns of its argument and the operator B is encoded (isomorphically) by the product $\boldsymbol{B} \otimes \boldsymbol{B}$.

It is easy to show that (4.11) requires $md^2 + d^2n + mn$ storage for m covariance matrices of size $d \times d$, while (4.10) takes $md^2 + dn + mn$. Computationally, also the first formulation is cheaper, so we prefer it. As to $\boldsymbol{A}$, we consider two choices:

1. $\boldsymbol{A} = \mathrm{Diag}(\alpha_1, \ldots, \alpha_n)$ where $\alpha_i \geq 0$; and
2. $\boldsymbol{A} = \sum_{j=1}^{k} \boldsymbol{\alpha}_j \boldsymbol{\alpha}_j^T$, a potentially low-rank (if $k < n$) SPD matrix.

Although diagonal $\boldsymbol{A}$ might appear to be simple, it is quite powerful. Indeed, with it GDL models SPD matrices as weighted sums of rank-one matrices since

$$\boldsymbol{X} \approx \boldsymbol{B}\boldsymbol{A}\boldsymbol{B}^T = \sum_{i=1}^{n} \alpha_i \boldsymbol{b}_i \boldsymbol{b}_i^T, \quad \text{where } \alpha_i = \boldsymbol{A}_{ii}, \tag{4.12}$$

which offers a rich yet computationally tractable model.

4.2.1.1 Stochastic Gradient Descent for GDL

Now we derive a stochastic-gradient descent procedure for approximately solving GDL. We use the convex function $\mathrm{sp}(\boldsymbol{A}) = \|\boldsymbol{A}\|_1$ for enforcing sparsity. Then, using representation (4.12) with diagonal $\boldsymbol{A}_i$, the GDL optimization problem (4.9) becomes

$$\min_{\boldsymbol{A}_1,\ldots,\boldsymbol{A}_N \geq 0, \boldsymbol{B}} \quad \tfrac{1}{2}\sum_{i=1}^{m} \|\boldsymbol{X}_i - \boldsymbol{B}\boldsymbol{A}_i\boldsymbol{B}^T\|_{\mathrm{F}}^2 + \sum_{i=1}^{m} \beta_i \|\boldsymbol{A}_i\|_1. \tag{4.13}$$

Problem (4.13) is nonconvex and difficult. However, for a fixed dictionary $\boldsymbol{B}$, it is individually convex in $(\boldsymbol{A}_1, \ldots, \boldsymbol{A}_m)$. It is thus amenable to the idea of alternating between updating $\boldsymbol{B}$ and optimizing over $(\boldsymbol{A}_1, \ldots, \boldsymbol{A}_m)$. But in many applications the number of input data points m is very large, so the alternating steps can easily become rather costly. Therefore, we follow a stochastic gradient approach that can scale to large data sets, as long as the stochastic gradients can be obtained efficiently.

To prevent degenerate solutions we also impose normalization constraints $\|\boldsymbol{b}_j\|_2 \leq 1$ on each column of matrix $\boldsymbol{B}$. We denote these requirements by the feasible set $\mathscr{B}$. We run stochastic gradient using K "mini-batches," for which we rewrite (4.13) as

$$\min_{\boldsymbol{B} \in \mathscr{B}} \quad \Phi(\boldsymbol{B}) := \sum_{b=1}^{K} \phi_b(\boldsymbol{B}), \tag{4.14}$$

where ϕ_b denotes the objective function for batch b. Let k_b be the size of batch b $(1 \le b \le K)$ containing the matrices $\left\{\boldsymbol{X}_{j(i)} | 1 \le i \le k_b\right\}$, where $j(i)$ is an appropriate index in $1, \ldots, m$. With this notation, the objective function for batch b is

$$\phi_b(\boldsymbol{B}) := \min_{\boldsymbol{A}_{j(1)},\ldots,\boldsymbol{A}_{j(k)} \ge 0} \quad \tfrac{1}{2} \sum\nolimits_{i=1}^{k_b} \|\boldsymbol{X}_{j(i)} - \boldsymbol{B}\boldsymbol{A}_{j(i)}\boldsymbol{B}^T\|_{\mathrm{F}}^2 + \beta_{j(i)} \|\boldsymbol{A}_{j(i)}\|_1. \tag{4.15}$$

We apply stochastic-gradient to (4.14), which performs the iteration

$$\boldsymbol{B}_{t+1} = \Pi_{\mathscr{B}}(\boldsymbol{B}_t - \eta_t \nabla_{\boldsymbol{B}} \phi_{b(t)}(\boldsymbol{B}_t)), \quad b(t) \in [1..K],\ t = 0, 1, \ldots, \tag{4.16}$$

where $\Pi_{\mathscr{B}}$ denotes orthogonal projection onto $\mathscr{B}$, i.e., normalizing each column of the dictionary matrix to have unit norm. Assuming (4.15) has a unique solution, the gradient $\nabla_{\boldsymbol{B}} \phi_{b(t)}$ is well defined. Specifically, let $(\boldsymbol{A}^*_{j(1)}, \ldots, \boldsymbol{A}^*_{j(k)})$ be the argmin of (4.15). Then, writing $b \equiv b(t)$, we have

$$\nabla_{\boldsymbol{B}} \phi_b(\boldsymbol{B}) = 2 \sum\nolimits_{i=1}^{k_b} \left(\boldsymbol{B}\boldsymbol{A}^*_{j(i)}\boldsymbol{B}^T - \boldsymbol{X}_{j(i)}\right)\boldsymbol{B}\boldsymbol{A}^*_{j(i)}. \tag{4.17}$$

The computationally intensive part is to compute (4.17), which we now consider.

4.2.1.2 Sparse Coding: Computing $\nabla \phi_b$

Observe that (4.15) is a sum of k_b independent problems, so it suffices to describe the computation for a subproblem of the form

$$\min_{\boldsymbol{A} \ge 0} \ f(\boldsymbol{A}) := \tfrac{1}{2} \|\boldsymbol{X} - \boldsymbol{B}\boldsymbol{A}\boldsymbol{B}^T\|_{\mathrm{F}}^2 + \beta \|\boldsymbol{A}\|_1. \tag{4.18}$$

Since $\boldsymbol{A} \ge 0$ and diagonal, problem (4.18) is nothing but a regularized nonnegative least-squares (NNLS) problem. There exist a variety of solvers for NNLS, for example, LBFGS-B [34], or Spectral Projected-Gradient (SPG) [8]. We prefer to use the latter, as it is not only simple, but also exhibits excellent empirical performance. In a nutshell, SPG minimizes a given objective in a closed and convex set via a sequence of steepest descent and projection steps. Each descent iteration uses a non-monotone line search for stepsize selection with the well-known Barzellai–Borwein secant algorithm.

In Sect. 4.3, we will apply this sparse coding scheme to the problem of nearest neighbor retrieval on covariance datasets. But, before proceeding to the experiments, we will elucidate a much richer and more powerful dictionary learning and sparse coding scheme on SPD matrices that leads to significantly better results on various applications; this scheme uses the natural Riemannian geometry for the sparse construction loss instead of the Euclidean distance as in (4.9).

4.2.2 Riemannian Dictionary Learning and Sparse Coding

Recall that we wish to compute a dictionary with "SPD atoms". We work on manifold $\mathbb{M}_n^d = \prod_{i=1}^n \mathbb{S}_+^d \subset \mathbb{R}^{d\times d\times n}$, which is the Cartesian product of n SPD manifolds. Our goals are (i) to learn a dictionary $\mathsf{B} \in \mathbb{M}_n^d$ in which each slice represents an SPD dictionary atom $\boldsymbol{B}_j \in \mathbb{S}_+^d$ $1 \le j \le n$; and (ii) to approximate each $\boldsymbol{X}_i$ as a sparse conic combination of atoms in B; i.e., $\boldsymbol{X}_i \sim \mathsf{B}(\boldsymbol{\alpha}_i)$ where $\alpha_i \in \mathbb{R}_+^n$ and $\mathsf{B}(\boldsymbol{\alpha}) := \sum_{i=1}^n \boldsymbol{\alpha}_i \boldsymbol{B}_i$. With this notation our *dictionary learning and sparse coding* (DLSC) problem is

$$\min_{\mathsf{B}\in\mathbb{M}_n^d,\boldsymbol{\alpha}\in\mathbb{R}_+^{n\times m}} \tfrac{1}{2}\sum\nolimits_{j=1}^m \mathrm{d_R}^2\left(\boldsymbol{X}_j, \mathsf{B}\boldsymbol{\alpha}_j\right) + \mathrm{Sp}(\boldsymbol{\alpha}_\mathrm{j}) + \Omega(\mathsf{B}), \tag{4.19}$$

where Sp and Ω regularize the codes $\boldsymbol{\alpha}_j$ and the dictionary tensor B respectively.

Formulation (4.19) is a direct SPD analog of the vector DL setup. Instead of learning a dictionary matrix for vectors, we learn a third-order tensor dictionary as our input is matricial data. We constraint the sparse codes to be non-negative to ensure that the linear combination $\mathsf{B}(\boldsymbol{\alpha})$ remains SPD. In contrast to usual DL problems where the dictionary learning and sparse coding subproblems are convex, problem (4.19) is much harder: it is neither convex in itself nor are its subproblems convex.

Pragmatically speaking, this lack of subproblem convexity is not too damaging: we just need a set of dictionary atoms that can sparse code the input data, and such a set can still be computed by performing an alternating minimization (actually, just descent here). We describe the details below.

4.2.2.1 Dictionary Learning Subproblem

Assuming that the sparse code vectors $\boldsymbol{\alpha}$ are available, the subproblem of updating the dictionary atoms can be separated from (4.19) and written as:

$$\begin{aligned}\min_{\mathsf{B}\in\mathbb{M}_n^d} \Psi(\mathsf{B}) &:= \tfrac{1}{2}\sum\nolimits_{j=1}^m \mathrm{d}_\mathrm{R}^2\left(\boldsymbol{X}_j, \mathsf{B}\alpha_j\right) + \Omega(\mathsf{B}),\\ &= \tfrac{1}{2}\sum\nolimits_{j=1}^m \left\|\mathrm{Log}\left(\boldsymbol{X}_\mathrm{j}^{-1/2}\left(\mathsf{B}\alpha_\mathrm{j}\right)\boldsymbol{X}_\mathrm{j}^{-1/2}\right)\right\|_F^2 + \Omega(\mathsf{B}).\end{aligned} \tag{4.20}$$

Due to its good empirical performance we choose $\Omega(\mathsf{B}) := \lambda_\mathsf{B}\sum_{i=1}^n \mathrm{Tr}(\boldsymbol{B}_i)$.

Riemannian CG

Due to the excellent empirical performance observed for the Riemannian Conjugate Gradient (RCG) algorithm [1, Chap. 8] for optimizing over the SPD atoms (in comparison to other first-order alternatives such as the steepest-descent, trust-region methods [2], etc.), we decided to use RCG. For completeness of our presentation, we provide below a brief review of this optimization technique.

For some non-linear function $\psi(x),\ x \in \mathbb{R}^n$, the CG method uses the following recurrence at step $k+1$

$$x_{k+1} = x_k + \gamma_k \xi_k, \tag{4.21}$$

where the descent direction ξ_k is

$$\xi_k = -\text{grad}\,\psi(x_k) + \mu_k \xi_{k-1}, \tag{4.22}$$

with grad $\psi(x_k)$ defining gradient of ψ at x_k ($\xi_0 = -\text{grad}\,\psi(x_0)$), and μ_k given by

$$\mu_k = \frac{(\text{grad}\,\psi(x_k))^T(\text{grad}\,\psi(x_k) - \text{grad}\,\psi(x_{k-1}))}{\text{grad}\,\psi(x_{k-1})^T \text{grad}\,\psi(x_{k-1})}, \tag{4.23}$$

The step-size γ_k in (4.21) is usually found via a line-search method [6]. When ψ is quadratic with a Hessian Q, the directions generated by (4.22) are Q-conjugate to previous directions of descent $\xi^0, \xi^1, \cdots, \xi^{k-1}$ (i.e., $\xi^{k-1}Q\xi^k = 0$); thereby (4.21) providing the exact minimizer of f in fewer than d iterations (d is the manifold dimension) [6, Sect. 1.6].

For $\mathsf{B} \in \mathbb{M}_n^d$ and referring back to (4.20), the recurrence in (4.21) uses the Riemannian retraction [1, Chap. 4] and the gradient grad $\Psi(\mathsf{B}_k)$ is the Riemannian gradient (here B_k represents the dictionary tensor at the k-th iteration). This leads to an important issue: the gradients grad $\Psi(\mathsf{B}_k)$ and grad $\Psi(\mathsf{B}_{k-1})$ belong to two different tangent spaces $T_{\mathsf{B}_k}\mathbb{M}$ and $T_{\mathsf{B}_{k-1}}\mathbb{M}$ respectively, and thus cannot be combined as in (4.23). Thus, following [1, Chap. 8] we resort to vector transport – a scheme to transport a tangent vector at $P \in \mathbb{M}$ to a point $\text{Exp}_P(S)$ where $S \in T_P\mathbb{M}$ and Exp_P is the exponential map with foot at P. The resulting formula for the direction update becomes

$$\xi_{\mathsf{B}_k} = -\text{grad}\,\Psi(\mathsf{B}_k) + \mu_k \mathfrak{T}_{\gamma_k\xi_{k-1}}(\xi_{k-1}), \tag{4.24}$$

where

$$\mu_k = \frac{\left\langle \text{grad}\,\Psi(\mathsf{B}_k), \text{grad}\,\Psi(\mathsf{B}_k) - \mathfrak{T}_{\gamma_k\xi^{k-1}}(\text{grad}\,\Psi(\mathsf{B}_{k-1}))\right\rangle}{\langle \text{grad}\,\Psi(\mathsf{B}_{k-1}), \text{grad}\,\Psi(\mathsf{B}_{k-1})\rangle}. \tag{4.25}$$

Here for $Z_1, Z_2 \in T_P\mathbb{M}$, the map $\mathfrak{T}_{Z_1}(Z_2)$ defines the vector transport given by:

$$\mathfrak{T}_{Z_1}(Z_2) = \frac{d}{dt}\text{Exp}_P(Z_1 + tZ_2)\bigg|_{t=0}. \tag{4.26}$$

It remains to derive an expression for the Riemannian gradient grad $\Psi(\mathsf{B})$.

Riemannian Gradient

Lemma 1 connects the Riemannian gradient to the Euclidean gradient of $\Psi(\mathsf{B})$.

Lemma 1 *For dictionary tensor* $\mathsf{B} \in \mathbb{M}_n^d$, *let* $\Psi(\mathsf{B})$ *be a differentiable function. Then, the Riemannian gradient* grad $\Psi(\mathsf{B})$ *satisfies:*

$$\langle \text{grad}\, \Psi(\mathsf{B}), \zeta \rangle_{\mathsf{B}} = \langle \nabla \Psi(\mathsf{B}), \zeta \rangle_I, \forall \zeta \in T_P \mathbb{M}_n^d, \tag{4.27}$$

where $\nabla \Psi(\mathsf{B})$ *is the Euclidean gradient of* $\Psi(\mathsf{B})$. *The Riemannian gradient for the i-th dictionary atom is given by* $\text{grad}_{\,i} \Psi(\mathsf{B}) = \boldsymbol{B}_i \nabla_{\boldsymbol{B}_i} \Psi(\mathsf{B}) \boldsymbol{B}_i$.

Proof See [1, Chap. 5]. The final result is obtained by substituting the definition of the Riemannian metric (4.5) to the LHS of (4.27).

The Euclidean gradient $\nabla \Psi(\mathsf{B})$ is obtained as follows: let $\boldsymbol{S}_j = \boldsymbol{X}_j^{-1/2}$ and $M_j(\mathsf{B}) := \mathsf{B}(\boldsymbol{\alpha}_j) = \sum_{i=1}^n \boldsymbol{\alpha}_j^i \boldsymbol{B}_i$. Then,

$$\Psi(\mathsf{B}) = \tfrac{1}{2} \sum_{j=1}^m \operatorname{Tr}(\operatorname{Log}(\boldsymbol{S}_j M_j(\mathsf{B}) \boldsymbol{S}_j)^2) + \lambda_{\mathsf{B}} \sum_{i=1}^n \operatorname{Tr}(\boldsymbol{B}_i). \tag{4.28}$$

The derivative $\nabla_{\boldsymbol{B}_i} \Psi(\mathsf{B})$ of (4.28) w.r.t. to atom $\boldsymbol{B}_i$ is:

$$\sum_{j=1}^m \alpha_j^i \big(\boldsymbol{S}_j \operatorname{Log}(M_j(\mathsf{B})) \big(M_j(\mathsf{B})\big)^{-1} \boldsymbol{S}_j\big) + \lambda_{\mathsf{B}} I. \tag{4.29}$$

4.2.2.2 Sparse Coding Subproblem

Referring back to (4.19), we now consider the sparse coding subproblem. Given a dictionary tensor B and a data matrix $\boldsymbol{X}_j \in \mathbb{S}_+^d$, this subproblem requires solving

$$\begin{aligned} \min_{\boldsymbol{\alpha}_j \geq 0} \quad & \phi(\boldsymbol{\alpha}_j) := \tfrac{1}{2} \mathrm{d}_{\mathrm{R}}^2 \left(\boldsymbol{X}_i, \mathsf{B}(\boldsymbol{\alpha}_j)\right) + \operatorname{Sp}(\boldsymbol{\alpha}_j) \\ & = \tfrac{1}{2} \left\| \operatorname{Log}\Big(\sum_{\mathrm{i}=1}^{\mathrm{n}} \alpha_{\mathrm{j}}^{\mathrm{i}} \boldsymbol{X}^{-1/2} \boldsymbol{B}_{\mathrm{j}} \boldsymbol{X}^{-1/2}\Big) \right\|_F^2 + \operatorname{Sp}(\boldsymbol{\alpha}_j), \end{aligned} \tag{4.30}$$

where α_j^i is the i-th component of $\boldsymbol{\alpha}_j$ and Sp is a sparsity inducing function. For simplicity, we use $\operatorname{Sp}(\boldsymbol{\alpha}) = \lambda \|\boldsymbol{\alpha}\|_1$, where $\lambda > 0$ is a regularization parameter. Since we are working with $\boldsymbol{\alpha} \geq 0$, we replace this penalty by $\lambda \sum_i \alpha_i$, which is differentiable.

Problem (4.30) measures reconstruction quality offered by a sparse non-negative linear combination of the atoms to a given input point $\boldsymbol{X}$. It will turn out (see experiments in Sect. 4.3) that the reconstructions obtained via this model actually lead to significant improvements in performance over sparse coding models that ignore SPD geometry. But this gain comes at a price: objective (4.30) is difficult to optimize, and remains difficult even if we take into account geodesic convexity of d_{R}.

While in practice this non-convexity does not seem to hurt our model, we digress below to show a surprising but intuitive constraint under which Problem (4.30) actually becomes convex. Although we do not exploit this observation to save on computation, we highlight it here due to its theoretical appeal.

Theorem 1 *([14]) The function* $\phi(\alpha) := \mathrm{d}_{\mathrm{R}}^2(\sum_i \alpha_i B_i, X)$ *is convex on the set*

$$\mathscr{A} := \{\alpha \mid \sum_i \alpha_i B_i \preceq X, \ \textit{and}\ \alpha \geq 0\}. \tag{4.31}$$

4.2.2.3 Optimizing Sparse Codes

We minimize (4.30) using a projected gradient method. Specifically, we run the iteration

$$\alpha^{k+1} \leftarrow \mathscr{P}[\alpha^k - \eta_k \nabla\phi(\alpha^k)], \qquad k = 0, 1, \ldots, \tag{4.32}$$

where $\mathscr{P}[\cdot]$ denotes the projection operator defined as

$$\mathscr{P}[\alpha] \equiv \alpha \mapsto \operatorname{argmin}_{\alpha'} \tfrac{1}{2}\|\alpha' - \alpha\|_2^2, \quad \text{s.t.}\,,\ \alpha' \in \mathscr{A}. \tag{4.33}$$

To implement iteration (4.32) we need to specify three components: (i) the stepsize η_k; (ii) the gradient $\nabla\phi(\alpha^k)$; and (iii) the projection (4.33). Lemma 2 shows how to compute the gradient. The projection task (4.33) is a special least-squares (dual) semidcfinite program (SDP), which can be solved using any SDP solver. However, in the interest of speed, we avoid the heavy computational burden imposed by an SDP, and drop the constraint $\alpha \in \mathscr{A}$. Although this sacrifices convexity, the resulting computation is vastly easier, and works well empirically. With this change, we simply have $\mathscr{P}[\alpha] = \max(0, \alpha)$.

It remains to specify how to obtain the stepsize η_k. There are several choices in the nonlinear programming literature [6], but most of them can be expensive in our setting. We wish to avoid expensive iterative algorithms for computing η_k, and thus choose to use Barzilai-Borwein stepsizes [5] that have closed forms and that often work remarkably well in practice [5, 44]. In particular, we use the Spectral Projected Gradient (SPG) method [9] by adapting a simplified implementation of [44].

Lemma 2 *Let $\boldsymbol{B}$, $\boldsymbol{C}$, and $\boldsymbol{X}$ be fixed SPD matrices. Consider the function* $f(x) := \mathrm{d}_{\mathrm{R}}^2(x\boldsymbol{B} + \boldsymbol{C}, \boldsymbol{X})$. *The derivative $f'(x)$ is given by*

$$f'(x) = 2\,\mathrm{Tr}(\log(\boldsymbol{S}(x\boldsymbol{B} + \boldsymbol{C})\boldsymbol{S})\boldsymbol{S}^{-1}(x\boldsymbol{B} + \boldsymbol{C})^{-1}\boldsymbol{B}\boldsymbol{S}), \qquad \textit{where } \boldsymbol{S} = \boldsymbol{X}^{-1/2}. \tag{4.34}$$

Proof Introduce the shorthand $\boldsymbol{M}(x) \equiv x\boldsymbol{B} + \boldsymbol{C}$. Using (4.5) we have

$$f(x) = \mathrm{Tr}([\log(\boldsymbol{S}\boldsymbol{M}(x)\boldsymbol{S})]^T[\log(\boldsymbol{S}\boldsymbol{M}(x)\boldsymbol{S})]),$$

The chain-rule of calculus then immediately yields the desired result

$$f'(x) = 2\,\mathrm{Tr}(\log(\boldsymbol{S}\boldsymbol{M}(x)\boldsymbol{S})(\boldsymbol{S}\boldsymbol{M}(x)\boldsymbol{S})^{-1}\boldsymbol{S}\boldsymbol{M}'(x)\boldsymbol{S}).$$

Writing $\boldsymbol{M}(\alpha_p) = \alpha_p\boldsymbol{B}_p + \sum_{i\neq p}\alpha_i\boldsymbol{B}_i$ and using Lemma 2 we obtain

$$\frac{\partial\phi(\alpha)}{\partial\alpha_p} = \mathrm{Tr}\big(\log\big(\boldsymbol{S}\boldsymbol{M}(\alpha_p)\boldsymbol{S}\big)\big(\boldsymbol{S}\boldsymbol{M}(\alpha_p)\boldsymbol{S}\big)^{-1}\boldsymbol{S}\boldsymbol{B}_p\boldsymbol{S}\big) + \lambda. \tag{4.35}$$

Computing (4.35) for all α is the dominant when running SPG. A a naïve implementation of (4.35) costs $O(nd^3)$, but with slight care this cost can be reduced $O(nd^2) + O(d^3)$ [14].

4.3 Applications of Sparse Coding

In this section, we describe an application of sparse coding for covariances, namely nearest neighbor (NN) retrieval. This is a fundamental task in several computer vision and machine learning applications in which the goal is to find a data point closest to a given query point within a large database.

4.3.1 Nearest Neighbors on Covariance Descriptors

Suppose we have obtained a sparse code matrix $\boldsymbol{A}$ using either GDL or Riemannian DLSC for an input matrix $\boldsymbol{X}$. Since, we use an overcomplete dictionary typically only a few dictionary atoms participate in the reconstruction of $\boldsymbol{X}$. Thus, with high probability dissimilar input points will obtain different sparse codes. In other words, suppose that we use a dictionary with n rank-one atoms and that only r of these matrices are used to obtain a reconstruction for a given input. Then, there are $\binom{n}{r}$ unique basis combinations possible. With appropriate choices of n and r, we will likely obtain a unique set of rank-one matrices that encode $\boldsymbol{X}$.

Using this intuition, we propose a sorted integer tuple representation to encode an input covariance matrix; the integers simply index the dictionary atoms used in the encoding. Formally, let $\boldsymbol{X} \in \mathbb{S}_+^d$ be the input, B an overcomplete dictionary, and $\boldsymbol{u}_i$ ($i \in [n]$) a unique identifier for the i-th atom of B. If $\boldsymbol{\alpha} = (\alpha_1, \alpha_2, \ldots, \alpha_n)^T$ is the coefficient vector corresponding to $\boldsymbol{X} \approx \mathsf{B}(\boldsymbol{\alpha})$, then the tuple $h(\boldsymbol{X}) = \langle \boldsymbol{u}_i, \ldots, \boldsymbol{u}_k \rangle$ is the *hash code* of $\boldsymbol{X}$. We choose only those identifiers for which the code α_j is larger than a threshold $\varepsilon \geq 0$.

In our case, we assume that the u_i's are just integers in $\{1, \ldots, n\}$ corresponding to the dimension index of the respective $\alpha_j > 0$, and that the hash code is thus a sorted tuple of these indices. For example, suppose $\alpha_1, \alpha_{10}, \alpha_{12}$ are the sparse coefficients that are greater than a threshold ε after the sparse coding of a given data point $\boldsymbol{X}$. Then, the hash code will be $h(\boldsymbol{X}) = \langle 1, 10, 12 \rangle$. The threshold ε helps select significant coefficients from the sparse coding, and makes the chosen code robust to noise. This coded representation enables the use of hash tables for fast locality sensitive hashing. Let us see how. Each column of the dictionary is identified by its index number; so each hash-key is a set of integers encoded as a character string. To tackle collisions in the hash buckets, the colliding input matrices are organized as a linked list. If the linked list gets too long, the data within a hash bucket can be further organized using a metric tree or any other efficient data structure. This idea of hashing is a direct adaptation of the scheme proposed in [12].

Given a query SPD matrix, we solve the sparse coding problem to first obtain the corresponding sparse coefficients. From these coefficients we compute the above hash code query the hash table. If there are several entries in a matching bucket, we run a linear scan using the AIRM distance (4.5) to find the best matches (the bucket can also be organized for faster than linear scans, if desired).

4.3.2 GDL Experiments

This section illustrates nearest neighbor search based upon our dictionary learning examples. We use the following datasets:

- **Face recognition.** The *FERET face dataset* [41, 42] contains facial appearances segregated into multiple classes. Each class has different views of the face of the same person for varying poses. We selected six images from each class. Inspired by the success of covariances created from Gabor filters for face recognition [33], we applied 40 Gabor filters on each image, later combining the filters into a covariance of size 40×40. We created a covariance dataset of approximately 10 K descriptors using this approach.
- **Texture classification.** Texture is an essential cue in many data mining applications like satellite imagery, industry inspection systems, etc. Thus, we used a combination of the *Brodatz dataset* [11] and the *Current dataset* [16] for creating a texture covariance dataset. Brodatz dataset contains approximately 111 texture classes, while Curret dataset contains 60 classes. To create the covariances data, we used the feature vector $F = [x, y, I, I_x, I_y]$, where the first two dimensions are the relative location of the pixel with respect to the texture patch, the third dimension encodes the grayscale intensity, and the last two dimensions capture the pixel gradients. Thus, each covariance is 5×5, and we created approximately 40 K such covariances.

Methods Compared. We compare against locality sensitive hashing (LSH) of vectorized covariances (VEC), hashing after log-Euclidean embedding (L2LSH), and kernelized LSH [30] using an RBF kernel using the AIRM distance.

4.3.2.1 GDL Experimental Setup

We implemented GDL in Matlab. For L2LSH, VEC, and HAM we used the C-implementation from the Caltech Toolbox.[1] Since the programs have different computational baselines, we cannot compare their retrieval speed. Rather, we show in Table 4.1 the average portion of each of the datasets scanned by GDL to find the nearest neighbor. The geodesic distance was used to resolve hash table collisions. As is seen from the table, the percentage coverage is low, which is exactly as desired.

[1] http://www.vision.caltech.edu/malaa/software/research/image-search/.

Table 4.1 Percentage of the database searched to find the nearest neighbor using sparse codes generated by GDL

Dataset	Faces	Texture
Avg. coverage (%)	3.54	6.26

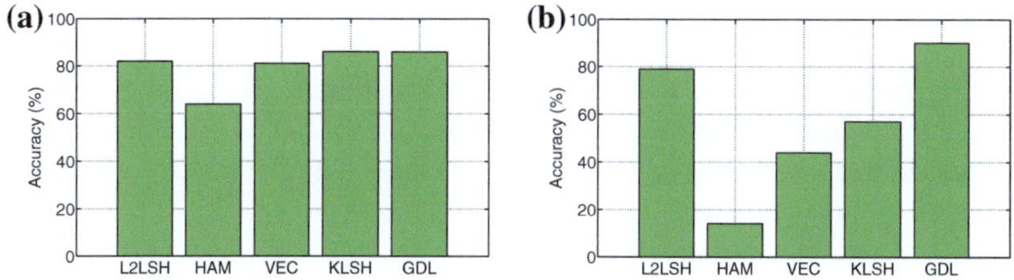

Fig. 4.2 Plots demonstrating nearest neighbor classification accuracy of GDL generated sparse codes compared to various standard techniques; **a** on faces dataset and **b** on texture dataset

Next, we substantiate the effectiveness of our NN retrieval scheme. To this end, we split each of the datasets into a gallery and query sets (approximately 5 % of the data for the latter). To compute the ground truth, we use a linear scan over the entire gallery set using geodesic distances to measure nearness. Since ensuring exact NN is hard, we restrict our search to Approximate Nearest Neighbors (ANN). Assume $\boldsymbol{Q}$ is a query point, $\boldsymbol{X}_{ls}$ is the exact NN found by a linear scan and $\boldsymbol{X}_{algo}$ is the neighbor returned by an NN algorithm. Using d_{AIRM} as the geodesic distance computed using AIRM distance, we classify an NN as correct if $\frac{d_{AIRM}(\boldsymbol{Q},\boldsymbol{X}_{ls})}{d_{AIRM}(\boldsymbol{Q},\boldsymbol{X}_{algo})} > \varepsilon$; we use $\varepsilon = 0.75$ below. Figure 4.2 shows the accuracy of different methods, where

$$\text{Accuracy} := \frac{\#\text{correct matches}}{\#\text{query size}}. \tag{4.36}$$

The plots indicate that GDL performs well across the datasets, while performance of the other methods varies. Vectorizing input matrices fails on all datasets, while KLSH performs reasonably well. We note, however, that KLSH needs to compute the kernel matrix for the query point against the *entire* dataset—this can drastically slow it down. On the face dataset, all methods had high accuracy, most probably because this dataset is noise free.

4.3.3 Riemannian Dictionary Learning Experiments

Next, we evaluate the Riemannian DLSC setup, and denote below dictionary learning by DL and sparse coding by SC. We compare our Riemannian (Riem) formulation against combinations of several state-of-the-art DLSC methods on SPD matrices, namely (i) log-Euclidean (LE) metric for DLSC [25], (ii) Frobenius norm (Frob) which discards the manifold structure, (iii) kernel methods such as the Stein-Kernel [48] proposed in [26], and the log-Euclidean kernel [32].

We experiment on data available from three standard computer vision applications: (i) 3D object recognition on the RGBD objects dataset [31]; (ii) texture recognition on the standard Brodatz dataset [38]; and (iii) person re-identification on the ETHZ people dataset [19]. We describe (i) and (iii) below.

- **Person re-identification task.** We use the benchmark ETHZ dataset [43] for evaluating people re-identification. This dataset consists of low-resolution images of tracked people from a real-world surveillance setup. The images are from 146 different individuals. There are about 5–356 images per person. There are a total of 8580 images in this dataset. Rather than detailing the results on several feature combinations, we describe here the feature combination that worked the best in our experiments. For this purpose, we used a validation set of 500 covariances and 10 true clusters from this dataset. The performance was evaluated using the Log-Euclidean SC setup with a dictionary learning via Log-Euclidean K-Means. We used a combination of nine features for each image as described below:

$$F_{ETHZ} = \left[x,\ I_r,\ I_g,\ I_b,\ Y_i,\ |I_x|,\ \left|I_y\right|,\ |\sin(\theta) + \cos(\theta)|,\ \left|H_y\right|\right],$$

  where x is the x-coordinate of a pixel location, I_r, I_g, I_b are the RGB color of a pixel, Y_i is the pixel intensity in the YCbCr color space, I_x, I_y are the gray scale pixel gradients, and H_y is the y-gradient of pixel hue. We also use the gradient angle $\theta = \tan^{-1}(I_y/I_x)$ in our feature set. Each image is resized to a fixed size 300×100, and divided into upper and lower parts. We compute two different region covariances for each part, which are combined as two block diagonal matrices to form a single covariance of size 18×18 for each appearance image.
- **3D Object Recognition.** The goal of this experiment is to recognize objects in 3D point clouds. We use the public RGB-D Object dataset [31], which consists of about 300 objects belonging to 51 categories and spread across $\sim$250K frames. We used approximately 15 K frames for our evaluation with approximately 250–350 frames devoted to every object seen from three different viewpoints (30, 45, and 60 degrees above the horizon). Following the procedure suggested in [22, Chap. 5], for every frame, the object was segmented out and 18 dimensional feature vectors generated for every 3D point in the cloud (and thus 18×18 covariance descriptors); the features we used are as follows:

$$F_{RGBD} = \left[x, y, z, I_r, I_g, I_b, I_x, I_y, I_{xx}, I_{yy}, I_{xy}, I_m, \delta_x, \delta_y, \delta_m, \nu_x, \nu_y, \nu_z\right], \quad (4.37)$$

  where the first three dimensions are the spatial coordinates, I_m is the magnitude of the intensity gradient, δ's represent gradients over the depth-maps, and ν represents the surface normal at the given 3D point.

4.3.3.1 Evaluation Techniques

We evaluate our algorithms for nearest neighbor (NN) retrieval against a gallery set via computing the Euclidean distances between sparse codes. We use the standard Recall@K accuracy defined as follows: Given a gallery $\mathscr{X}$ and a query set $\mathscr{Q}$. Recall@K computes the average accuracy when retrieving K nearest neighbors from $\mathscr{X}$ for each instance in $\mathscr{Q}$. Suppose G_K^q is the set of ground truth class labels associated with the qth query, and S_K^q is the set of labels associated with the K neighbors found by some algorithm, then

$$\text{Recall@K} := \frac{1}{|Q|} \sum_{q \in Q} \frac{|G_K^q \cap S_K^q|}{|G_K^q|}. \tag{4.38}$$

All the experiments used fivefold cross-validation in which 80 % of the datasets were used for training the dictionary, 10 % for generating the gallery set, and the rest as queries. The size of the dictionary was considered to be twice the number of classes in the respective dataset. This scheme was considered for all the comparison methods as well. Our DLSC scheme was implemented in MATLAB. We used the Manopt optimization toolbox [10] for implementing the CG method for our DL subproblem. We found that initializing the dictionary learning setup using K-Means clustering (using the Karcher mean algorithm [40]) led to faster convergence of CG.

4.3.3.2 Results

We compare below performance of our Riem-DL and Riem-SC against several prior DLSC schemes on the three datasets described above. In particular, we compare (i) Riemannian geometric methods such as log-Euclidean (LE-DL + LE-SC), (ii) Kernelized methods using the Stein kernel (Kernel-Stein-DL and kernel-Stein-SC), (iii) Euclidean DLSC (Frob-DL + Frob-SC), and using a dictionary generated by random sampling the dataset followed by sparse coding using our Riemannian method (Random-DL + Riem-SC). In Fig. 4.4, we show the performance comparison for the task of K-NN where K is increased from 1 to 25.

A commonly adopted alternative to dictionary learning is to approximate the dictionary using centroids of a K-Means clustering of the dataset. Such a method is faster than Riemannian DL, and also demonstrate reasonable performance [13, 46]. Thus, an important experiment is to ensure that learning the dictionary actually provides superior performance compared to the *ad hoc* clustering setup. In Fig. 4.3, we plot the K-NN retrieval when we use a clustering scheme to generate the dictionary.

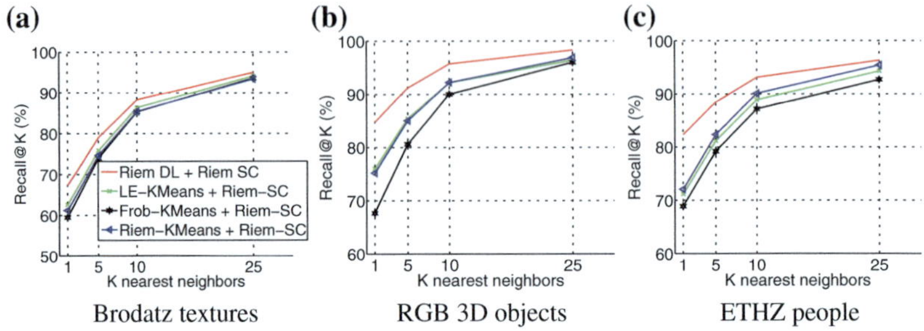

Fig. 4.3 Results of Nearest neighbor recall@K accuracy against increasing number of retrieved points (K). Comparisons of Riem-DL and Riem-SC against other DL learning schemes based on clustering, while using our Riem-SC for sparse coding

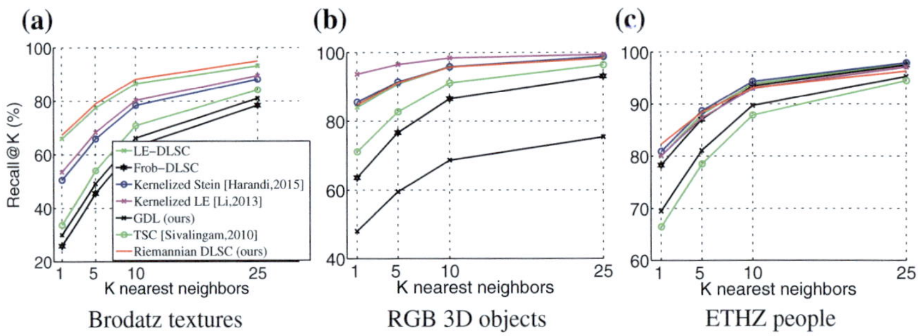

Fig. 4.4 Results of k-nearest neighbor retrieval accuracy against an increasing number of retrieved points (K). Comparisons of Riem-DL and Riem-SC against other dictionary learning and sparse coding combinations

4.3.3.3 Discussion of Results

With regard to Fig. 4.4, we found that the performance of different methods is diverse across datasets. For example, the log-euclidean DLSC variant (LE-DL+LE-SC) is generally seen to show good performance, but its performance is inferior when the number of data instances per class is small (as in the ETHZ people dataset). The kernelized DLSC method (Kernel-Stein-DL) [27] performs favorably on most datasets. Note that this is the current state of the art sparse coding algorithm for SPD matrix valued data and has shown to be better than alternatives such as [28]. From the plots in Fig. 4.4, it is clear that our Riemannian sparse coding demonstrates competitive (or even better on the texture dataset) to this scheme and to other kernelized schemes such as the kernelized-LE scheme [32]. Moreover, given that there is no need to construct any kernel matrix in our scheme, it is more scalable to very large datasets in comparison to these other kernelized alternatives.

The most surprising of the results that we found was for Frob-DL. It is generally assumed that using Frobenius distance for comparing SPD matrices leads to poor accuracy, a view echoed by Fig. 4.4a and b. However, when the matrices are ill-conditioned, taking the logarithm (as in the LE-DL scheme) of these matrices results in amplifying the influence of the smaller eigenvalues, which is essentially noise. When learning a dictionary, the atoms will be learned to reconstruct this noise against the signal, thus leading to inferior performance than for FrobDL (which do not use the logarithm0. In comparison to all the compared methods, Riem-DL+Riem-SC was found to produce consistent and competitive performance, substantiating the usefulness of our model. While running the experiments, we found that the initialization of our DL sub-problem (using Riemannian K-Means) played an important role in achieving this superior performance.

We further compare Riem-DL against alternative DL schemes via clustering in Fig. 4.3. We see that learning the dictionary using Riem-DL demonstrates the best performance against the next best and efficient alternative of using LE-KMeans as was done in [13]. Using Frob-KMeans or using a random dictionary are generally seen to have inferior performance compared to other learning methods.

4.3.4 GDL Versus Riemannian Sparse Coding

Finally, we compare sparse coding via our GDL model and the Riem-SC setup. We use the Brodatz and RGB-D Object recognition datasets. Tables 4.2 and 4.3 show the results. As is clear from the table, the Riemannian approach leads to much higher accuracy. However, GDL using the diagonal sparse coefficient formulation is generally seen to be much faster to solve than the Riemannian setup.

Comparison of the average classification accuracy using a linear SVM.

Table 4.2 Brodatz texture dataset

Method	Accuracy (%)
Frob-SC	32.3 (4.4)
TSC [46]	35.6 (7.1)
GDL	43.7 (6.3)
Riem-SC	**53.9** (3.4)

Table 4.3 RGB-D object recognition

Method	Accuracy (%)
Frob-SC	80.3 (1.1)
TSC [46]	72.8 (2.1)
GDL	61.9 (0.4)
Riem-SC	**84.0** (0.6)

4.4 Conclusion and Future Work

In this chapter, we reviewed the steps for constructing covariance descriptors, followed by a brief exposition of SPD matrix geometry motivated by the design of novel machine learning models on these descriptors. We covered the concrete problems of dictionary learning and sparse coding, and noted two approaches: (i) a framework that uses Euclidean embedding of SPD matrices for sparse coding; and (ii) a Riemannian geometric approach. Our experiments demonstrated that designing machine learning algorithms for SPD matrices that respect the Riemannian geometry fares significantly better than using Euclidean embedding.

That said, Riemannian optimization algorithms are usually computationally more expensive. This is mainly due to the need for operating in the tangent space of the SPD manifold, which involves matrix exponentials and logarithms that require $O(d^3)$ flops. Designing faster Riemannian machine learning algorithms is a challenge that needs to be addressed for these algorithms to be more widely accepted.

Acknowledgments AC is funded by the Australian Research Council Centre of Excellence for Robotic Vision (number CE140100016). SS acknowledges support from NSF grant IIS-1409802.

References

1. P.A. Absil, R. Mahony, R. Sepulchre, *Optimization Algorithms on Matrix Manifolds*. (Princeton University Press, Princeton, 2009)
2. P.A. Absil, C.G. Baker, K.A. Gallivan, Trust-region methods on riemannian manifolds. Found. Comput. Math. **7**(3), 303–330 (2007)
3. D.C. Alexander, C. Pierpaoli, P.J. Basser, J.C. Gee, Spatial transformations of diffusion tensor magnetic resonance images. IEEE Trans. Med. Imaging **20**(11), 1131–1139 (2001)
4. V. Arsigny, P. Fillard, X. Pennec, N. Ayache, Log-Euclidean metrics for fast and simple calculus on diffusion tensors. Magn. Reson. Med. **56**(2), 411–421 (2006)
5. J. Barzilai, J.M. Borwein, Two-point step size gradient methods. IMA J. Num. Analy. **8**(1), 141–148 (1988)
6. D.P. Bertsekas, Nonlinear Programming, 2nd edn. (Athena Scientific, Belmont, 1999)
7. R. Bhatia, *Positive Definite Matrices*. (Princeton University Press, Princeton, 2007)
8. E. Birgin, J. Martínez, M. Raydan, Nonmonotone spectral projected gradient methods on convex sets. SIAM J. Optim. **10**(4), 1196–1211 (2000)
9. E.G. Birgin, J.M. Martínez, M. Raydan, Algorithm 813: SPG-Software for Convex-constrained Optimization. ACM Trans. Math. Softw. **27**, 340–349 (2001)
10. N. Boumal, B. Mishra, P.A. Absil, R. Sepulchre, Manopt, a matlab toolbox for optimization on manifolds. J. Mach. Learn. Res. **15**(1), 1455–1459 (2014)
11. P. Brodatz, *Textures: a photographic album for artists and designers*, vol. 66. (Dover, New York, 1966)
12. A. Cherian, Nearest neighbors using compact sparse codes,in *Proceedings of the International Conference on Machine Learning*, pp. 1053–1061 (2014)
13. A. Cherian, S. Sra, Riemannian sparse coding for positive definite matrices, in *Proceedings of the European Conference on Computer Vision*. Springer (2014)
14. A. Cherian, S. Sra, Riemannian dictionary learning and sparse coding for positive definite matrices. (2015). arXiv preprint arXiv:1507.02772

15. A. Cherian, S. Sra, A. Banerjee, N. Papanikolopoulos, Jensen-bregman logdet divergence with application to efficient similarity search for covariance matrices. IEEE Trans. Pattern Anal. Mach. Intell. **35**(9), 2161–2174 (2013)
16. K. Dana, B. Van Ginneken, S. Nayar, J. Koenderink, Reflectance and texture of real-world surfaces. ACM Trans. Graph. (TOG) **18**(1), 1–34 (1999)
17. L. Dodero, H.Q. Minh, M.S. Biagio, V. Murino, D. Sona, Kernel-based classification for brain connectivity graphs on the Riemannian manifold of positive definite matrices, in *Proceedings of the International Symposium on Biomedical Imaging*, pp. 42–45. IEEE (2015)
18. M. Elad, M. Aharon, Image denoising via learned dictionaries and sparse representation, in *Proceedings of the EEE Computer Society Conference on Computer Vision and Pattern Recognition*, vol. 1, pp. 895–900. IEEE (2006)
19. A. Ess, B. Leibe, L.V. Gool, Depth and appearance for mobile scene analysis, in *Proceedings of the International Conference on Computer Vision*. IEEE (2007)
20. U. Fano, Description of states in quantum mechanics by density matrix and operator techniques. Rev. Mod. Phys. **29**(1), 74–93 (1957)
21. M. Faraki, M. Harandi, Bag of riemannian words for virus classification. Case Studies in Intelligent Computing: Achievements and Trends. pp. 271–284 (2014)
22. D.A. Fehr, *Covariance Based Point Cloud Descriptors for Object Detection and Classification*. University Of Minnesota, Minneapolis (2013)
23. D. Fehr, A. Cherian, R. Sivalingam, S. Nickolay, V. Morellas, N. Papanikolopoulos, Compact covariance descriptors in 3d point clouds for object recognition, in *Proceedings of the IEEE International Conference on Robotics and Automation*, pp. 1793–1798. IEEE (2012)
24. L. Ferro-Famil, E. Pottier, J. Lee, Unsupervised classification of multifrequency and fully polarimetric SAR images based on the H/A/Alpha-Wishart classifier. IEEE Trans. Geosci. Remote Sens. **39**(11), 2332–2342 (2001)
25. K. Guo, P. Ishwar, J. Konrad, Action recognition using sparse representation on covariance manifolds of optical flow, in *Proceedings of the Advanced Video and Signal Based Surveillance*. IEEE (2010)
26. M.T. Harandi, R. Hartley, B. Lovell, C. Sanderson, Sparse coding on symmetric positive definite manifolds using bregman divergences. IEEE Trans. Neural Netw. Learn. Syst. PP(99), 1–1 (2015)
27. M. Harandi, M. Salzmann, Riemannian coding and dictionary learning: Kernels to the rescue, in *Proceedings of the IEEE Conference on Computer Vision and Pattern Recognition* (CVPR), pp. 3926–3935 (2015)
28. J. Ho, Y. Xie, B. Vemuri, On a nonlinear generalization of sparse coding and dictionary learning, in *Proceedings of the International Conference on Machine Learning* (2013)
29. S. Jayasumana, R. Hartley, M. Salzmann, H. Li, M. Harandi, Kernel methods on the Riemannian manifold of symmetric positive definite matrices, in *Proceedings of the Computer Vision and Pattern Recognition*. IEEE (2013)
30. B. Kulis, K. Grauman, Kernelized locality-sensitive hashing for scalable image search. ICCV, October **1**, 3 (2009)
31. K. Lai, L. Bo, X. Ren, D. Fox, A large-scale hierarchical multi-view RGB-D object dataset, in *Proceedings of the International Conference on Robotics and Automation* (2011)
32. P. Li, Q. Wang, W. Zuo, L. Zhang, Log-euclidean kernels for sparse representation and dictionary learning, in *Proceedings of the International Conference on Computer Vision*. IEEE (2013)
33. C. Liu, Gabor-based kernel PCA with fractional power polynomial models for face recognition. Pattern Anal. Mach. Intell. **26**(5), 572–581 (2004)
34. D. Liu, J. Nocedal, On the limited memory BFGS method for large scale optimization. Math. program. **45**(1), 503–528 (1989)
35. B. Ma, Y. Su, F. Jurie, BiCov: a novel image representation for person re-identification and face verification, in *Proceedings of the British Machine Vision Conference* (2012)
36. J. Mairal, F. Bach, J. Ponce, Sparse modeling for image and vision processing (2014). arXiv preprint arXiv:1411.3230

37. S. Marčelja, Mathematical description of the responses of simple cortical cells*. JOSA **70**(11), 1297–1300 (1980)
38. T. Ojala, M. Pietikäinen, D. Harwood, A comparative study of texture measures with classification based on featured distributions. Pattern Recogn. **29**(1), 51–59 (1996)
39. Y. Pang, Y. Yuan, X. Li, Gabor-based region covariance matrices for face recognition. IEEE Trans. Circuits Syst. Video Technol. **18**(7), 989–993 (2008)
40. X. Pennec, P. Fillard, N. Ayache, A Riemannian framework for tensor computing. Int. J. Comput. Vis. **66**(1), 41–66 (2006)
41. P. Phillips, H. Wechsler, J. Huang, P. Rauss, The FERET database and evaluation procedure for face-recognition algorithms. Image Vis. Comput. **16**(5), 295–306 (1998)
42. P. Phillips, H. Moon, S. Rizvi, P. Rauss, The FERET evaluation methodology for face-recognition algorithms. Pattern Anal. Mach. Intell. **22**(10), 1090–1104 (2000)
43. W. Schwartz, L. Davis, Learning Discriminative Appearance-Based Models Using Partial Least Squares, in *Proceedings of the XXII Brazilian Symposium on Computer Graphics and Image Processing* (2009)
44. M. Schmidt, E. van den Berg, M. Friedlander, K. Murphy, Optimizing costly functions with simple constraints: a limited-memory projected Quasi-Newton algorithm, in *Proceedings of the International Conference on Artificial Intelligence and Statistics* (2009)
45. Y. Shinohara, T. Masuko, M. Akamine, Covariance clustering on Riemannian manifolds for acoustic model compression, in *proceedings of the International Conference on Acoustics, Speech and Signal Processing* (2010)
46. R. Sivalingam, D. Boley, V. Morellas, N. Papanikolopoulos, Tensor sparse coding for region covariances, in *Proceedings of the European Conference on Computer Vision*. Springer (2010)
47. G. Somasundaram, A. Cherian, V. Morellas, N. Papanikolopoulos, Action recognition using global spatio-temporal features derived from sparse representations. Comput. Vis. Image Underst. **123**, 1–13 (2014)
48. S. Sra, Positive Definite Matrices and the S-Divergence, in Proceedings of the American Mathematical Society (2015). arXiv:1110.1773v4
49. S. Sra, A. Cherian, Generalized dictionary learning for symmetric positive definite matrices with application to nearest neighbor retrieval, in *Proceedings of the European Conference on Machine Learning*. Springer (2011)
50. J. Su, A. Srivastava, F. de Souza, S. Sarkar, Rate-invariant analysis of trajectories on riemannian manifolds with application in visual speech recognition, in *Proceedings of the Computer Vision and Pattern Recognition*, pp. 620–627. IEEE (2014)
51. D. Tosato, M. Farenzena, M. Spera, V. Murino, M. Cristani, Multi-class classification on Riemannian manifolds for video surveillance, in *Proceedings of the European Conference on Computer Vision* (2010)
52. O. Tuzel, F. Porikli, P. Meer.: Covariance Tracking using Model Update Based on Lie Algebra in *Proceedings of the IEEE Computer Society Conference on Computer Vision and Pattern Recognition* (2006)
53. O. Tuzel, F. Porikli, P. Meer.: Region covariance: a fast descriptor for detection and classification. in *Proceedings of the European Conference on Computer Vision* (2006)
54. R. Wang, H. Guo, L.S. Davis, Q. Dai, Covariance discriminative learning: A natural and efficient approach to image set classification, in *Proceedings of the Computer Vision and Pattern Recognition*. IEEE (2012)
55. W. Zheng, H. Tang, Z. Lin, T.S. Huang, Emotion recognition from arbitrary view facial images, in *Proceedings of the European Conference on Computer Vision*, pp. 490–503. Springer (2010)

Chapter 5
From Covariance Matrices to Covariance Operators: Data Representation from Finite to Infinite-Dimensional Settings

Hà Quang Minh and Vittorio Murino

Abstract This chapter presents some of the recent developments in the generalization of the data representation framework using finite-dimensional covariance matrices to infinite-dimensional covariance operators in Reproducing Kernel Hilbert Spaces (RKHS). We show that the proper mathematical setting for covariance operators is the infinite-dimensional Riemannian manifold of positive definite Hilbert–Schmidt operators, which are the generalization of symmetric, positive definite (SPD) matrices. We then give the closed form formulas for the affine-invariant and Log-Hilbert–Schmidt distances between RKHS covariance operators on this manifold, which generalize the affine-invariant and Log-Euclidean distances, respectively, between SPD matrices. The Log-Hilbert–Schmidt distance in particular can be used to design a two-layer kernel machine, which can be applied directly to a practical application, such as image classification. Experimental results are provided to illustrate the power of this new paradigm for data representation.

5.1 Introduction

Symmetric Positive Definite (SPD) matrices, in particular covariance matrices, play an important role in many areas of mathematics, statistics, machine learning, and their applications. In practice, the applications of SPD matrices are numerous, including brain imaging [3, 12, 34], kernel learning [19] in machine learning, object detection [39, 40] and image retrieval [11] in computer vision, and radar signal processing [5, 15].

In the field of computer vision and image processing, covariance matrices have recently been utilized as a powerful image representation approach, which is

H.Q. Minh (✉) · V. Murino
Pattern Analysis and Computer Vision (PAVIS),
Istituto Italiano di Tecnologia (IIT), Genova, 16163, Italy
e-mail: minh.haquang@iit.it

V. Murino
e-mail: vittorio.murino@iit.it

H.Q. Minh and V. Murino (eds.), *Algorithmic Advances in Riemannian Geometry and Applications*, Advances in Computer Vision and Pattern Recognition, DOI 10.1007/978-3-319-45026-1_5

commonly called *covariance descriptor*. In this approach, an image is compactly represented by a covariance matrix encoding correlations between different features extracted from that image. This representation has been demonstrated to work very well in practice and consequently, covariance descriptors have been applied with success to many computer vision tasks, including tracking [33], object detection and classification [39, 40], and image retrieval [11]. A more detailed discussion of the covariance matrix representation can be found in the chapter by Cherian and Sra in this volume.

Riemannian geometric framework for covariance matrices. Covariance matrices, properly regularized if necessary, are examples of SPD matrices. In the following, we denote by $\mathrm{Sym}^{++}(n)$ the set of all $n \times n$ SPD matrices. A key mathematical property of $\mathrm{Sym}^{++}(n)$ is that it is not a vector subspace of Euclidean space under the standard matrix addition and scalar multiplication operations. Instead, it is an open convex cone, since it is only closed under positive scalar multiplication, and at the same time admits a differentiable manifold structure. Consequently, in general, the optimal measure of similarity between covariance matrices is not the Euclidean distance, but a metric that captures the geometry of $\mathrm{Sym}^{++}(n)$. Among the most widely used metrics for $\mathrm{Sym}^{++}(n)$ is the classical *affine-invariant Riemannian metric* [6, 7, 23, 30, 31, 40], under which $\mathrm{Sym}^{++}(n)$ becomes a Riemannian manifold with nonpositive curvature. Another commonly used Riemannian metric for $\mathrm{Sym}^{++}(n)$ is the recently introduced Log-Euclidean metric [3, 4], which is *bi-invariant* and under which the manifold is flat. Compared to the affine-invariant metric, the Log-Euclidean metric is faster to compute, especially on large datasets, and can be used to define many positive definite kernels, such as the Gaussian kernel, allowing kernel methods to be applied directly on the manifold [17, 24].

Positive definite kernels and covariance operators. While they have been shown to be effective in many applications, one major limitation of covariance matrices is that they only capture *linear* correlations between input features. In order to encode *nonlinear* correlations, we generalize the covariance matrix representation framework to the infinite-dimensional setting by the use of positive definite kernels defined on the original input features. Intuitively, from the viewpoint of kernel methods in machine learning [37], each positive definite kernel, such as the Gaussian kernel, induces a feature map that nonlinearly maps each input point into a high (generally infinite) dimensional feature space. We then represent each image by an infinite-dimensional covariance operator, which can be thought as the covariance matrix of the infinite-dimensional features in the feature space. Since the high-dimensional feature maps are nonlinear, the resulting covariance operators thus encode the nonlinear correlations between the original input features. A key property of this framework, as is common for kernel methods, is that the infinite-dimensional feature maps and the corresponding covariance operators are all *implicit*, and all necessary computations are carried out via the Gram matrices associated with the given kernels.

Infinite-dimensional Riemannian manifold setting for covariance operators. Having represented each image by a covariance operator, we need to define a notion of distances between these operators. Instead of the finite-dimensional manifold setting for covariance matrices, in the infinite-dimensional setting, regularized covariance operators lie on an infinite-dimensional Hilbert manifold. This is the manifold of positive definite unitized Hilbert–Schmidt operators, which are scalar perturbations of Hilbert–Schmidt operators on a Hilbert space and which are infinite-dimensional generalizations of SPD matrices. On this manifold, the generalization of the affine-invariant Riemannian metric on $\mathrm{Sym}^{++}(n)$ was recently carried out by [1, 21, 22] from a purely mathematical viewpoint. For the case of RKHS covariance operators, the explicit formulas for the affine-invariant distance, in terms of the Gram matrices, were obtained in [26]. The generalization of the Log-Euclidean metric, called the *Log-Hilbert–Schmidt metric*, was formulated by [28], including the explicit formulas for the distances between RKHS covariance operators. As with the Log-Euclidean metric, the Log-Hilbert–Schmidt metric can be used to define many positive definite kernels, such as the Gaussian kernel, allowing kernel methods to be applied on top of the infinite-dimensional manifold and effectively creating a two-layer kernel machine.

Differences between the finite and infinite-dimensional settings. In [32], in the context of functional data analysis, the authors discussed the difficulty of generalizing the affine-invariant and Log-Euclidean metrics to the infinite-dimensional setting and proposed several other metrics instead. As we analyze in [26, 28] and below, this difficulty is due to the fundamental differences between the finite and infinite-dimensional cases. The reason is that many concepts, such as *principal matrix logarithm, determinant, and norm*, all involve infinite sums and products and therefore are well-defined only on specific classes of infinite-dimensional operators. In particular, the infinite-dimensional distance formulas are *not* the limits of the finite-dimensional ones as the dimension approaches infinity.

The aim of this chapter. In the present chapter, we first show how to generalize the data representation framework by covariance matrices to RKHS covariance operators. We then report on the recent development in mathematical theory of infinite-dimensional positive definite operators [1, 21, 22, 28], which successfully resolves the problems of extending the affine-invariant and Log-Euclidean metrics to the infinite-dimensional setting. We then show how this theory can be applied to compute distances between RKHS covariance operators. In particular, we describe in detail the two-layer kernel machine which arises from the Log-Hilbert–Schmidt distance between RKHS operators, which can be used in a practical application such as image classification.

Related work. In the literature on kernel methods in machine learning, it is well-known that RKHS covariance operators defined on nonlinear features, which are obtained by mapping the original input data into a high-dimensional feature space, can better capture input correlations than covariance matrices defined on the original input data, see e.g. KernelPCA [36]. However, the use of RKHS covariance operators for data representation is quite recent and has its origin in computer vision [14, 16, 41]. The main problem with the approaches in [14, 16, 41] is that they lack the theoretical foundation provided by the mathematical theory of infinite-dimensional operators and infinite-dimensional manifolds. As such, they are necessarily heuristic and many results obtained are only valid in the finite-dimensional setting.

Organization. We start by recalling the data representation framework using covariance matrices in Sect. 5.2. Then in Sect. 5.3, we show how this framework generalizes to RKHS covariance operators which are induced by positive definite kernels and their associated feature maps. In Sect. 5.4, we give the closed form formulas for the Hilbert–Schmidt, affine-invariant, and Log-Hilbert–Schmidt distances between RKHS covariance operators. The two-layer kernel machine resulting from the Log-Hilbert–Schmidt distance is described in Sect. 5.5, with experimental results illustrating its power in Sect. 5.6. Mathematical proofs are given in the Appendix.

5.2 Covariance Matrices for Data Representation

Before presenting covariance operators for data representation, we first recall how covariance matrices are employed as a form of image representation. For each image, at every pixel (or a subset of the pixels), we extract an image feature vector consisting of n features, for example intensity, gradient, and colors. Suppose that we perform feature extraction at m pixels, each one giving a feature vector $x_i \in \mathbb{R}^n, i = 1, \ldots, m$, we then obtain a data matrix of size $n \times m$, given by

$$\mathbf{x} = [x_1, \ldots, x_m], \tag{5.1}$$

with each column consisting of image features extracted at one pixel. The $n \times n$ covariance matrix

$$C_{\mathbf{x}} = \frac{1}{m}\mathbf{x}J_m\mathbf{x}^T = \frac{1}{m}\sum_{j=1}^{m}(x_j - \mu)(x_j - \mu)^T, \tag{5.2}$$

then encodes *linear correlations* between all the different extracted features and is used as the representation for the image. Here J_m is the centering matrix, defined by $J_m = I_m - \frac{1}{m}\mathbf{1}_m\mathbf{1}_m^T$, where $\mathbf{1}_m = (1, \ldots, 1)^T \in \mathbb{R}^m$ and $\mu = \frac{1}{m}\sum_{j=1}^{m} x_j \in \mathbb{R}^n$ denotes the mean column of $\mathbf{x}$. In general, $C_{\mathbf{x}}$ is a symmetric, positive *semi-definite* matrix.

In a practical application, such as classification, we need to have a similarity measure between images. By representing images as covariance matrices, this means that we need to compute distances between covariance matrices. Let A and B be two symmetric, positive semi-definite matrices. A straightforward distance between A and B is the Euclidean distance, given by

$$d_E(A, B) = ||A - B||_F. \tag{5.3}$$

Here $|| \; ||_F$ denotes the Frobenius norm, which, for $A = (a_{ij})_{i,j=1}^n$, is defined by

$$||A||_F^2 = \mathrm{tr}(A^TA) = \sum_{i,j=1}^{n} a_{ij}^2. \tag{5.4}$$

It is clear from the definition of the Frobenius norm that the distance $||A - B||_F$ depends only on the entries of $A - B$, without taking into account any structure of A and B. Furthermore, the set of symmetric, positive semi-definite matrices is not a vector subspace of Euclidean space under the standard matrix addition and scalar multiplication operations, but a convex cone, since it is only closed under *positive* scalar multiplication. By simply vectorizing A and B, the Euclidean distance $||A - B||_F$ reflects neither the positivity of A and B nor the convex cone structure of the set of positive matrices.

We note that, *empirically*, by a simple regularization, the regularized covariance matrix $(C_{\mathbf{x}} + \gamma I)$ for any constant $\gamma > 0$ is an element of $\mathrm{Sym}^{++}(n)$, which has been studied extensively, both mathematically and computationally. Thus, we can apply to the set of regularized covariance matrices any distance on $\mathrm{Sym}^{++}(n)$ that reflects its intrinsic geometry as a set of SPD matrices. The regularization $(C_{\mathbf{x}} + \gamma I)$ is called *diagonal loading* in the literature (see [2, 13] for more general forms of regularizations). We show in Sect. 5.4 below that for infinite-dimensional covariance operators, this form of regularization is always necessary, both *theoretically* and *empirically*.

Let $\mathrm{Sym}^{++}(n)$ denote the set of SPD matrices of size $n \times n$. Let $A, B \in \mathrm{Sym}^{++}(n)$ be arbitrary. We now review three distances that exploit the geometry of $\mathrm{Sym}^{++}(n)$, namely the affine-invariant distance, Log-Euclidean distance, and Bregman divergences.

Affine-invariant metric. In the first approach, the set $\mathrm{Sym}^{++}(n)$ is equipped with a Riemannian metric, the so-called *affine-invariant metric* [6, 7, 23, 30, 31]. For each $P \in \mathrm{Sym}^{++}(n)$, the tangent space at P is $T_P(\mathrm{Sym}^{++}(n)) \cong \mathrm{Sym}(n)$, the space of symmetric matrices of size $n \times n$. The affine-invariant metric is defined by the following inner product on the tangent space at P

$$\langle A, B \rangle_P = \langle P^{-1/2} A P^{-1/2}, P^{-1/2} B P^{-1/2} \rangle_F, \quad \forall P \in \mathrm{Sym}^{++}(n), A, B \in \mathrm{Sym}(n). \tag{5.5}$$

Under the affine-invariant metric, $\mathrm{Sym}^{++}(n)$ becomes a Riemannian manifold with *nonpositive sectional curvature*. The affine-invariant geodesic distance between A and B is given by

$$d_{\mathrm{aiE}}(A, B) = || \log(A^{-1/2} B A^{-1/2}) ||_F, \tag{5.6}$$

where log denotes the principal matrix logarithm.

Log-Euclidean metric. In the second approach, the set $\mathrm{Sym}^{++}(n)$ is equipped with a *bi-invariant* Riemannian metric, the so-called Log-Euclidean metric [4]. This metric arises from the following commutative Lie group multiplication on $\mathrm{Sym}^{++}(n)$

$$\begin{aligned} \odot : \mathrm{Sym}^{++}(n) \times \mathrm{Sym}^{++}(n) &\to \mathrm{Sym}^{++}(n), \\ A \odot B &= \exp(\log(A) + \log(B)). \end{aligned} \tag{5.7}$$

Under the Log-Euclidean metric, the geodesic distance between A and B is given by

$$d_{\mathrm{logE}}(A,B) = ||\log(A) - \log(B)||_F. \tag{5.8}$$

Along with the group operation $\odot$, one can also define the scalar multiplication

$$\begin{aligned}\circledast : \mathbb{R} \times \mathrm{Sym}^{++}(n) &\to \mathrm{Sym}^{++}(n),\\ \lambda \circledast A = \exp(\lambda \log(A)) &= A^{\lambda}, \quad \lambda \in \mathbb{R}.\end{aligned} \tag{5.9}$$

Endowed with the commutative group multiplication $\odot$ and the scalar multiplication $\circledast$, $(\mathrm{Sym}^{++}, \odot, \circledast)$ becomes a vector space [4]. Furthermore, we can endow this vector space with the *Log-Euclidean inner product*.

$$\langle A, B\rangle_{\mathrm{logE}} = \langle \log(A), \log(B)\rangle_F = \mathrm{tr}[\log(A)\log(B)]. \tag{5.10}$$

along with the corresponding *Log-Euclidean norm*

$$||A||^2_{\mathrm{logE}} = \langle \log(A), \log(A)\rangle_F = \mathrm{tr}[\log^2(A)], \tag{5.11}$$

giving us the inner product space

$$(\mathrm{Sym}^{++}(n), \odot, \circledast, \langle\ ,\ \rangle_{\mathrm{logE}}). \tag{5.12}$$

This inner product space structure was first discussed in [24]. The Log-Euclidean distance in Eq. (5.8) is then expressed as

$$d_{\mathrm{logE}}(A,B) = ||\log(A) - \log(B)||_F = ||A \odot B^{-1}||_{\mathrm{logE}}. \tag{5.13}$$

With this viewpoint, it follows that $\mathrm{Sym}^{++}(n)$ under the Log-Euclidean metric is flat, that is it has *zero sectional curvature*. Furthermore, the map

$$\begin{aligned}\log : (\mathrm{Sym}^{++}(n), \odot, \circledast, \langle\ ,\ \rangle_{\mathrm{logE}}) &\to (\mathrm{Sym}(n), +, \cdot, \langle\ ,\ \rangle_F)\\ A &\to \log(A).\end{aligned} \tag{5.14}$$

is an isometrical isomorphism between inner product spaces, where $(+, \cdot)$ denote the standard matrix addition and scalar multiplication operations, respectively.

Log-Euclidean versus Euclidean. The previous discussion shows that the Log-Euclidean metric essentially flattens $\mathrm{Sym}^{++}(n)$ via the map $A \to \log(A)$. However, the vector space operations $(\odot, \circledast)$ are *not* the Euclidean space operations $(+, \cdot)$ and $(\mathrm{Sym}^{++}(n), \odot, \circledast, \langle\ ,\ \rangle_{\mathrm{logE}})$ is *not* a vector subspace of Euclidean space. Furthermore, $(\mathrm{Sym}^{++}(n), ||\ ||_E)$ is an incomplete metric space, whereas, since $||\ ||_{\mathrm{logE}}$ is an inner product distance, the metric space $(\mathrm{Sym}^{++}(n), ||\ ||_{\mathrm{logE}})$ is complete, which is a desirable property when dealing with converging sequences of SPD matrices.

One can also clearly see that the SPD property of the matrices A and B is encoded by the principal matrix logarithms in the distance formula $||\log(A) - \log(B)||_F$ (if A has a negative eigenvalue, for example, its principal matrix logarithm is *not* even defined). This is in strong contrast to the Euclidean distance formula $||A - B||_F$, which depends only on the entries of $A - B$ and therefore does *not* reflect any inherent structure in A and B.

Kernel methods with the Log-Euclidean metric. For the purposes of kernel methods in machine learning and applications, since $(\mathrm{Sym}^{++}(n), \odot, \circledast, \langle\ ,\ \rangle_{\mathrm{logE}})$ is an inner product space, one can define positive definite kernels on $\mathrm{Sym}^{++}(n)$ using the inner product $\langle\ ,\ \rangle_{\mathrm{logE}}$ and the corresponding norm $||\ ||_{\mathrm{logE}}$. This enables us to apply kernel methods directly on $\mathrm{Sym}^{++}(n)$, as is done in [17, 18, 24]. In particular, we have the following result.

Proposition 1 *The following kernels* $K : \mathrm{Sym}^{++}(n) \times \mathrm{Sym}^{++}(n) \to \mathbb{R}$ *are positive definite*

$$K(A, B) = (c + \langle A, B\rangle_{\mathrm{logE}})^d = (c + \langle \log(A), \log(B)\rangle_F)^d, \quad c \geq 0, d \in \mathbb{N}. \tag{5.15}$$

$$\begin{aligned} K(A, B) &= \exp\left(-||A \odot B^{-1}||^p_{\mathrm{logE}}\right), \quad \sigma \neq 0, \quad 0 < p \leq 2, \\ &= \exp\left(-\frac{||\log(A) - \log(B)||^p_F}{\sigma^2}\right). \end{aligned} \tag{5.16}$$

Remark 1 The proofs of Proposition 1 and all subsequent propositions are given in the Appendix. The kernel K in Eq. (5.16) in particular generalizes the results in [17, 18, 24], which show that K is positive definite for $p = 2$.

Bregman divergences. In the third approach, one defines distance-like functions based on the convex cone structure of $\mathrm{Sym}^{++}(n)$. One well-known example of this approach is the Stein divergence, defined by [38]

$$d^2_{\mathrm{stein}}(A, B) = \log \frac{\det(\frac{A+B}{2})}{\sqrt{\det(A)\det(B)}}. \tag{5.17}$$

The Bregman divergences do not arise from Riemannian metrics on $\mathrm{Sym}^{++}(n)$ and, apart from special cases such as d_{stein} in Eq. (5.17), they are generally *not* metric distances. However, they can be computed efficiently and have been shown to work well in diverse applications [11, 19].

In this chapter, we show how to generalize both the affine-invariant distance in Eq. (5.6) and the Log-Euclidean distance in Eq. (5.8) to the infinite-dimensional setting, in particular to RKHS covariance operators. The generalization of the Bregman divergences to the infinite-dimensional setting will be presented in a separate work.

5.3 Infinite-Dimensional Covariance Operators

Having reviewed the data representation framework using finite-dimensional covariance matrices, we now present infinite-dimensional covariance operators in RKHS and show how they can be used as a form of data representation. This framework is grounded in the setting of positive definite kernels and their associated RKHS and feature maps, which we discuss first.

5.3.1 *Positive Definite Kernels, Reproducing Kernel Hilbert Spaces, and Feature Maps*

Positive definite kernels. Let $\mathscr{X}$ be an arbitrary non-empty set. A function $K : \mathscr{X} \times \mathscr{X} \to \mathbb{R}$ is said to be a positive definite kernel if it is symmetric and satisfies

$$\sum_{i,j=1}^{N} a_i a_j K(x_i, x_j) \geq 0 \tag{5.18}$$

for any set of points $\mathbf{x} = \{x_i\}_{i=1}^N$ in $\mathscr{X}$ and any set of real numbers $\{a_i\}_{i=1}^N$. In other words, the $N \times N$ matrix $K[\mathbf{x}]$ defined by $(K[\mathbf{x}])_{ij} = K(x_i, x_j)$ is symmetric, positive semi-definite.

Examples of commonly used positive definite kernels include the Gaussian kernel $K(x, y) = \exp\left(-\frac{||x-y||^2}{\sigma^2}\right)$, $\sigma \neq 0$, and polynomial kernels $K(x, y) = (\langle x, y\rangle + c)^d$, $c \geq 0$, $d \in \mathbb{N}$, for $x, y \in \mathbb{R}^n$, $n \in \mathbb{N}$.

Reproducing kernel Hilbert spaces (RKHS). Each positive definite kernel K corresponds to a unique Hilbert space of functions on $\mathscr{X}$ as follows. For each $x \in \mathscr{X}$, there corresponds a function $K_x : \mathscr{X} \to \mathbb{R}$ defined by $K_x(y) = K(x, y)$. Consider the set $\mathscr{H}_0$ of all linear combinations of functions of the form K_x, $x \in \mathscr{X}$, that is

$$\mathscr{H}_0 = \left\{ \sum_{j=1}^{N} a_j K_{x_j} \ : \ a_j \in \mathbb{R}, x_j \in \mathscr{X}, N \in \mathbb{N} \right\}. \tag{5.19}$$

On $\mathscr{H}_0$, we define the following inner product

$$\langle \sum_{i=1}^{N} a_i K_{x_i}, \sum_{j=1}^{M} b_j K_{y_j} \rangle_{\mathscr{H}_K} = \sum_{i=1}^{N} \sum_{j=1}^{M} a_i b_j K(x_i, y_j). \tag{5.20}$$

This inner product is well-defined by the assumption that K is a positive definite kernel, making $\mathscr{H}_0$ an inner product space. Let $\mathscr{H}_K$ be the Hilbert completion of $\mathscr{H}_0$, obtained by adding the limits of all the Cauchy sequences in $\mathscr{H}_0$, then $\mathscr{H}_K$

is a Hilbert space of functions on $\mathscr{X}$, called the *reproducing kernel Hilbert space* (RKHS) induced by the kernel K.

The terminology RKHS comes from the *reproducing property*, which states that for all $f \in \mathscr{H}_K$ and all $x \in \mathscr{X}$,

$$f(x) = \langle f, K_x \rangle_{\mathscr{H}_K}. \tag{5.21}$$

Feature maps. A very useful and intuitive geometrical view of positive definite kernels is that of *feature maps*, which comes from machine learning and pattern recognition. In this viewpoint, a function $K : \mathscr{X} \times \mathscr{X} \to \mathbb{R}$ is a positive definite kernel if and only if there exists a Hilbert space $\mathscr{H}$, called *feature space*, and a map $\Phi : \mathscr{X} \to \mathscr{H}$, called *feature map*, such that

$$K(x, y) = \langle \Phi(x), \Phi(y) \rangle_{\mathscr{H}} \quad \forall x, y \in \mathscr{H}. \tag{5.22}$$

As the simplest example, consider the quadratic polynomial kernel $K : \mathbb{R}^2 \times \mathbb{R}^2 \to \mathbb{R}$ defined by $K(x, y) = \langle x, y \rangle^2 = (x_1 y_1 + x_2 y_2)^2$. It can be readily verified, via a simple algebraic calculation, that this kernel possesses the 3-dimensional feature map $\Phi : \mathbb{R}^2 \to \mathbb{R}^3$, defined by

$$\Phi(x) = (x_1^2, \sqrt{2} x_1 x_2, x_2^2) \in \mathbb{R}^3.$$

For a general positive definite kernel K, from the definition of RKHS above, it follows that the RKHS $\mathscr{H}_K$ induced by K is a feature space associated with K, with corresponding feature map $\Phi : \mathscr{X} \to \mathscr{H}_K$, defined by

$$\Phi(x) = K_x \quad \forall x \in \mathscr{X}, \tag{5.23}$$

which is called the *canonical feature map* [27] associated with K. If $\mathscr{X} \subset \mathbb{R}^n$ is a set with non-empty interior, then $\dim(\mathscr{H}_K) = \infty$ (see [25]), so that the feature map Φ is infinite-dimensional. We refer to [27] for a more detailed discussion of feature maps, including their equivalence to the canonical feature map, and many other examples.

The feature map viewpoint is particularly useful from an algorithmic perspective, since it allows one to transform any *linear* algorithm, which is expressed in terms of the inner product $\langle x, y \rangle$ of input examples in Euclidean space, into a *nonlinear* algorithm, simply by replacing $\langle x, y \rangle$ with $\langle \Phi(x), \Phi(y) \rangle_{\mathscr{H}_K} = K(x, y)$ for some nonlinear kernel K.

For our present purposes, feature maps enable us to generalize covariance matrices, which encode *linear correlations* between input features, to covariance operators in RKHS, which encode *nonlinear correlations* between input features.

5.3.2 Covariance Operators in RKHS and Data Representation

With the kernels and feature maps in Sect. 5.3.1, we now define RKHS covariance operators using these feature maps and show how they are employed for image representation. This framework is a generalization of the covariance matrix representation described in Sect. 5.2.

As in Sect. 5.2, for each image, let $\mathbf{x} = [x_1, \ldots, x_m]$ be the data matrix of size $n \times m$, with each column being the vector of features $x_i \in \mathbb{R}^n$ sampled at pixel i, $1 \leq i \leq m$. Now let $K : \mathbb{R}^n \times \mathbb{R}^n \to \mathbb{R}$ be a positive definite kernel, such as the Gaussian kernel, which induces implicitly a feature map $\Phi : \mathbb{R}^n \to \mathscr{H}_K$, where $\mathscr{H}_K$ is the RKHS induced by K. The map Φ gives us the matrix of features in $\mathscr{H}_K$

$$\Phi(\mathbf{x}) = [\Phi(x_1), \ldots, \Phi(x_m)], \tag{5.24}$$

which can be viewed informally as a (potentially infinite) matrix of size $\dim(\mathscr{H}_K) \times m$. Formally, it is a bounded linear operator $\Phi(\mathbf{x}) : \mathbb{R}^m \to \mathscr{H}_K$, defined by

$$\Phi(\mathbf{x})\mathbf{b} = \sum_{i=1}^{m} b_i \Phi(x_i), \tag{5.25}$$

with corresponding adjoint operator $\Phi(\mathbf{x})^* : \mathscr{H}_K \to \mathbb{R}^m$. The operator $\Phi(\mathbf{x})$ gives rise to the RKHS covariance operator

$$C_{\Phi(\mathbf{x})} = \frac{1}{m}\Phi(\mathbf{x})J_m\Phi(\mathbf{x})^* : \mathscr{H}_K \to \mathscr{H}_K, \tag{5.26}$$

which can be viewed informally as a (potentially infinite) matrix of size $\dim(\mathscr{H}_K) \times \dim(\mathscr{H}_K)$. The covariance operator $C_{\Phi(\mathbf{x})}$ is now the *representation* for the image and encodes, for a nonlinear kernel K, *nonlinear correlations* between all the different extracted features.

Remark 2 We say that the covariance operator representation is a generalization of the covariance matrix representation, since for the linear kernel $K(x, y) = \langle x, y\rangle$ on $\mathbb{R}^n \times \mathbb{R}^n$, we have $\Phi(x) = x$ and $C_{\Phi(\mathbf{x})} = C_{\mathbf{x}}$.

A crucial feature of the RKHS covariance operator representation is that it is *implicit*, that is neither the matrix of features $\Phi(\mathbf{x})$ nor the covariance operator $C_{\Phi(\mathbf{x})}$ is ever computed. Instead, all the necessary computations involving $\Phi(\mathbf{x})$ and $C_{\Phi(\mathbf{x})}$ are done via the Gram matrices of the kernel K on the original data matrix $\mathbf{x}$. We show that this is indeed the case for the distances between the covariance operators.

5.4 Distances Between RKHS Covariance Operators

Having described the image representation framework by RKHS covariance operators, we now describe the distances between covariance operators. These distances can then be directly employed in a practical application, e.g. image classification.

Since covariance operators are Hilbert–Schmidt operators, a natural distance between them is the Hilbert–Schmidt distance, which is the infinite-dimensional generalization of the Euclidean distance given by the Frobenius norm $|| \ ||_F$. However, just like the Euclidean distance, the Hilbert–Schmidt distance does *not* capture the *positivity* of covariance operators. In order to do so, as with $\mathrm{Sym}^{++}(n)$, we need to consider the manifold setting of covariance operators.

As a generalization from the finite-dimensional setting, it can be shown [22] that regularized covariance operators lie on an *infinite-dimensional* Hilbert manifold, namely the manifold of *positive definite unitized Hilbert–Schmidt operators* on a separable Hilbert space $\mathscr{H}$. Each point on this manifold has the form $A + \gamma I > 0$, $\gamma > 0$, where A is a Hilbert–Schmidt operator on $\mathscr{H}$. As we now show, both the affine-invariant distance in Eq. (5.6) and the Log-Euclidean distance in Eq. (5.8) admit a generalization on this manifold. However, there are several *key differences* between the finite and infinite-dimensional settings:

1. In the finite-dimensional case, the regularization $(C_{\mathbf{x}} + \gamma I)$ is often necessary *empirically* since in general $C_{\mathbf{x}}$ is not guaranteed to be positive definite. In contrast, when $\dim(\mathscr{H}) = \infty$, the regularization form $(A + \gamma I)$ is *always* needed, both *theoretically* and *empirically*, even if A is strictly positive. This is because $\log(A)$ is unbounded and we must always consider $\log(A + \gamma I)$. We explain this in detail in Sect. 5.4.2.1.
2. When $\dim(\mathscr{H}) = \infty$, the identity operator I is not Hilbert–Schmidt and therefore the Hilbert–Schmidt norm of $\log(A + \gamma I)$ is generally infinite. Furthermore, the distance between any two different multiples of I would be infinite. This problem is resolved by the introduction of the *extended Hilbert–Schmidt inner product*. We explain this in detail in Sect. 5.4.2.2.

In general, the distance formulas for the finite and infinite-dimensional cases are different and the infinite-dimensional formulas are generally *not* the limits of the finite-dimensional ones as the dimension approaches infinity. For RKHS covariance operators, all three distances admit closed forms in terms of Gram matrices.

5.4.1 Hilbert–Schmidt Distance

We first consider the generalization of the Frobenius norm in Eq. (5.4) to the separable Hilbert space setting. We recall that a bounded linear operator $A : \mathscr{H} \to \mathscr{H}$ is said to be a Hilbert–Schmidt operator if

$$||A||_{\mathrm{HS}}^2 = \mathrm{tr}(A^*A) = \sum_{k=1}^{\infty} ||Ae_k||^2 < \infty, \tag{5.27}$$

for any countable orthonormal basis $\{e_k\}_{k\in\mathbb{N}}$ in $\mathscr{H}$. $||\ ||_{\mathrm{HS}}$ is called the Hilbert–Schmidt norm, which is the infinite-dimensional version of the Frobenius norm in Eq. (5.4).

Let $\mathrm{HS}(\mathscr{H})$ denote the class of all Hilbert–Schmidt operators on $\mathscr{H}$. The Hilbert–Schmidt norm corresponds to the Hilbert–Schmidt inner product on $\mathrm{HS}(\mathscr{H})$, which is defined by

$$\langle A, B\rangle_{\mathrm{HS}} = \mathrm{tr}(A^*B) = \sum_{k=1}^{\infty} \langle Ae_k, Be_k\rangle, \quad A, B \in \mathrm{HS}(\mathscr{H}). \tag{5.28}$$

For a self-adjoint operator $A \in \mathrm{HS}(\mathscr{H})$, A is compact and hence possesses a countable spectrum $\{\lambda_k\}_{k=1}^{\infty}$, with $\lim_{k\to\infty} \lambda_k = 0$, and

$$||A||_{\mathrm{HS}}^2 = \sum_{k=1}^{\infty} \lambda_k^2 < \infty. \tag{5.29}$$

It is clear then that if $\dim(\mathscr{H}) = \infty$, then the identity operator I is not Hilbert–Schmidt, since obviously

$$||I||_{\mathrm{HS}} = \infty.$$

We explain the consequence of this fact on the infinite-dimensional generalization of the affine-invariant and Log-Euclidean distances in Sect. 5.4.2.2.

For two RKHS covariance operators $C_{\Phi(\mathbf{x})}$ and $C_{\Phi(\mathbf{y})}$, their Hilbert–Schmidt distance is expressed explicitly in terms of Gram matrices, as follows. Let $K[\mathbf{x}]$, $K[\mathbf{y}]$, $K[\mathbf{x},\mathbf{y}]$ denote the $m \times m$ matrices defined by

$$(K[\mathbf{x}])_{ij} = K(x_i, x_j), \quad (K[\mathbf{y}])_{ij} = K(y_i, y_j), \quad (K[\mathbf{x},\mathbf{y}])_{ij} = K(x_i, y_j). \tag{5.30}$$

By definition of feature maps, we have $K(x, y) = \langle \Phi(x), \Phi(y)\rangle_{\mathscr{H}_K}$ $\forall (x, y) \in \mathscr{X} \times \mathscr{X}$, so that the Gram matrices and the feature maps are closely related as follows

$$K[\mathbf{x}] = \Phi(\mathbf{x})^*\Phi(\mathbf{x}), \quad K[\mathbf{y}] = \Phi(\mathbf{y})^*\Phi(\mathbf{y}), \quad K[\mathbf{x},\mathbf{y}] = \Phi(\mathbf{x})^*\Phi(\mathbf{y}). \tag{5.31}$$

Lemma 1 *The Hilbert–Schmidt distance between two RKHS covariance operators* $C_{\Phi(\mathbf{x})}$ *and* $C_{\Phi(\mathbf{y})}$ *is given by*

$$\begin{aligned} ||C_{\Phi(\mathbf{x})} - C_{\Phi(\mathbf{y})}||_{\mathrm{HS}}^2 &= \frac{1}{m^2}\langle J_m K[\mathbf{x}], K[\mathbf{x}]J_m\rangle_F - \frac{2}{m^2}\langle J_m K[\mathbf{x},\mathbf{y}], K[\mathbf{x},\mathbf{y}]J_m\rangle_F \\ &+ \frac{1}{m^2}\langle J_m K[\mathbf{y}], K[\mathbf{y}]J_m\rangle_F. \end{aligned} \tag{5.32}$$

Remark 3 For the linear kernel $K(x, y) = \langle x, y\rangle_{\mathbb{R}^n}$, we have $\Phi(\mathbf{x}) = \mathbf{x}$, $\Phi(\mathbf{y}) = \mathbf{y}$ and thus Eq. (5.32) gives us the Euclidean distance $||C_{\mathbf{x}} - C_{\mathbf{y}}||_F$.

5.4.2 Riemannian Distances Between Covariance Operators

We now show how the affine-invariant and Log-Euclidean distances in Eqs. (5.6) and (5.8), respectively, can be generalized to the infinite-dimensional settings, specifically to self-adjoint, positive Hilbert–Schmidt operators on a separable Hilbert space $\mathscr{H}$. These generalizations were carried out recently in [21, 22], for the affine-invariant metric, and in [28], for the Log-Euclidean metric. As a special case of these formulations, we obtain the respective distances between infinite-dimensional covariance operators on Hilbert spaces, which assume explicit forms in the RKHS setting [26, 28].

Looking at Eqs. (5.6) and (5.8), we see that generalizing them to the infinite-dimensional setting requires the following two steps

1. Generalization of the set $\mathrm{Sym}^{++}(n)$ of all $n \times n$ SPD matrices and the corresponding generalization for the principal matrix logarithm.
2. Generalization of the Frobenius norm $||\ ||_F$.

We show next that the first step leads us to the concept of *positive definite operators* and the second to the concept of *extended Hilbert–Schmidt norm.*

5.4.2.1 Positive Definite Operators

We first consider the bounded operators $A : \mathscr{H} \to \mathscr{H}$ that generalize $n \times n$ SPD matrices and the corresponding generalization for the principal matrix logarithm. To this end, we first recall the definition of the principal matrix logarithm for an SPD matrix A of size $n \times n$. Let $\{\lambda_k\}_{k=1}^n$ be the eigenvalues of A, which are all positive, with corresponding normalized eigenvectors $\{\mathbf{u}_k\}_{k=1}^n$. Then A admits the spectral decomposition

$$A = \sum_{k=1}^{n} \lambda_k \mathbf{u}_k \mathbf{u}_k^T.$$

The principal matrix logarithm of A is then given by

$$\log(A) = \sum_{k=1}^{n} \log(\lambda_k) \mathbf{u}_k \mathbf{u}_k^T. \tag{5.33}$$

To generalize this formula to the Hilbert space setting, let us assume that $A : \mathscr{H} \to \mathscr{H}$ is a self-adjoint, compact operator, so that it possesses a countable spectrum

$\{\lambda_k\}_{k=1}^{\infty}$, with corresponding normalized eigenvectors $\{\mathbf{u}_k\}_{k=1}^{\infty}$. The eigenvalues λ_k's are all real and satisfy $\lim_{k\to\infty}\lambda_k = 0$. We assume next that A is *strictly positive*, that is

$$\langle x, Ax\rangle > 0 \quad \forall x \in \mathscr{H}, x \neq 0. \tag{5.34}$$

Then the eigenvalues $\{\lambda_k\}_{k=1}^{\infty}$ of A are all strictly positive. However, in contrast to the case $\dim(\mathscr{H}) < \infty$, when $\dim(\mathscr{H}) = \infty$, strict positivity is *not* a sufficient condition for $\log(A)$ to be well-defined. To see why, consider the following direct generalization of the principal matrix logarithm in Eq. (5.33),

$$\log(A) = \sum_{k=1}^{\infty} (\log\lambda_k)\mathbf{u}_k \otimes \mathbf{u}_k : \mathscr{H} \to \mathscr{H}, \tag{5.35}$$

where $\mathbf{u}_k \otimes \mathbf{u}_k : \mathscr{H} \to \mathscr{H}$ is a rank-one operator defined by $(\mathbf{u}_k \otimes \mathbf{u}_k)w = \langle \mathbf{u}_k, w\rangle \mathbf{u}_k$, which directly generalizes the rank-one matrix $\mathbf{u}_k\mathbf{u}_k^T$. Thus

$$\log(A)w = \sum_{k=1}^{\infty} (\log\lambda_k)\langle \mathbf{u}_k, w\rangle \mathbf{u}_k \quad \forall w \in \mathscr{H}. \tag{5.36}$$

However, since $\lim_{k\to\infty}\log\lambda_k = -\infty$, this operator is *unbounded*. Thus in particular, if A is a covariance operator, then $\log(A)$ is unbounded.

This problem can be resolved via regularization as follows. Instead of considering $\log(A)$, we consider $\log(A + \gamma I)$, $\gamma > 0$, and we see immediately that

$$\log(A + \gamma I) = \sum_{k=1}^{\infty} [\log(\lambda_k + \gamma)]\mathbf{u}_k \otimes \mathbf{u}_k : \mathscr{H} \to \mathscr{H}, \tag{5.37}$$

is a bounded operator $\forall \gamma > 0$. The operator $A + \gamma I$ is an example of a *positive definite operator*, that is it is a member of the set

$$\mathbb{P}(\mathscr{H}) = \{B \in \mathscr{L}(\mathscr{H}) : \exists M_B > 0 \text{ such that } \langle x, Bx\rangle \geq M_B||x||^2 \ \ \forall x \in \mathscr{H}\}. \tag{5.38}$$

Clearly, if $B \in \mathbb{P}(\mathscr{H})$, then the eigenvalues of B, if they exist, are all bounded below by the constant $M_B > 0$. Thus in the infinite-dimensional setting, positive definiteness is a stronger requirement than strict positivity. In fact, it can be shown that

$$B \text{ positive definite} \Longleftrightarrow B \text{ strictly positive and invertible.} \tag{5.39}$$

Subsequently, we use the notation $B > 0$ for $B \in \mathbb{P}(\mathscr{H})$.

5.4.2.2 Extended Hilbert–Schmidt Inner Product and Norm

We now consider operators of the form $A + \gamma I > 0$, where A is a self-adjoint, compact operators, so that $\log(A + \gamma I)$, as given by Eq. (5.37) is well-defined and bounded. To generalize Eq. (5.8), we then need to have

$$||\log(A + \gamma I)||^2_{\mathrm{HS}} = \sum_{k=1}^{\infty}[\log(\lambda_k + \gamma)]^2 < \infty. \tag{5.40}$$

We show below that this is also sufficient for the generalization of Eq. (5.6). For $\gamma = 1$, this holds if and only if $A \in \mathrm{HS}(\mathscr{H})$, as shown by the following.

Proposition 2 *Assume that* $A + I > 0$, *with* A *being a self-adjoint, compact operator. Then* $\log(A + I) \in \mathrm{HS}(\mathscr{H})$ *if and only if* $A \in \mathrm{HS}(\mathscr{H})$.

However, for $\gamma \neq 1$, $\gamma > 0$, $\log(A + \gamma I)$ *cannot* be a Hilbert–Schmidt operator for any compact operator A with $A + \gamma I > 0$, as shown in the following.

Proposition 3 *Assume that* $A + \gamma I > 0$, $\gamma > 0$, $\gamma \neq 1$, *with* A *being a self-adjoint, compact operator. Then* $\log(A + \gamma I) \notin \mathrm{HS}(\mathscr{H})$.

The result given in Proposition 3 is due to the fact that $||I||_{\mathrm{HS}} = \infty$, as can be viewed via the decomposition

$$\log(A + \gamma I) = \log\left(\frac{A}{\gamma} + I\right) + \log(\gamma)I. \tag{5.41}$$

On the right handside, the first term $\log\left(\frac{A}{\gamma} + I\right)$ is Hilbert–Schmidt if and only if A is Hilbert–Schmidt, as guaranteed by Proposition 2. However, for $\gamma \neq 1$, the second term $\log(\gamma)I$ *cannot* be Hilbert–Schmidt, since $||I||_{\mathrm{HS}} = \infty$ for $\dim(\mathscr{H}) = \infty$.

Thus, taken together, Propositions 2 and 3 show that to generalize Eqs. (5.6) and (5.8), we need to consider operators of the form $A + \gamma I > 0$, where A is Hilbert–Schmidt, and at the same time, extend the definition of the Hilbert–Schmidt inner product and norm so that the norm of the identity operator I is finite.

The desired extension is called the *extended Hilbert–Schmidt inner product* $\langle\ ,\ \rangle_{\mathrm{eHS}}$ [21, 22], which is defined by

$$\langle A + \gamma I, B + \mu I\rangle_{\mathrm{eHS}} = \langle A, B\rangle_{\mathrm{HS}} + \gamma\mu. \tag{5.42}$$

Under the extended Hilbert–Schmidt inner product, the scalar operators γI are orthogonal to the Hilbert–Schmidt operators. The corresponding *extended Hilbert–Schmidt norm* is then given by

$$||A + \gamma I||^2_{\mathrm{eHS}} = ||A||^2_{\mathrm{HS}} + \gamma^2. \tag{5.43}$$

One can see that this is a form of compactification, which gives $||I||_{\text{eHS}} = 1$, in contrast to the infinite Hilbert–Schmidt norm $||I||_{\text{HS}} = \infty$. Thus instead of the Hilbert space of self-adjoint Hilbert–Schmidt operators, we consider the Hilbert space of self-adjoint *extended (or unitized) Hilbert–Schmidt operators*

$$\mathscr{H}_{\mathbb{R}} = \{A + \gamma I \ : \ A^* = A, \ A \in \text{HS}(\mathscr{H}), \ \gamma \in \mathbb{R}\}, \tag{5.44}$$

under the extended Hilbert–Schmidt inner product. By the decomposition given in Eq. (5.41), we immediately obtain the following.

Proposition 4 *Assume that* $A + \gamma I > 0$ *where* $\gamma > 0$ *and* A *is a self-adjoint, compact operator. Then*

$$\log(A + \gamma I) \in \mathscr{H}_{\mathbb{R}} \Longleftrightarrow A + \gamma I > 0, \ A \in \text{HS}(\mathscr{H}), \ A^* = A. \tag{5.45}$$

Furthermore, when $\dim(\mathscr{H}) = \infty$,

$$|| \log(A + \gamma I)||^2_{\text{eHS}} = \left\| \log\left(\frac{A}{\gamma} + I\right)\right\|^2_{\text{HS}} + (\log \gamma)^2. \tag{5.46}$$

5.4.2.3 The Hilbert Manifold of Positive Definite Unitized Hilbert–Schmidt Operators

In summary, to generalize Eqs. (5.6) and (5.8) to the Hilbert space setting, we need to consider operators of the form $A + \gamma I > 0$, where A is self-adjoint, Hilbert–Schmidt, so that $\log(A + \gamma I)$ is well-defined and bounded, along with the extended Hilbert–Schmidt norm $|| \ ||_{\text{eHS}}$, so that $|| \log(A + \gamma I)||_{\text{eHS}}$ is finite. We have thus arrived at the following generalization of $\text{Sym}^{++}(n)$

$$\Sigma(\mathscr{H}) = \mathbb{P}(\mathscr{H}) \cap \mathscr{H}_{\mathbb{R}} = \{A + \gamma I > 0 \ : \ A^* = A, \ A \in \text{HS}(\mathscr{H}), \ \gamma \in \mathbb{R}\}, \tag{5.47}$$

which was first introduced by [21, 22]. This is an infinite-dimensional Hilbert manifold, with the tangent space at each point $T_P(\Sigma(\mathscr{H})) \cong \mathscr{H}_{\mathbb{R}} \ \forall P \in \Sigma(\mathscr{H})$. By Proposition 4,

$$A + \gamma I \in \Sigma(\mathscr{H}) \Longleftrightarrow \log(A + \gamma I) \in \mathscr{H}_{\mathbb{R}}. \tag{5.48}$$

5.4.3 The Affine-Invariant Distance

The affine-invariant Riemannian metric was introduced on the Hilbert manifold $\Sigma(\mathscr{H})$ by [21, 22]. Under this metric, the geodesic distance between any two positive definite operators $(A+\gamma I), (B+\mu I) \in \Sigma(\mathscr{H})$ is given by

$$d_{\mathrm{aiHS}}[(A+\gamma I), (B+\mu I)] = ||\log[(A+\gamma I)^{-1/2}(B+\mu I)(A+\gamma I)^{-1/2}]||_{\mathrm{eHS}}. \tag{5.49}$$

The following result confirms that the distance $d_{\mathrm{aiHS}}[(A+\gamma I), (B+\mu I)]$ is always finite for any pair of operators $(A+\gamma I), (B+\mu I) \in \Sigma(\mathscr{H})$.

Proposition 5 *For any two operators* $(A+\gamma I), (B+\mu I) \in \Sigma(\mathscr{H})$, *we can write* $(A+\gamma I)^{-1/2}(B+\mu I)(A+\gamma I)^{-1/2} = Z + \nu I > 0$ *for* $\nu = \frac{\mu}{\gamma}$ *and* $Z = (A+\gamma I)^{-1/2} B(A+\gamma I)^{-1/2} - \frac{\mu}{\gamma}A(A+\gamma I)^{-1}$ *satisfying* $Z = Z^*$, $Z \in \mathrm{HS}(\mathscr{H})$. *Thus the affine-invariant geodesic distance*

$$d_{\mathrm{aiHS}}[(A+\gamma I), (B+\mu I)] = ||\log(Z+\nu I)||_{\mathrm{eHS}} \tag{5.50}$$

is always finite. Furthermore, when $\dim(\mathscr{H}) = \infty$,

$$d^2_{\mathrm{aiHS}}[(A+\gamma I), (B+\mu I)] = \left\| \log\left(\frac{Z}{\nu} + I\right) \right\|^2_{\mathrm{HS}} + (\log \nu)^2. \tag{5.51}$$

In the RKHS setting, the affine-invariant distance between regularized RKHS covariance operators $d_{\mathrm{aiHS}}[(C_{\Phi(\mathbf{x})}+\gamma I), (C_{\Phi(\mathbf{y})}+\mu I)]$ admits a closed form, which was given by [26], as follows.

Theorem 1 ([26]) *Assume that* $\dim(\mathscr{H}_K) = \infty$. *Let* $\gamma > 0$, $\mu > 0$. *Then*

$$d^2_{\mathrm{aiHS}}[(C_{\Phi(\mathbf{x})}+\gamma I), (C_{\Phi(\mathbf{y})}+\mu I)] = \mathrm{tr}\left\{\log\left[\begin{pmatrix} C_{11} & C_{12} & C_{13} \\ C_{21} & C_{22} & C_{23} \\ C_{11} & C_{12} & C_{13} \end{pmatrix} + I_{3m}\right]\right\}^2 + \left(\log\frac{\gamma}{\mu}\right)^2, \tag{5.52}$$

where the $m \times m$ *matrices* C_{ij}, $i, j = 1, 2, 3$, *are given by*

$$C_{11} = \frac{1}{\mu m} J_m K[\mathbf{y}] J_m,$$

$$C_{12} = -\frac{1}{\sqrt{\gamma\mu} m} J_m K[\mathbf{y}, \mathbf{x}] J_m \left(I_m + \frac{1}{\gamma m} J_m K[\mathbf{x}] J_m\right)^{-1},$$

$$C_{13} = -\frac{1}{\gamma\mu m^2} J_m K[\mathbf{y}, \mathbf{x}] J_m \left(I_m + \frac{1}{\gamma m} J_m K[\mathbf{x}] J_m\right)^{-1} J_m K[\mathbf{x}, \mathbf{y}] J_m,$$

$$C_{21} = \frac{1}{\sqrt{\gamma\mu}m} J_m K[\mathbf{x}, \mathbf{y}] J_m,$$

$$C_{22} = -\frac{1}{\gamma m} J_m K[\mathbf{x}] J_m \left(I_m + \frac{1}{\gamma m} J_m K[\mathbf{x}] J_m\right)^{-1},$$

$$C_{23} = -\frac{1}{\gamma m} J_m K[\mathbf{x}] J_m \left(I_m + \frac{1}{\gamma m} J_m K[\mathbf{x}] J_m\right)^{-1} \frac{1}{\sqrt{\gamma\mu}m} J_m K[\mathbf{x}, \mathbf{y}] J_m.$$

Theorem 2 ([26]) *Assume that* $\dim(\mathcal{H}_K) < \infty$. *Let* $\gamma > 0$, $\mu > 0$. *Then*

$$d^2_{\mathrm{aiHS}}[(C_{\Phi(\mathbf{x})} + \gamma I), (C_{\Phi(\mathbf{y})} + \mu I)] = \mathrm{tr}\left\{\log\left[\begin{pmatrix} C_{11} & C_{12} & C_{13} \\ C_{21} & C_{22} & C_{23} \\ C_{11} & C_{12} & C_{13} \end{pmatrix} + I_{3m}\right]\right\}^2$$

$$-2\left(\log\frac{\gamma}{\mu}\right)\mathrm{tr}\left\{\log\left[\begin{pmatrix} C_{11} & C_{12} & C_{13} \\ C_{21} & C_{22} & C_{23} \\ C_{11} & C_{12} & C_{13} \end{pmatrix} + I_{3m}\right]\right\} + \left(\log\frac{\gamma}{\mu}\right)^2 \dim(\mathcal{H}_K), \quad (5.53)$$

where the $m \times m$ *matrices* C_{ij}*'s,* $i, j = 1, 2, 3$, *are as in Theorem 1.*

We see that the formula for the affine-invariant distance for the case $\dim(\mathcal{H}_K) = \infty$ is generally different from that for the case $\dim(\mathcal{H}_K) < \infty$, except when $\gamma = \mu$, in which case they are identical. One can see that for $m \in \mathbb{N}$ fixed, $\gamma \neq \mu$, the right hand side of Eq. (5.53) approaches infinity as $\dim(\mathcal{H}_K) \to \infty$. Thus for $\gamma \neq \mu$, one *cannot* approximate the infinite-dimensional distance in Eq. (5.52) by the finite-dimensional distance in Eq. (5.53).

5.4.3.1 Log-Hilbert–Schmidt Distance

Similar to the affine-invariance distance in Eq. (5.49), the generalization of the Log-Euclidean distance [4] is the Log-Hilbert–Schmidt distance

$$d_{\mathrm{logHS}}[(A + \gamma I), (B + \mu I)] = ||\log(A + \gamma I) - \log(B + \mu I)||_{\mathrm{eHS}}, \quad (5.54)$$

which was recently formulated by [28]. The well-definedness of this distance for any pair of operators $(A + \gamma I), (B + \mu I) \in \Sigma(\mathcal{H})$ is confirmed by the following result.

Proposition 6 *For any pair of operators* $(A + \gamma I), (B + \mu I) \in \Sigma(\mathcal{H})$, *the distance* $d_{\mathrm{logHS}}[(A + \gamma I), (B + \mu I)] = ||\log(A + \gamma I) - \log(B + \mu I)||_{\mathrm{eHS}}$ *is always finite. Furthermore, when* $\dim(\mathcal{H}) = \infty$,

$$||\log(A + \gamma I) - \log(B + \mu I)||^2_{\mathrm{eHS}} = \left\|\log\left(\frac{A}{\gamma} + I\right) - \log\left(\frac{B}{\mu} + I\right)\right\|^2_{\mathrm{HS}} + \left(\log\frac{\gamma}{\mu}\right)^2.$$

As in the case of the affine-invariant distance, in the case of regularized RKHS covariance operators, the Log-HS distance

$$d_{\text{logHS}}[C_{\Phi(\mathbf{x})} + \gamma I, C_{\Phi(\mathbf{y})} + \mu I] = ||\log(C_{\Phi(\mathbf{x})} + \gamma I) - \log(C_{\Phi(\mathbf{y})} + \mu I)||_{\text{eHS}} \quad (5.55)$$

also admits an explicit form, expressed via the Gram matrices corresponding to $\mathbf{x}$ and $\mathbf{y}$ [28]. To state this explicit form, we first define the following operators

$$A = \frac{1}{\sqrt{\gamma m}}\Phi(\mathbf{x})J_m : \mathbb{R}^m \to \mathscr{H}_K, \quad B = \frac{1}{\sqrt{\mu m}}\Phi(\mathbf{y})J_m : \mathbb{R}^m \to \mathscr{H}_K, \quad (5.56)$$

so that

$$A^*A = \frac{1}{\gamma m}J_m K[\mathbf{x}]J_m, \quad B^*B = \frac{1}{\mu m}J_m K[\mathbf{y}]J_m, \quad A^*B = \frac{1}{\sqrt{\gamma \mu} m}J_m K[\mathbf{x}, \mathbf{y}]J_m. \quad (5.57)$$

Let N_A and N_B be the numbers of nonzero eigenvalues of A^*A and B^*B, respectively. Let Σ_A and Σ_B be the diagonal matrices of size $N_A \times N_A$ and $N_B \times N_B$, and U_A and U_B be the matrices of size $m \times N_A$ and $m \times N_B$, respectively, which are obtained from the spectral decompositions

$$\frac{1}{\gamma m}J_m K[\mathbf{x}]J_m = U_A \Sigma_A U_A^T, \quad \frac{1}{\mu m}J_m K[\mathbf{y}]J_m = U_B \Sigma_B U_B^T. \quad (5.58)$$

Let $\circ$ denote the Hadamard (element-wise) matrix product and define

$$C_{AB} = \mathbf{1}_{N_A}^T \log(I_{N_A} + \Sigma_A)\Sigma_A^{-1}(U_A^T A^* B U_B \circ U_A^T A^* B U_B)\Sigma_B^{-1}\log(I_{N_B} + \Sigma_B)\mathbf{1}_{N_B}. \quad (5.59)$$

In terms of the quantities just defined, the Log-HS distance can be expressed as follows. As in the case of the affine-invariant distance, the distance formulas are different for the cases $\dim(\mathscr{H}_K) = \infty$ and $\dim(\mathscr{H}_K) < \infty$, with the finite-dimensional distance approaching infinity as $\dim(\mathscr{H}_K) \to \infty$.

Theorem 3 ([28]) *Assume that* $\dim(\mathscr{H}_K) = \infty$. *Let* $\gamma > 0$, $\mu > 0$. *Then*

$$d^2_{\text{logHS}}[(C_{\Phi(\mathbf{x})} + \gamma I), (C_{\Phi(\mathbf{y})} + \mu I)] = \text{tr}[\log(I_{N_A} + \Sigma_A)]^2 + \text{tr}[\log(I_{N_B} + \Sigma_B)]^2 - 2C_{AB} + (\log\gamma - \log\mu)^2. \quad (5.60)$$

Theorem 4 ([28]) *Assume that* $\dim(\mathscr{H}_K) < \infty$. *Let* $\gamma > 0$, $\mu > 0$. *Then*

$$\begin{aligned} d^2_{\text{logHS}}[(C_{\Phi(\mathbf{x})} + \gamma I), (C_{\Phi(\mathbf{y})} + \mu I)] &= \text{tr}[\log(I_{N_A} + \Sigma_A)]^2 + \text{tr}[\log(I_{N_B} + \Sigma_B)]^2 - 2C_{AB} \\ &+ 2\left(\log\frac{\gamma}{\mu}\right)(\text{tr}[\log(I_{N_A} + \Sigma_A)] - \text{tr}[\log(I_{N_B} + \Sigma_B)]) \\ &+ (\log\gamma - \log\mu)^2 \dim(\mathscr{H}_K). \end{aligned} \quad (5.61)$$

Remark 4 In the case of the linear kernel $K(x, y) = \langle x, y \rangle$, $x, y \in \mathbb{R}^n$, Theorem 4 gives the Log-Euclidean distance $|| \log(C_{\mathbf{x}} + \gamma I) - \log(C_{\mathbf{y}} + \mu I)||$.

Remark 5 We showed in [28] that the two operations $\odot$ and $\circledast$ on $\mathrm{Sym}^{++}(n)$ as defined in Sect. 5.2 can both be generalized to the Hilbert manifold $\Sigma(\mathscr{H})$, so that $(\Sigma(\mathscr{H}), \odot, \circledast)$ is a vector space. This vector space can be endowed with the *Log-Hilbert–Schmidt inner product*, defined by

$$\langle A + \gamma I, B + \mu I \rangle_{\mathrm{logHS}} = \langle \log(A + \gamma I), \log(B + \mu I) \rangle_{\mathrm{eHS}}. \tag{5.62}$$

With this inner product, the space $(\Sigma(\mathscr{H}), \odot, \circledast, \langle\ ,\ \rangle_{\mathrm{logHS}})$ is a Hilbert space and the distance in this Hilbert space is precisely the Log-Hilbert–Schmidt distance, see [28] for detail.

5.5 Two-Layer Kernel Machines with RKHS Covariance Operators

Having presented the explicit formulas for the affine-invariant and Log-Hilbert–Schmidt distances between RKHS covariance operators, we now show how the Log-Hilbert–Schmidt distance in particular can be used to design a two-layer kernel machine for machine learning, with an application in image classification.

5.5.1 The Interplay Between Positive Definite Kernels and Riemannian Manifolds

The geometric framework for RKHS covariance operators that we have just described reveals a close link between positive definite kernels and Riemannian manifolds, as follows.

Kernels giving rise to Manifolds. Let $\mathscr{X}$ be any non-empty set. Each positive definite kernel defined on $\mathscr{X} \times \mathscr{X}$ gives rise to a set of RKHS covariance operators, each of the form $C_{\Phi(\mathbf{x})}$, where $\mathbf{x}$ is a data matrix sampled from $\mathscr{X}$ according to a probability distribution. The corresponding set of regularized RKHS covariance operators $(C_{\Phi(\mathbf{x})} + \gamma I)$, $\gamma > 0$, forms a subset of the Hilbert manifold of positive definite Hilbert–Schmidt operators.

For the case of the Log-Hilbert–Schmidt distance, we have the link in the other direction as well.

Distances on Manifolds giving rise to Kernels. Since the Log-Hilbert–Schmidt distance is a Hilbert space distance, it can be used to define many positive definite kernels on $\Sigma(\mathscr{H}) \times \Sigma(\mathscr{H})$. The following result naturally generalizes Proposition 1 to the infinite-dimensional setting.

Proposition 7 ([28]) *The following kernels $K : \Sigma(\mathscr{H}) \times \Sigma(\mathscr{H}) \to \mathbb{R}$ are positive definite*

$$K[(A+\gamma I), (B+\mu I)] = (c + \langle \log(A+\gamma I), \log(B+\mu I)\rangle_{\mathrm{eHS}})^d, \quad c \geq 0, \; d \in \mathbb{N}, \tag{5.63}$$

$$K[(A+\gamma I), (B+\mu I)] = \exp\left(-\frac{||\log(A+\gamma I) - \log(B+\mu I)||^p_{\mathrm{eHS}}}{\sigma^2}\right), \tag{5.64}$$

for $0 < p \leq 2$.

5.5.2 Two-Layer Kernel Machines

The interplay between positive definite kernels and Riemannian manifolds as we described above allows us to design a two-layer kernel machine by utilizing the Log-Hilbert–Schmidt distance as follows.

1. In the first layer, a positive definite kernel, such as the Gaussian kernel, is applied to the original features extracted from each image, giving an *implicit* covariance operator representing that image. Using the Log-Hilbert–Schmidt distance, we then compute the pairwise distances between all the images.
2. In the second layer, using the pairwise Log-Hilbert–Schmidt distances obtained in the first layer, we define a new positive definite kernel, such as another Gaussian kernel. We can then apply any kernel method, such as SVM, using this kernel.

Remark 6 The approach in [17, 24], which applies kernel methods on top of the Log-Euclidean distance, is a special case our our framework, where the kernel in the first layer is linear (which is equivalent to *not* having the first layer).

5.6 Experiments in Image Classification

In this section, we report empirical results on the task of image classification using the two-layer kernel machine with the Log-Hilbert–Schmidt distance, as described in Sect. 5.5. The results obtained are substantially better than those obtained using the corresponding one-layer kernel machine with the Log-Euclidean distance. These thus clearly demonstrate the superior power and effectiveness of the covariance operator representation framework compared to the covariance matrix representation. The results presented here were first reported in [28].

We recall from our previous discussion that each image is represented by a covariance operator as follows. At every pixel (or a subset of pixels) of the image, we extract n low-level features, such as intensity and colors, giving us a low-level feature vector in $\mathbb{R}^n$. Sampling at m pixels in the image gives us a data matrix $\mathbf{x}$ of size $n \times m$. By applying a positive definite kernel, defined on $\mathbb{R}^n \times \mathbb{R}^n$, to the low-level feature vectors, we obtain *implicitly* a matrix of features $\Phi(\mathbf{x})$, as defined in Eq. (5.24), and the corresponding covariance operator $C_{\Phi(\mathbf{x})}$, as defined in Eq. (5.26). The image is then represented by the covariance operator $C_{\Phi(\mathbf{x})}$. In the current experiments, we used the Gaussian kernel and the resulting covariance operator is called *Gaussian-COV*. The distance between two images is the distance between the corresponding covariance operators, which in this case is the Log-Hilbert–Schmidt distance, given by Eq. (5.60) when $\dim(\mathscr{H}_K) = \infty$, e.g. for the Gaussian kernel, and Eq. (5.61) when $\dim(\mathscr{H}_K) < \infty$, e.g. for the polynomial kernels.

Given a set of images, we then have a corresponding set of covariance operators and a matrix of pairwise Log-Hilbert–Schmidt distances between these operators. In the following experiments, the task of image classification was carried out by applying Gaussian Support Vector Machine (SVM) on top of this distance matrix, using LIBSVM [10]. Thus this corresponds to a two-layer kernel machine *Gaussian-Gaussian* involving two Gaussian kernels, with the first Gaussian kernel defined on the low-level features and the second Gaussian kernel defined using the Log-Hilbert–Schmidt distances between the covariance operators of those features. For the sake of comparison, we also evaluated the kernel machine *Gaussian-Laplacian*, with the second kernel being the Laplacian kernel, which corresponds to $p = 1$ in Eq. (5.64).

In comparison, with the covariance matrix representation, one represents the image by the covariance matrix $C_{\mathbf{x}}$ defined directly on the data matrix $\mathbf{x}$ of low-level features, which we call *linearCOV*, since it is precisely the covariance operator obtained with the linear kernel. Given a set of images, we then obtain a set of corresponding covariance matrices and a matrix of pairwise Log-Euclidean distances between these covariance matrices. One can then carry out the task of image classification by applying Gaussian SVM on top of this distance matrix. Thus this corresponds to the two-layer kernel machine *linear-Gaussian*, which is equivalent to the one-layer kernel machine *Gaussian* on top of the Log-Euclidean distances, since the first layer in this case, being linear, has no effect.

Texture classification. The first dataset used is the Kylberg texture dataset [20], which contains 28 texture classes of different natural and man-made surfaces, with each class consisting of 160 images. The experimental protocols are the same as those in [16] and are as follows. Each image is resized to a dimension of 128×128, with $m = 1024$ observations computed on a coarse grid (i.e., every 4 pixels in the horizontal and vertical direction). At each pixel, 5 low-level features are extracted: $\mathbf{F}(x, y) = \left[I_{x,y}, |I_x|, |I_y|, |I_{xx}|, |I_{yy}|\right]$, where I, I_x, I_y, I_{xx} and I_{yy}, are the intensity, first- and second-order derivatives of the texture image. We randomly selected 5 images in each class for training and used the remaining ones as testing data, repeating the entire procedure 10 times.

Material classification. The second dataset used is the KTH-TIPS2b dataset [9], which contains images of 11 materials captured under 4 different illuminations, in 3 poses, and at 9 scales. The total number of images per class is 108. The same experimental protocols as used for the previous dataset [16] are employed, where at each pixel 23 low-level dense features are extracted: $\mathbf{F}(x, y) =$

Table 5.1 Classification accuracies over the 3 datasets. The accuracies shown are the mean accuracy across all the different splits for each dataset, along with the standard deviation. Here *Log-HS* denotes SVM with the Gaussian kernel on top of the Log-Hilbert–Schmidt distances, *Log-HS*$_\Delta$ denotes SVM with the Laplacian kernel on top of the Log-Hilbert–Schmidt distances, and *Log-E* denotes SVM with the Gaussian kernel on top of the Log-Euclidean distances

	Methods	Kylberg texture	KTH-TIPS2b	KTH-TIPS2b (RGB)	Fish
Gaussian COV	*Log-HS*	**92.58** %(**±1.23**)	**81.91** %(**±3.3**)	**79.94** %(**±4.6**)	**56.74** %(**±2.87**)
	Log-HS$_\Delta$	92.56 %(±1.26)	81.50 %(±3.90)	77.53 %(±5.2)	56.43 %(±3.02)
linear COV	*Log-E*	87.49 %(±1.54)	74.11 %(±7.41)	74.13 %(±6.1)	42.70 %(±3.45)

$\left[R_{x,y}, G_{x,y}, B_{x,y}, \left|G^{0,0}_{x,y}\right|, \ldots, \left|G^{4,5}_{x,y}\right|\right]$, where $R_{x,y}, G_{x,y}, B_{x,y}$ are the color intensities and $\left|G^{o,s}_{x,y}\right|$ are the 20 Gabor filters at 4 orientations and 5 scales. The experiment is repeated across 4 splits of the dataset.

Fish recognition. The third dataset used is the Fish Recognition dataset [8], which consists of 27,370 fish images belonging to 23 different classes. The number of images per class ranges from 21 to 12,112, with a medium resolution of roughly 150×120 pixels. The same experimental protocols are employed, where at each pixel the 3 color intensities are extracted: $\mathbf{F}(x, y) = \left[R_{x,y}, G_{x,y}, B_{x,y}\right]$. We randomly selected 5 images from each class for training and 15 for testing, repeating the entire procedure 10 times.

Results and discussion. Table 5.1 shows the classification accuracies obtained on the three tested datasets. As can be seen, the Log-Hilbert–Schmidt distance with the GaussianCOV displays significant improvements over the Log-Euclidean distance with the linearCOV. This strong improvement in performance is as expected, since, as we have discussed previously, covariance operators, by capturing nonlinear correlations between input features, offer a more general, more powerful, and more expressive data representation than covariance matrices.

5.7 Discussion, Conclusion, and Future Work

In this chapter, we have reviewed some of the recent progress in the generalization of the data representation framework using finite-dimensional covariance matrices to infinite-dimensional RKHS covariance operators, which are induced by positive definite kernels on the original data. In particular, we treated covariance operators in the setting of the infinite-dimensional manifold of positive definite operators, which is the generalization of the Riemannian manifold setting for SPD matrices. We presented the affine-invariant and Log-Hilbert–Schmidt distances on this manifold, which are generalizations of the affine-invariant and Log-Euclidean distances, respectively, between SPD matrices. For RKHS covariance operators, these distances admit closed form expressions via the corresponding Gram matrices and thus can be employed directly in a practical algorithm, such as image classification.

The Log-Hilbert–Schmidt distance, in particular, can be used to define new positive definite kernels, giving rise to a two-layer kernel machine. Experiments on the task of image classification have demonstrated that results obtained using the infinite-dimensional covariance operator representation significantly outperform those obtained using the finite-dimensional covariance matrix representation.

There are several ongoing and potential future research directions for the data representation framework using covariance operators. *On the methodological side*, one challenge faced by the framework is that both the affine-invariant and Log-Hilbert–Schmidt distances between covariance operators are computationally intensive on large scale datasets. One way to tackle this computational complexity for large scale applications is by approximating the infinite-dimensional covariance operators using approximate kernel feature maps. This has recently been carried out by [14], which effectively computed an approximate version of the affine-invariant distance, and [29], which computed approximate Log-Hilbert–Schmidt distances, both using Fourier feature maps. It would be interesting and fruitful to explore other approximations and computation schemes as well. *On the application side*, given the numerous applications of covariance matrices in diverse domains, ranging from statistics to machine learning to brain imaging, we expect that the covariance operator framework will find many more applications beyond those that we have presented or surveyed in this chapter.

Appendix

Proofs of Mathematical Results

Proof (**of Proposition** 1) For the first kernel, we have the property that the sum and product of positive definite kernels are also positive definite. Thus from the positivity of the inner product $\langle A, B\rangle_F$, it follows that $K(A,B) = (c + \langle A, B\rangle_{\mathrm{logE}})^d$ is positive definite, as in the Euclidean setting.

For the second kernel, since $(\mathrm{Sym}^{++}(n), \odot, \circledast, \langle\ ,\ \rangle_{\mathrm{logE}})$ is an inner product space, it follows that the kernel

$$K(A,B) = \exp(-d^p_{\mathrm{logE}}(A,B)/\sigma^2) = \exp(-||\log(A) - \log(B)||^p_F/\sigma^2)$$

is positive definite for $0 < p \leq 2$ by a classical result due to Schoenberg on positive definite functions and the imbeddability of metric spaces into Hilbert spaces (see [35], Theorem 1 and Corollary 1).

Proof (**of Lemma** 1) Recall that we have $\Phi(\mathbf{x})^*\Phi(\mathbf{x}) = K[\mathbf{x}]$, $\Phi(\mathbf{y})^*\Phi(\mathbf{y}) = K[\mathbf{y}]$, $\Phi(\mathbf{x})^*\Phi(\mathbf{y}) = K[\mathbf{x},\mathbf{y}]$. By definition of the Hilbert–Schmidt norm and property of the trace operation, we have

$$
\begin{aligned}
||C_{\Phi(\mathbf{x})} - C_{\Phi(\mathbf{y})}||^2_{\mathrm{HS}} &= \left\| \frac{1}{m}\Phi(\mathbf{x})J_m\Phi(\mathbf{x})^* - \frac{1}{m}\Phi(\mathbf{y})J_m\Phi(\mathbf{y})^* \right\|^2_{\mathrm{HS}} \\
&= \frac{1}{m^2}||\Phi(\mathbf{x})J_m\Phi(\mathbf{x})^*||^2_{\mathrm{HS}} - \frac{2}{m^2}\langle \Phi(\mathbf{x})J_m\Phi(\mathbf{x})^*, \Phi(\mathbf{y})J_m\Phi(\mathbf{y})^*\rangle_{\mathrm{HS}} \\
&+ \frac{1}{m^2}||\Phi(\mathbf{y})J_m\Phi(\mathbf{y})^*||^2_{\mathrm{HS}} \\
&= \frac{1}{m^2}\mathrm{tr}[\Phi(\mathbf{x})J_m\Phi(\mathbf{x})^*\Phi(\mathbf{x})J_m\Phi(\mathbf{x})^*] - \frac{2}{m^2}\mathrm{tr}[\Phi(\mathbf{x})J_m\Phi(\mathbf{x})^*\Phi(\mathbf{y})J_m\Phi(\mathbf{y})^*] \\
&+ \frac{1}{m^2}\mathrm{tr}[\Phi(\mathbf{y})J_m\Phi(\mathbf{y})^*\Phi(\mathbf{y})J_m\Phi(\mathbf{y})^*] \\
&= \frac{1}{m^2}\mathrm{tr}[(K[\mathbf{x}]J_m)^2 - 2K[\mathbf{y},\mathbf{x}]J_mK[\mathbf{x},\mathbf{y}]J_m + (K[\mathbf{y}]J_m)^2] \\
&= \frac{1}{m^2}[\langle J_mK[\mathbf{x}], K[\mathbf{x}]J_m\rangle_F - 2\langle J_mK[\mathbf{x},\mathbf{y}], K[\mathbf{x},\mathbf{y}]J_m\rangle_F + \langle J_mK[\mathbf{y}], K[\mathbf{y}]J_m\rangle_F].
\end{aligned}
$$

This completes the proof of the lemma. □

Lemma 2 *Let B be a constant with $0 < B < 1$. Then for all $|x| \leq B$,*

$$
|\log(1+x)| \leq \frac{1}{1-B}|x|. \tag{5.65}
$$

Proof For $x \geq 0$, we have the well-known inequality $0 \leq \log(1+x) \leq x$, so clearly $0 \leq \log(1+x) < \frac{1}{1-B}x$. Consider now the case $-B \leq x \leq 0$. Let

$$
f(x) = \log(1+x) - \frac{1}{1-B}x.
$$

We have

$$
f^{'}(x) = \frac{1}{1+x} - \frac{1}{1-B} \leq 0,
$$

with $f^{'}(-B) = 0$. Thus the function f is decreasing on $[-B, 0]$ and reaches its minimum at $x = 0$, which is $f(0) = 0$. Hence we have for all $-1 < -B \leq x \leq 0$

$$
0 \geq \log(1+x) \geq \frac{1}{1-B}x \Rightarrow |\log(1+x)| \leq \frac{1}{1-B}|x|,
$$

as we claimed. □

Proof (**of Propositions** 2 **and** 3) We first show that for an operator $A + \gamma I > 0$, where A is self-adjoint, compact, the operator $\log(A + \gamma I) \notin \mathrm{HS}(\mathscr{H})$ if $\gamma \neq 1$. Since A is compact, it has a countable spectrum $\{\lambda_k\}_{k\in\mathbb{N}}$, with $\lim_{k\to\infty}\lambda_k = 0$, so that $\lim_{k\to\infty}\log(\lambda_k + \gamma) = \log(\gamma)$. Thus if $\gamma \neq 1$, so that $\log\gamma \neq 0$, we have

$$||\log(A+\gamma I)||^2_{\text{HS}} = \sum_{k=1}^{\infty}[\log(\lambda_k+\gamma)]^2 = \infty.$$

Hence $\log(A+\gamma I) \notin \text{HS}(\mathscr{H})$ if $\gamma \neq 1$.

Assume now that $\gamma = 1$. We show that $\log(A+I) \in \text{HS}(\mathscr{H})$ if and only if $A \in \text{HS}(\mathscr{H})$. For the first direction, assume that $B = \log(A+I) \in \text{HS}(\mathscr{H})$. By definition, we have $A + I = \exp(B) \Longleftrightarrow A = \exp(B) - I = \sum_{k=1}^{\infty}\frac{B^k}{k!}$, with

$$||A||_{\text{HS}} = \left\|\sum_{k=1}^{\infty}\frac{B^k}{k!}\right\|_{\text{HS}} \leq \sum_{k=1}^{\infty}\frac{||B||^k_{\text{HS}}}{k!} = \exp(||B||_{\text{HS}}) - 1 < \infty.$$

This shows that $A \in \text{HS}$. Conversely, assume $A \in \text{HS}(\mathscr{H})$, so that

$$||A||^2_{\text{HS}} = \sum_{k=1}^{\infty}\lambda_k^2 < \infty,$$

and that $A + I > 0$, so that $\log(A+I)$ is well-defined and bounded, with eigenvalues $\{\log(\lambda_k+1)\}_{k=1}^{\infty}$. Since $\lim_{k\to\infty}\lambda_k = 0$, for any constant $0 < \varepsilon < 1$, there exists $N = N(\varepsilon)$ such that $|\lambda_k| < \varepsilon\ \forall k \geq N$. By Lemma 2, we have

$$\begin{aligned} ||\log(A+I)||^2_{\text{HS}} &= \sum_{k=1}^{\infty}[\log(\lambda_k+1)]^2 = \sum_{k=1}^{N-1}[\log(\lambda_k+1)]^2 + \sum_{k=N}^{\infty}[\log(\lambda_k+1)]^2 \\ &\leq \sum_{k=1}^{N-1}[\log(\lambda_k+1)]^2 + \frac{1}{1-\varepsilon}\sum_{k=N}^{\infty}\lambda_k^2 < \infty. \end{aligned}$$

This shows that $\log(A+I) \in \text{HS}(\mathscr{H})$, which completes the proof. □

Proof (**Proof of Proposition 4**) Since the identity operator I commutes with any operator A, we have the decomposition

$$\log(A+\gamma I) = \log\left(\frac{A}{\gamma} + I\right) + (\log\gamma)I.$$

We first note that the operator $\log\left(\frac{A}{\gamma}+I\right)$ is compact, since it possesses a countable set of eigenvalues $\{\log(\frac{\lambda_k}{\gamma}+1)\}_{k\in\mathbb{N}}$ satisfying $\lim_{k\to\infty}\log(\frac{\lambda_k}{\gamma}+1) = 0$.

If A is Hilbert–Schmidt, then by Proposition 2, we have $\log\left(\frac{A}{\gamma}+I\right) \in \text{HS}(\mathscr{H})$, and thus $\log(A+\gamma I) \in \mathscr{H}_{\mathbb{R}}$. By definition of the extended Hilbert–Schmidt norm,

$$||\log(A+\gamma I)||^2_{\text{eHS}} = \left\|\log\left(\frac{A}{\gamma}+I\right)\right\|^2_{\text{HS}} + (\log\gamma)^2 < \infty.$$

Conversely, if $\log(A+\gamma I) \in \mathscr{H}_{\mathbb{R}}$, then together with the fact that $\log\left(\frac{A}{\gamma}+I\right)$ is compact, the above decomposition shows that we must have $\log\left(\frac{A}{\gamma}+I\right) \in \mathrm{HS}(\mathscr{H})$ and hence $A \in \mathrm{HS}(\mathscr{H})$ by Proposition 2. □

Proof (**of Proposition** 5) Since $(A+\gamma I) > 0$, $(B+\mu I) > 0$, it is straightforward to see that $(A+\gamma I)^{-1/2}(B+\mu I)(A+\gamma I)^{-1/2} > 0$. Using the identity

$$(A+\gamma I)^{-1} = \frac{1}{\gamma}I - \frac{A}{\gamma}(A+\gamma I)^{-1},$$

we obtain

$$\begin{aligned}
&(A+\gamma I)^{-1/2}(B+\mu I)(A+\gamma I)^{-1/2} \\
&= \frac{\mu}{\gamma}I + (A+\gamma I)^{-1/2}B(A+\gamma I)^{-1/2} - \frac{\mu}{\gamma}A(A+\gamma I)^{-1} = Z + \nu I,
\end{aligned}$$

where $\nu = \frac{\mu}{\gamma}$ and $Z = (A+\gamma I)^{-1/2}B(A+\gamma I)^{-1/2} - \frac{\mu}{\gamma}A(A+\gamma I)^{-1}$. It is clear that $Z = Z^*$ and that $Z \in \mathrm{HS}(\mathscr{H})$, since $\mathrm{HS}(\mathscr{H})$ is a two-sided ideal in $\mathscr{L}(\mathscr{H})$. It follows that $\log(Z+\gamma I) \in \mathscr{H}_{\mathbb{R}}$ by Proposition 4. Thus the geodesic distance

$$\begin{aligned}
d_{\mathrm{aiHS}}[(A+\gamma I), (B+\mu I)] &= ||\log[(A+\gamma I)^{-1/2}(B+\mu I)(A+\gamma I)^{-1/2}]||_{\mathrm{eHS}} \\
&= ||\log(Z+\nu I)||_{\mathrm{eHS}}
\end{aligned}$$

is always finite. Furthermore, by Proposition 2, $\log(\frac{Z}{\nu}+I) \in \mathrm{HS}(\mathscr{H})$ and thus by definition of the extended Hilbert–Schmidt norm, when $\dim(\mathscr{H}) = \infty$,

$$d^2_{\mathrm{aiHS}}[(A+\gamma I), (B+\mu I)] = ||\log(Z+\nu I)||^2_{\mathrm{eHS}} = ||\log\left(\frac{Z}{\nu}+I\right)||^2_{\mathrm{HS}} + (\log\nu)^2.$$

This completes the proof. □

Proof (**of Proposition** 6) By Proposition 4, $(A+\gamma I), (B+\mu I) \in \Sigma(\mathscr{H}) \Longleftrightarrow \log(A+\gamma I), \log(B+\mu I) \in \mathscr{H}_{\mathbb{R}}$. It follows that $[\log(A+\gamma I) - \log(B+\mu I)] \in \mathscr{H}_{\mathbb{R}}$, so that $||\log(A+\gamma I) - \log(B+\mu I)||_{\mathrm{eHS}}$ is always finite.

Furthermore, by Proposition 2, $\log\left(\frac{A}{\gamma}+I\right)$, $\log\left(\frac{B}{\mu}+I\right) \in \mathrm{HS}(\mathscr{H})$ and by definition of the extended Hilbert–Schmidt norm, when $\dim(\mathscr{H}) = \infty$,

$$\begin{aligned}
||\log(A+\gamma I) - \log(B+\mu I)||^2_{\mathrm{eHS}} &= \left\|\log\left(\frac{A}{\gamma}+I\right) - \log\left(\frac{B}{\mu}+I\right) + \left(\log\frac{\gamma}{\mu}\right)I\right\|^2_{\mathrm{eHS}} \\
&= \left\|\log\left(\frac{A}{\gamma}+I\right) - \log\left(\frac{B}{\mu}+I\right)\right\|^2_{\mathrm{HS}} + \left(\log\frac{\gamma}{\mu}\right)^2.
\end{aligned}$$

This completes the proof. □

References

1. E. Andruchow, A. Varela, Non positively curved metric in the space of positive definite infinite matrices. Revista de la Union Matematica Argentina **48**(1), 7–15 (2007)
2. V.I. Arsenin, A.N. Tikhonov, *Solutions of Ill-Posed Problems* (Winston, Washington, 1977)
3. V. Arsigny, P. Fillard, X. Pennec, N. Ayache, Fast and simple calculus on tensors in the Log-Euclidean framework, *Medical Image Computing and Computer-Assisted Intervention–MICCAI 2005* (Springer, New York, 2005), pp. 115–122
4. V. Arsigny, P. Fillard, X. Pennec, N. Ayache, Geometric means in a novel vector space structure on symmetric positive-definite matrices. SIAM J. Matrix Anal. Appl. **29**(1), 328–347 (2007)
5. F. Barbaresco, Information geometry of covariance matrix: Cartan-Siegel homogeneous bounded domains, Mostow/Berger fibration and Frechet median, *Matrix Information Geometry* (Springer, New York, 2013), pp. 199–255
6. R. Bhatia, *Positive Definite Matrices* (Princeton University Press, Princeton, 2007)
7. D.A. Bini, B. Iannazzo, Computing the Karcher mean of symmetric positive definite matrices. Linear Algebra Appl. **438**(4), 1700–1710 (2013)
8. B.J. Boom, J. He, S. Palazzo, P.X. Huang, C. Beyan, H.-M. Chou, F.-P. Lin, C. Spampinato, R.B. Fisher, A research tool for long-term and continuous analysis of fish assemblage in coral-reefs using underwater camera footage. Ecol. Inf. **23**, 83–97 (2014)
9. B. Caputo, E. Hayman, P. Mallikarjuna, Class-specific material categorisation, in *ICCV* (2005), pp. 1597–1604
10. C.-C. Chang, C.-J. Lin, LIBSVM: a library for support vector machines. ACM Trans. Intell. Syst. Technol. **2**(3), 27:1–27:27 (2011)
11. A. Cherian, S. Sra, A. Banerjee, N. Papanikolopoulos, Jensen-Bregman LogDet divergence with application to efficient similarity search for covariance matrices. TPAMI **35**(9), 2161–2174 (2013)
12. I.L. Dryden, A. Koloydenko, D. Zhou, Non-Euclidean statistics for covariance matrices, with applications to diffusion tensor imaging. Ann. Appl. Stat. **3**, 1102–1123 (2009)
13. H.W. Engl, M. Hanke, A. Neubauer, *Regularization of Inverse Problems*, vol. 375, Mathematics and Its Applications (Springer, New York, 1996)
14. M. Faraki, M. Harandi, F. Porikli, Approximate infinite-dimensional region covariance descriptors for image classification, in *IEEE International Conference on Acoustics, Speech and Signal Processing (ICASSP)* (2015)
15. P. Formont, J.-P. Ovarlez, F. Pascal, On the use of matrix information geometry for polarimetric SAR image classification, *Matrix Information Geometry* (Springer, New York, 2013), pp. 257–276
16. M. Harandi, M. Salzmann, F. Porikli, Bregman divergences for infinite dimensional covariance matrices, in *CVPR* (2014)
17. S. Jayasumana, R. Hartley, M. Salzmann, H. Li, M. Harandi, Kernel methods on the Riemannian manifold of symmetric positive definite matrices, in *CVPR* (2013)
18. S. Jayasumana, R. Hartley, M. Salzmann, H. Li, M. Harandi, Kernel methods on Riemannian manifolds with Gaussian RBF kernels. IEEE Trans. Pattern Anal. Mach. Intell. **37**(12), 2464–2477 (2015)
19. B. Kulis, M.A. Sustik, I.S. Dhillon, Low-rank kernel learning with Bregman matrix divergences. J. Mach. Learn. Res. **10**, 341–376 (2009)
20. G. Kylberg, The Kylberg texture dataset v. 1.0. External report (Blue series) 35, Centre for Image Analysis, Swedish University of Agricultural Sciences and Uppsala University (2011)
21. G. Larotonda, *Geodesic Convexity, Symmetric Spaces and Hilbert-Schmidt Operators*. Ph.D. thesis, Universidad Nacional de General Sarmiento, Buenos Aires, Argentina (2005)
22. G. Larotonda, Nonpositive curvature: a geometrical approach to Hilbert-Schmidt operators. Differ. Geom. Appl. **25**, 679–700 (2007)
23. J.D. Lawson, Y. Lim, The geometric mean, matrices, metrics, and more. Am. Math. Monthly **108**(9), 797–812 (2001)

24. P. Li, Q. Wang, W. Zuo, L. Zhang, Log-Euclidean kernels for sparse representation and dictionary learning, in *ICCV* (2013)
25. H.Q. Minh, Some properties of Gaussian reproducing kernel Hilbert spaces and their implications for function approximation and learning theory. Constr. Approx. **32**, 307–338 (2010)
26. H.Q. Minh, Affine-invariant Riemannian distance between infinite-dimensional covariance operators, in *Geometric Science of Information*, vol. 9389, Lecture Notes in Computer Science, ed. by F. Nielsen, F. Barbaresco (Springer International Publishing, Switzerland, 2015), pp. 30–38
27. H.Q. Minh, P. Niyogi, Y. Yao, Mercer's theorem, feature maps, and smoothing, in *Proceedings of 19th Annual Conference on Learning Theory* (Springer, Pittsburg, 2006)
28. H.Q. Minh, M. San Biagio, V. Murino, Log-Hilbert-Schmidt metric between positive definite operators on Hilbert spaces, in *Advances in Neural Information Processing Systems 27 (NIPS 2014)* (2014), pp. 388–396
29. H.Q. Minh, M. San Biagio, L. Bazzani, V. Murino, Approximate Log-Hilbert-Schmidt distances between covariance operators for image classification, in *IEEE Conference on Computer Vision and Pattern Recognition (CVPR)* (2016)
30. G.D. Mostow, Some new decomposition theorems for semi-simple groups. Mem. Am. Math. Soc. **14**, 31–54 (1955)
31. X. Pennec, P. Fillard, N. Ayache, A Riemannian framework for tensor computing. Int. J. Comput. Vis. **66**(1), 41–66 (2006)
32. D. Pigoli, J. Aston, I.L. Dryden, P. Secchi, Distances and inference for covariance operators. Biometrika **101**(2), 409–422 (2014)
33. F. Porikli, O. Tuzel, P. Meer, Covariance tracking using model update based on Lie algebra, in *CVPR*, vol. 1 (IEEE, 2006), pp. 728–735
34. A. Qiu, A. Lee, M. Tan, M.K. Chung, Manifold learning on brain functional networks in aging. Med. Image Anal. **20**(1), 52–60 (2015)
35. I.J. Schoenberg, Metric spaces and positive definite functions. Trans. Am. Math. Soc. **44**, 522–536 (1938)
36. B. Schölkopf, A. Smola, K.-R. Müller, Nonlinear component analysis as a kernel eigenvalue problem. Neural Comput. **10**(5), 1299 (1998)
37. J. Shawe-Taylor, N. Cristianini, *Kernel Methods for Pattern Analysis* (Cambridge University Press, Cambridge, 2004)
38. S. Sra, A new metric on the manifold of kernel matrices with application to matrix geometric means. Adv. Neural Inf. Process. Syst. **1**, 144–152 (2012)
39. D. Tosato, M. Spera, M. Cristani, V. Murino, Characterizing humans on Riemannian manifolds. TPAMI **35**(8), 1972–1984 (2013)
40. O. Tuzel, F. Porikli, P. Meer, Pedestrian detection via classification on Riemannian manifolds. TPAMI **30**(10), 1713–1727 (2008)
41. S.K. Zhou, R. Chellappa, From sample similarity to ensemble similarity: probabilistic distance measures in reproducing kernel Hilbert space. TPAMI **28**(6), 917–929 (2006)

Chapter 6
Dictionary Learning on Grassmann Manifolds

Mehrtash Harandi, Richard Hartley, Mathieu Salzmann and Jochen Trumpf

Abstract Sparse representations have recently led to notable results in various visual recognition tasks. In a separate line of research, Riemannian manifolds have been shown useful for dealing with features and models that do not lie in Euclidean spaces. With the aim of building a bridge between the two realms, we address the problem of sparse coding and dictionary learning in Grassmann manifolds, i.e, the space of linear subspaces. To this end, we introduce algorithms for sparse coding and dictionary learning by embedding Grassmann manifolds into the space of symmetric matrices. Furthermore, to handle nonlinearity in data, we propose positive definite kernels on Grassmann manifolds and make use of them to perform coding and dictionary learning.

6.1 Introduction

In the past decade, sparsity has become a popular term in neuroscience, information theory, signal processing, and related areas [7, 11, 12, 33, 46]. Through sparse representation and compressive sensing, it is possible to represent natural signals like images using only a few nonzero coefficients of a suitable basis. In computer vision, sparse and overcomplete image representations were first introduced for modeling the spatial receptive fields of simple cells in the human visual system by [33]. The

M. Harandi (✉) · R. Hartley · J. Trumpf
College of Engineering and Computer Science, Australian National University, Canberra, ACT 2601, Australia
e-mail: mehrtash.harandi@anu.edu.au

R. Hartley
e-mail: richard.hartley@anu.edu.au

J. Trumpf
e-mail: jochen.trumpf@anu.edu.au

M. Salzmann
CVLab, EPFL, Lausanne, Switzerland
e-mail: mathieu.salzmann@epfl.ch

H.Q. Minh and V. Murino (eds.), *Algorithmic Advances in Riemannian Geometry and Applications*, Advances in Computer Vision and Pattern Recognition, DOI 10.1007/978-3-319-45026-1_6

linear decomposition of a signal using a few atoms of a dictionary has been shown to deliver notable results for various visual inference tasks, such as face recognition [46, 47], image classification [30, 48], subspace clustering [13] and image restoration [31] to name a few. While significant steps have been taken to develop the theory of the sparse coding and dictionary learning in Euclidean spaces, similar problems on non-Euclidean geometry have received comparatively little attention [8, 20, 22, 26]. This chapter discusses techniques to sparsely represent p-dimensional linear subspaces in $\mathbf{R}^d$ using a combination of linear subspaces.

Linear subspaces can be considered as the core of many inference algorithms in computer vision and machine learning. Examples include but not limited to modeling the reflectance function of Lambertian objects [4, 34], video analysis [9, 14, 18, 21, 41, 42], chromatic noise filtering [39], domain adaptation [16, 17], and object tracking [37]. Our main motivation here is to develop new methods for analyzing video data and image sets. This is inspired by the success of sparse signal modeling and related topics that suggest natural signals like images (and hence video and image sets as our concern here) can be efficiently approximated by superposition of atoms of a dictionary. We generalize the traditional notion of coding, which operates on vectors, to coding on subspaces. Coding with subspaces can then be seamlessly used for categorizing video data. Toward this, we first provide efficient solutions to the following two fundamental problems on Grassmann manifolds: (see Fig. 6.1 for a conceptual illustration):

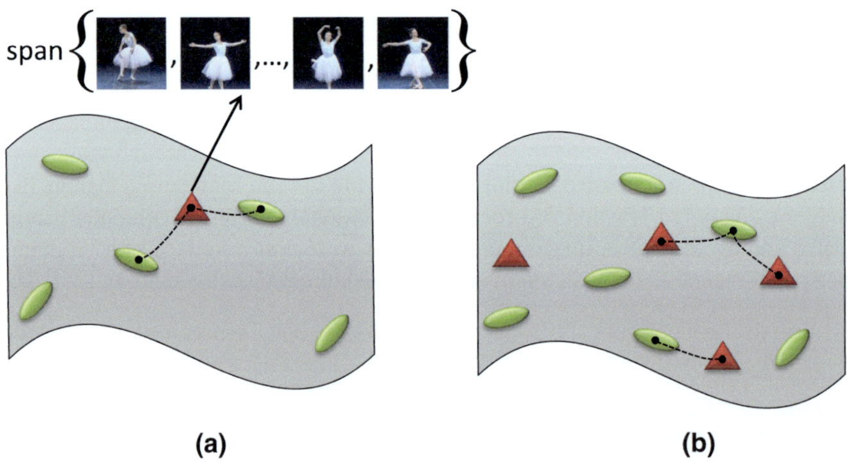

Fig. 6.1 A conceptual diagram of the problems addressed in this work. A video or an image set can be modeled by a linear subspace, which can be represented as a point on a Grassmann manifold. **a Sparse coding on a Grassmann manifold**. Given a dictionary (*green ellipses*) and a query signal (*red triangle*) on the Grassmann manifold, we are interested in estimating the query signal by a sparse combination of atoms while taking into account the geometry of the manifold (e.g, curvature). **b Dictionary learning on a Grassmann manifold**. Given a set of observations (*green ellipses*) on a Grassmann manifold, we are interested in determining a dictionary (*red triangles*) to describe the observations sparsely, while taking into account the geometry. This figure is best seen in color

1. **Coding**. Given a subspace $\mathscr{X}$ and a set $\mathbb{D} = \{\mathscr{D}_i\}_{i=1}^N$ with N elements (also known as atoms), where $\mathscr{X}$ and $\mathscr{D}_i$ are linear subspaces, how can $\mathscr{X}$ be approximated by a combination of atoms in $\mathbb{D}$?
2. **Dictionary learning**. Given a set of subspaces $\{\mathscr{X}_i\}_{i=1}^m$, how can a smaller set of subspaces $\mathbb{D} = \{\mathscr{D}_i\}_{i=1}^N$ be learned to represent $\{\mathscr{X}_i\}_{i=1}^m$ accurately?

Later, we tackle the problem of coding and dictionary learning on Grassmannian by embedding the manifold in Reproducing Kernel Hilbert Spaces (RKHS). To this end, we introduce a family of positive definite kernels on Grassmannian and make use of them to recast the coding problem in kernel spaces.

6.2 Problem Statement

In this section, we formulate the problem of coding and dictionary learning on the Grassmannian. Throughout this chapter, bold capital letters denote matrices (e.g X) and bold lowercase letters denote column vectors (e.g, x). The notation x_i (respectively $X_{i,j}$) is used to demonstrate the element in position i of the vector x (respectively (i,j) of the matrix X). $1_d \in \mathbf{R}^d$ and $0_d \in \mathbf{R}^d$ are vectors of ones and zeros. $\mathbf{I}_d$ is the $d \times d$ identity matrix. $\|x\|_1 = \sum_i |x_i|$ and $\|x\| = \sqrt{x^T x}$ denote the ℓ_1 and ℓ_2 norms, respectively, with T indicating transposition. $\|X\|_F = \sqrt{\operatorname{Tr}\left(X^T X\right)}$ designates the Frobenius norm, with $\operatorname{Tr}(\cdot)$ computing the matrix trace.

In vector spaces, by *coding* we mean the general notion of representing a vector x (the *query*) as some combination of other vectors d_i belonging to a *dictionary*. Typically, x is expressed as a linear combination $x = \sum_{j=1}^N y_j d_j$, or else as an *affine combination* in which the coefficients y_j satisfy the additional constraint $\sum_{j=1}^N y_j = 1$. (This constraint may also be written as $1^T y = 1$.)

In *sparse coding* one seeks to express the query in terms of a small number of dictionary elements. Given, a query $x \in \mathbf{R}^d$ and a dictionary $\mathbb{D}$ of size N, i.e, $\mathbb{D}_{d\times N} = \{d_1, d_2, \ldots, d_N\}$ with atoms $d_i \in \mathbf{R}^d$, the problem of coding x can be formulated as solving the minimization problem:

$$\ell_E(x, \mathbb{D}) \triangleq \min_y \left\| x - \sum\nolimits_{j=1}^N y_j d_j \right\|_2^2 + \lambda f(y). \tag{6.1}$$

The domain of y may be the whole of $\mathbf{R}^N$, so that the sum runs over all linear combinations of dictionary elements (or *atoms*), or alternatively, the extra constraint $1^T y = 1$ may be specified, to restrict to affine combinations.

The idea here is to (approximately) reconstruct the query x by a combination of dictionary atoms while forcing the coefficients of combination, i.e, y, to have some structure. The quantity $\ell_E(x, \mathbb{D})$ can be thought of as a coding cost combining the squared residual coding error, reflected in the energy term $\|\cdot\|_2^2$ in (6.1), along with a penalty term $f(y)$, which encourages some structure such as sparsity. The function

$f : \mathbf{R}^N \to \mathbf{R}$ could be the ℓ_1 norm, as in the Lasso problem [40], or some form of locality as proposed in [43].

The problem of dictionary learning is to determine $\mathbb{D}$ given a finite set of observations $\{x_i\}_{i=1}^m$, $x \in \mathbf{R}^d$, by minimizing the total coding cost for all observations, namely

$$h(\mathbb{D}) \triangleq \sum_{i=1}^{m} \ell_E(x_i, \mathbb{D}) \,, \tag{6.2}$$

while enforcing certain constraints on $\mathbb{D}$ to be satisfied to avoid trivial solutions. A "good" dictionary has a small residual coding error for all observations x_i while producing codes $y_i \in \mathbf{R}^N$ with the desired structure. For example, in the case of sparse coding, the ℓ_1 norm is usually taken as $f(\cdot)$ to obtain the most common form of dictionary learning in the literature. More specifically, the sparse dictionary learning problem may be written in full as that of jointly minimizing the total coding cost over all choices of coefficients and dictionary:

$$\min_{\{y_i\}_{i=1}^m, \mathbb{D}} \sum_{i=1}^{m} \Big\| x_i - \sum_{j=1}^{N} y_{ij} d_j \Big\|_2^2 + \lambda \sum_{i=1}^{m} \|y_i\|_1 . \tag{6.3}$$

A common approach to solving this is to alternate between the two sets of variables, $\mathbb{D}$ and $\{y_i\}_{i=1}^m$, as proposed for example by [2] (see [12] for a detailed treatment). Minimizing (6.3) over sparse codes y_i while dictionary $\mathbb{D}$ is fixed is a convex problem. Similarly, minimizing the overall problem over $\mathbb{D}$ with fixed $\{y_i\}_{i=1}^m$ is convex as well.

In generalizing the coding problem to a more general space $\mathcal{M}$, (e.g, Riemannian manifolds), one may write (6.1) as

$$\ell_{\mathcal{M}}(\mathcal{X}, \mathbb{D}) \triangleq \min_{y} \Big(d_{\mathcal{M}}\big(\mathcal{X},\, C(y, \mathbb{D})\big)^2 + \lambda f(y) \Big). \tag{6.4}$$

Here $\mathcal{X}$ and $\mathbb{D} = \{\mathcal{D}_j\}_{j=1}^N$ are points in the space $\mathcal{M}$, while $d_{\mathcal{M}}(\cdot,\cdot)$ is some distance metric and $C : \mathbf{R}^N \times \mathcal{M}^N \to \mathcal{M}$ is an *encoding function*, assigning an element of $\mathcal{M}$ to every choice of coefficients and dictionary. Note that (6.1) is a special case of this, in which $C(y, \mathbb{D})$ represents linear or affine combination, and $d_{\mathcal{M}}(\cdot,\cdot)$ is the Euclidean distance metric. To define the coding, one need only specify the metric $d_{\mathcal{M}}(\cdot,\cdot)$ to be used and the encoding function $C(\cdot,\cdot)$. Although this formulation may apply to a wide range of spaces, here we shall be concerned chiefly with coding on Grassmann manifolds.

A seemingly straightforward method for coding and dictionary learning is through embedding manifolds into Euclidean spaces via a fixed tangent space (the concepts related to differential geometry, such as tangent spaces will be shortly defined). The

Algorithm 1: Log-Euclidean sparse coding on Grassmann manifolds.

Input: Grassmann dictionary $\{\mathcal{D}_i\}_{i=1}^N$, $\mathcal{D}_i \in \mathcal{G}(p,d)$; the query sample $\mathcal{X} \in \mathcal{G}(p,d)$.
Output: The sparse code y^*.

Initialization.

for $i \leftarrow 1$ **to** N **do**
  $d_i \leftarrow \log_{\mathcal{P}}(\mathcal{D}_i)$;
end
$A \leftarrow [d_1|d_2|\cdots|d_N]$;

Processing.

$x \leftarrow \log_{\mathcal{P}}(\mathcal{X})$;
$y^* \leftarrow \arg\min_y \left\|x - A^T y\right\|_2^2 + \lambda \left\|y\right\|_1$;

embedding function in this case would be $\log_{\mathcal{P}}(\cdot)$, where $\mathcal{P}$ is some default base point.[1]

By mapping points in the manifold $\mathcal{M}$ to the tangent space, the problem at hand is transformed to its Euclidean counterpart. For example in the case of sparse coding, the encoding cost may be defined as follows:

$$\ell_{\mathcal{M}}(\mathcal{X}, \mathbb{D}) \triangleq \min_y \left\|\log_{\mathcal{P}}(\mathcal{X}) - \sum_{j=1}^N y_j \log_{\mathcal{P}}(\mathcal{D}_j)\right\|_{\mathcal{P}}^2 + \lambda\|y\|_1 \tag{6.5}$$

where the notation $\|\cdot\|_{\mathcal{P}}$ reminds us that the norm is in the tangent space at $\mathcal{X}$. We shall refer to this straightforward approach as *Log-Euclidean* sparse coding (the corresponding steps for Grassmann manifolds in Algorithm 1), following the terminology used in [3]. Since on a tangent space only distances to the base point are equal to true geodesic distances, the Log-Euclidean solution does not take into account the true structure of the underlying Riemannian manifold. Moreover, the solution is dependent upon the particular point $\mathcal{P}$ used as a base point.

Another approach is to measure the loss of $\mathcal{X}$ with respect to the dictionary $\mathbb{D}$, i.e, $\ell_{\mathcal{M}}(\mathcal{X}, \mathbb{D})$, by working in the tangent space of $\mathcal{X}$, rather than a fixed point $\mathcal{P}$ [8, 26]. The loss in this case becomes

$$\ell_{\mathcal{M}}(\mathcal{X}, \mathbb{D}) \triangleq \min_{\substack{y\in\mathbf{R}^N \\ 1^T y=1}} \left\|\sum_{j=1}^N y_j \log_{\mathcal{X}}(\mathcal{D}_j)\right\|_{\mathcal{X}}^2 + \|y\|_1 \tag{6.6}$$

[1]The function that maps each vector $y \in T_{\mathcal{P}}\mathcal{M}$ to a point $\mathcal{X}$ of the manifold that is reached after a unit time by the geodesic starting at $\mathcal{P}$ with this tangent vector is called the *exponential map*. For complete manifolds, this map is defined in the whole tangent space $T_{\mathcal{P}}\mathcal{M}$. The *logarithm map* is the inverse of the exponential map, i.e, $y = \log_{\mathcal{P}}(\mathcal{X})$ is the smallest vector y such that $\mathcal{X} = \exp_{\mathcal{P}}(y)$.

Following [26], given a set of training data $\{\mathcal{X}_i\}_{i=1}^m$, $\mathcal{X}_i \in \mathcal{M}$, the problem of dictionary learning can be written as

$$\min_{\{y_i\}_{i=1}^m, \mathbb{D}} \sum_{i=1}^m \left\| \sum_{j=1}^N y_{ij} \log_{\mathcal{X}_i}(\mathcal{D}_j) \right\|^2 + \lambda \sum_{i=1}^m \|y_i\|_1 \tag{6.7}$$
$$\text{s.t.} \quad 1^T y_i = 1, \; i = 1, 2, \ldots, m.$$

Similar to the Euclidean case, the problem in (6.7) can be solved by iterative optimization over $\{y_i\}_{i=1}^m$ and $\mathbb{D}$. Computing the sparse codes $\{y_i\}_{i=1}^m$ is done by solving (6.6). To update $\mathbb{D}$, [26] proposed a gradient descent approach along geodesics. That is, the update of $\mathcal{D}_r$ at time t while $\{y_i\}_{i=1}^m$ and $\mathcal{D}_j, j \neq r$ are kept fixed has the form

$$\mathcal{D}_r^{(t)} = \exp_{\mathcal{D}_r^{(t-1)}}(-\eta \Delta). \tag{6.8}$$

In Eq. (6.8) η is a step size and the tangent vector $\Delta \in T_{\mathcal{D}_r}(\mathcal{M})$ represents the direction of maximum ascent. That is $\Delta = \operatorname{grad} \mathcal{J}(\mathcal{D}_r)$,[2] where

$$\mathcal{J} = \sum_{i=1}^m \left\| \sum_{j=1}^N y_{ij} \log_{\mathcal{X}_i}(\mathcal{D}_j) \right\|^2. \tag{6.9}$$

Here is where the difficulty arises. Since the logarithm map does not have a closed-form expression on Grassmann manifolds, an analytic expression for Δ in Eq. (6.8) cannot be sought for the case of interest in this work, i.e, Grassmann manifolds. Having this in mind, we will describe various techniques to coding and dictionary learning specialized for Grassmann manifolds.

6.3 Background Theory

This section overviews Grassmann geometry and provides the groundwork for techniques described in following sections. Since the term "manifold" itself is often used in computer vision in a somewhat loose sense, we emphasize that the word is used in this chapter in its strict mathematical sense.

One most easily interprets the Grassmann manifolds in the more general context of group actions, to be described first. Consider a transitive (left) group action of a group G on a set S. The result of applying a group element g to a point $x \in S$ is written as gx. Choose a specific point $x_0 \in S$ and consider its *stabilizer* $\operatorname{Stab}(x_0) =$

[2] On an abstract Riemannian manifold $\mathcal{M}$, the gradient of a smooth real function f at a point $x \in \mathcal{M}$, denoted by $\operatorname{grad} f(x)$, is the element of $T_x(\mathcal{M})$ satisfying $\langle \operatorname{grad} f(x), \zeta \rangle_x = Df_x[\zeta]$ for all $\zeta \in T_x(\mathcal{M})$. Here, $Df_x[\zeta]$ denotes the directional derivative of f at x in the direction of ζ. The interested reader is referred to [1] for more details on how the gradient of a function on Grassmann manifolds can be computed.

$\{g \in G \mid gx_0 = x_0\}$. The stabilizer is a subgroup of G, which we will denote by H, and there is a one-to-one correspondence between the left cosets gH and the elements of S, whereby a point $x \in S$ is associated with the coset $\{g \in G \mid gx_0 = x\}$. The set of all left cosets is denoted by G/H, which we identify with the set S under this identification.

The (real, unoriented) *Grassmannian* $\mathscr{G}(p, d)$, where $0 < p \leq d$, is the set of all p-dimensional linear subspaces (we shall usually call them p-planes, or simply planes) of the real vector space $\mathbf{R}^d$. A geometric visualization of a point in the Grassmannian is a p-plane through the origin of d-dimensional Euclidean space. From this geometric picture, it is obvious that the *orthogonal group* $O(d)$ of all real orthogonal transforms of $\mathbf{R}^d$ acts on $\mathscr{G}(p, d)$, as any element of $O(d)$ transforms a p-plane through the origin to a p-plane through the origin. The action is transitive, since any plane can be reached in this way from any given one. The set of elements of $O(d)$ that transforms a given p-plane $\mathscr{X}$ to itself, the stabilizer $\mathrm{Stab}(\mathscr{X}) = \{U \in O(d) \mid U\mathscr{X} = \mathscr{X}\}$, is a subgroup of $O(d)$ and isomorphic to the product $O(p) \times O(d-p)$, where the factor $O(p)$ corresponds to in-plane transformations and the factor $O(d-p)$ corresponds to transformations that leave all points of the plane fixed. The Grassmannian can hence be identified with the coset space $O(d)/(O(p) \times O(d-p))$.

To make the above discussion more concrete, in terms of matrices, identify $O(d)$ as the group of orthogonal $d \times d$ matrices, a Lie group of dimension $d \times (d+1)/2$. In addition, think of p-planes in $\mathbf{R}^d$ as represented by *orthogonal Stiefel matrices*, that is, by rectangular $d \times p$-matrices X with orthonormal columns that form a basis of the plane. Since a given plane has many different orthogonal bases, two such matrices X and X' represent the same plane if and only if there exists a matrix $V \in O(p)$ such that $X' = XV$. This defines an equivalence relation between matrices X and X', and the planes may be identified with the equivalence classes of $d \times p$ matrices under this equivalence relation. The set of all such equivalence classes constitute the Grassmannian $\mathscr{G}(p, d)$.

A matrix U in $O(d)$ acts on a plane with representative Stiefel matrix X by left multiplication: $X \mapsto UX$. One immediately verifies that if X and X' represent the same plane, then so do UX and UX', so this defines a transitive left group action of $O(d)$ on $\mathscr{G}(p, d)$. Of particular interest is the element of the Grassmannian,

$$\mathscr{X}_0 = \mathrm{Span}(X_0) = \mathrm{Span}\begin{bmatrix} I_p \\ 0 \end{bmatrix},$$

where I_p is the $p \times p$ identity matrix. A transformation $U \in O(d)$ acts by left matrix multiplication and it is immediate that

$$\mathrm{Stab}(\mathscr{X}_0) = \left\{ \begin{bmatrix} U_1 & 0 \\ 0 & U_2 \end{bmatrix} \in O(d) \;\middle|\; U_1 \in O(p),\ U_2 \in O(d-p) \right\}, \tag{6.10}$$

showing again that $\mathrm{Stab}(\mathscr{X}_0) \simeq O(p) \times O(d-p)$.

Since the action of $O(d)$ is transitive on $\mathcal{G}(p, d)$, the elements of $\mathcal{G}(p, d)$ (planes) are in one-to-one correspondence with the set of left cosets of $\text{Stab}(\mathcal{X}_0)$ in $O(d)$. We think of the Grassmannian as the coset space $O(d)/(O(d-p) \times O(p))$, where $O(d-p) \times O(p)$ is identified with the block-diagonal subgroup in Eq. (6.10).

In this way, it is seen that the Grassmann manifolds form a special case of coset spaces G/H, in which $G = O(d)$ and H is the subgroup $O(d-p) \times O(p)$. In the Grassmann case, the group $G = O(d)$ is a matrix Lie group, and hence has the topological structure of a compact manifold. In this situation, G/H inherits a topology, a smooth manifold structure, and indeed a Riemannian metric from $G = O(d)$.

We continue the discussion denoting by G the matrix Lie group $O(d)$ and $H = \text{Stab}\,\mathcal{X}_0$, shown in Eq. (6.10), but the reader may bear in mind that the discussion holds equally well in the case of a general compact (sometimes also non-compact) Lie group G with Lie subgroup H.[3] This topic is treated in Chap. 21 of [29], which the reader may consult to fill in details.

The natural projection $\pi : G \to G/H \simeq \mathcal{G}(p, d)$ can now be used to equip the Grassmannian with quotient structures. For example, using the standard Lie group topology on G, the quotient as a quotient topology (the strongest topology such that π is continuous). Using the differential structure of the Lie group G, the quotient inherits a differential structure (the unique one that makes π a smooth submersion). With this differential structure, the action of G on G/H is smooth. With this smooth structure, the quotient space G/H is a manifold, according to the quotient manifold theorem (see Theorem 21.10 in [29]). Thus, the Grassmannian (or *Grassmann manifold*) is thus a *homogeneous space* of G. Its dimension is $p(d-p) = \dim(O(d)) - (\dim(O(d-p)) + \dim(O(p)))$.[4]

Tangent Space and Riemannian Metric

The homogeneous space structure of the Grassmannian can be used to equip it with a Riemannian metric, starting from a bi-invariant Riemannian metric on $G = O(d)$.

To this end, think of G as embedded in $\mathbf{R}^{d\times d}$ and equip the tangent space $T_I G$ at the identity with the Euclidean inner product inherited from that embedding. This defines a biinvariant Riemannian metric on G through right (or left) translation. For subgroup H of G define the Riemannian metric on the quotient G/H by

$$\langle X, Y \rangle_{UH} = \langle \tilde{X}, \tilde{Y} \rangle_U$$

where $U \in G$ and $X, Y \in T_{UH} G/H$. Further, $\tilde{X}$ and $\tilde{Y}$ in $T_U(G)$ are the horizontal lifts of X and Y with respect to π and the Riemannian metric on the group, that is, $\pi_U^*(\tilde{X}) = X$ and $\pi_U^*(\tilde{Y}) = Y$, and both $\tilde{X}$ and $\tilde{Y}$ are orthogonal to the kernel of π_U^* with respect to the inner product $\langle ., . \rangle_U$. It is easily verified that the above construction for (left) homogeneous spaces is well defined as long as the Riemannian metric on

[3] Another situation where this applies in Computer Vision is the study of the *essential manifold*, which may be envisaged as the coset space of $SO(3) \times SO(3)$ modulo a subgroup isomorphic to $SO(2)$. For details see [25].

[4] $O(d)$ has dimension $d(d-1)/2$, since its Lie algebra is the set of $n \times n$ skew-symmetric matrices.

the Lie group is right invariant under the action of the stabilizer. This is trivially the case for a biinvariant metric.

The tangent space to G at $U \in G$ is $T_U G = \{U\Omega \in \mathbf{R}^{d\times d} \mid \Omega \in \mathfrak{so}(d)\}$, where $\mathfrak{so}(d) = \{\Omega \in \mathbf{R}^{d\times d} \mid \Omega = -\Omega^\top\}$, the set of real skew-symmetric $d \times d$-matrices. The bi-invariant Riemannian metric on G inherited from the embedding of $T_I G$ in Euclidean d-space is given by

$$\langle U\Omega_1, U\Omega_2\rangle_U = -\operatorname{Tr}\Omega_1\Omega_2,$$

that is, by (left translations of) the Frobenius inner product. The tangent space to the Grassmannian $\mathscr{G}(p,d) \simeq G/H$ at UH is then the quotient

$$T_U G/UT_I H = \left\{U\left(\begin{bmatrix} 0 & -\Omega_{21}^\top \\ \Omega_{21} & 0\end{bmatrix} + T_I H\right) \mid \Omega_{21} \in \mathbf{R}^{(d-p)\times p}\right\}$$

where

$$T_I H = \left\{\begin{bmatrix}\Omega_1 & 0 \\ 0 & \Omega_2\end{bmatrix} \in \mathbf{R}^{d\times d} \mid \Omega_1 \in \mathfrak{so}(p),\ \Omega_2 \in \mathfrak{so}(d-p)\right\}.$$

The *horizontal subspace* of $T_U G$ formed by all horizontal lifts of tangent vectors to the Grassmannian $\mathscr{G}(p,d) \simeq G/H$ at UH is

$$V_U^\perp(G) = \left\{U\begin{bmatrix} 0 & -\Omega_{21}^\top \\ \Omega_{21} & 0\end{bmatrix} \in \mathbf{R}^{d\times d} \mid \Omega_{21} \in \mathbf{R}^{(d-p)\times p}\right\}.$$

This is most easily seen at $U = I$ where the elements of $V_I^\perp(G)$ are obviously perpendicular to the *vertical subspace* $V_I(G) = \operatorname{Ker}\pi_I^* = T_I H$ with respect to the Frobenius inner product. The normal Riemannian metric on the Grassmannian is then given by

$$\begin{aligned}\left\langle U\left(\begin{bmatrix} 0 & -\Omega_1^\top \\ \Omega_1 & 0\end{bmatrix} + T_I H\right), U\left(\begin{bmatrix} 0 & -\Omega_2^\top \\ \Omega_2 & 0\end{bmatrix} + T_I H\right)\right\rangle_{UH} \\ = \operatorname{Tr}\Omega_1^\top\Omega_2 + \operatorname{Tr}\Omega_1\Omega_2^\top \\ = 2\operatorname{Tr}\Omega_1\Omega_2^\top\end{aligned}$$

in terms of the horizontal lifts.

Projective Representation of Grassmann

An alternative representation of the Grassmannian $\mathscr{G}(p,d)$ is not as a quotient space of G, but as a subset of $\operatorname{Sym}(d)$, the vector space of all real symmetric $d \times d$-matrices. In this representation, a p-plane $\mathscr{X}$ is represented by the symmetric projection operator $P\colon \mathbf{R}^d \to \mathbf{R}^d$ with image $\operatorname{Im}(P) = \mathscr{X}$. In terms of matrix representations, P is a rank p real symmetric $d \times d$ matrix with $P^2 = P$ and $\operatorname{colspan}(P) = \mathscr{X}$. For example,

$$\mathscr{X}_0 = \operatorname{colspan}(P_0) = \operatorname{colspan}\begin{bmatrix} I_p & 0 \\ 0 & 0\end{bmatrix}.$$

In general, if $\mathscr{X}$ is represented by the orthogonal Stiefel matrix V then it is also represented by the symmetric matrix $P = VV^\top$. We denote the set of rank p real symmetric $d \times d$ matrix with $P^2 = P$ by $\mathscr{PG}(d, p)$ and obtain a bijection $\mathscr{PG}(d, p) \to \mathscr{G}(p, d)$ via $P \mapsto \text{colspan}(P)$.

The natural inclusion map

$$i \colon \mathscr{G}(p, d) \simeq \mathscr{PG}(d, p) \hookrightarrow \text{Sym}(d)$$

can now be used to equip the Grassmannian with subspace structures. For example, the Grassmannian inherits a subspace topology (the coarsest topology such that i is continuous) and a differential structure (the unique one that makes i a smooth embedding) from the standard topology resp. differential structure on $\text{Sym}(d)$. It turns out that both of these coincide with the respective quotient structures constructed above. The Grassmannian $\mathscr{G}(p, d)$ can hence equally be thought of as a homogeneous space of G or as an embedded submanifold of $\text{Sym}(d)$.

Since $\text{Sym}(d)$, as a linear subspace of $\mathbf{R}^{d \times d}$, carries a natural inner product, namely the restriction of the Frobenius inner product on $\mathbf{R}^{d \times d}$ to $\text{Sym}(d)$, each tangent space $T_P \mathscr{PG}(d, p)$ at a point $P \in \mathscr{PG}(d, p) \subset \text{Sym}(d)$ inherits this inner product via the inclusion $T_P \mathscr{PG}(d, p) \subset T_P \text{Sym}(d) \simeq \text{Sym}(d)$. This provides another construction for a Riemannian metric on $\mathscr{G}(p, d) \simeq \mathscr{PG}(d, p)$.

To make this construction more concrete, note that the tangent space

$$T_P \mathscr{PG}(d, p) = \{[P, \Omega] \mid \Omega \in \mathfrak{so}(d)\},$$

where $[P, \Omega] = P\Omega - \Omega P$ is the matrix commutator [24, Theorem 2.1]. In particular,

$$T_{P_0} \mathscr{PG}(d, p) = \left\{ \begin{bmatrix} 0 & \Omega_{12} \\ \Omega_{12}^\top & 0 \end{bmatrix} \in \text{Sym}(d) \,\middle|\, \Omega_{12} \in \mathbf{R}^{p \times (d-p)} \right\}.$$

Now observe that $O(d)$ acts transitively on $\mathscr{PG}(d, p)$ by conjugation since $UPU^\top \in \mathscr{PG}(d, p)$ for every $U \in O(d)$ and every $P \in \mathscr{PG}(d, p)$, and every point in $\mathscr{PG}(d, p)$ can be reached thus from a given one. Note that this action of $O(d)$ is different to the action by left multiplication that has been used to define the above homogeneous space structure of $\mathscr{G}(p, d)$. Nevertheless, it can be used to more explicitly describe the tangent space

$$T_{UP_0U^\top} \mathscr{PG}(d, p) = \left\{ U \begin{bmatrix} 0 & \Omega_{12} \\ \Omega_{12}^\top & 0 \end{bmatrix} U^\top \in \text{Sym}(d) \,\middle|\, \Omega_{12} \in \mathbf{R}^{p \times (d-p)} \right\}$$

at an arbitrary point $P = UP_0U^\top \in \mathscr{PG}(d, p)$. Note further that the first p columns of UP_0 form an orthogonal Stiefel matrix V (equal to the first p columns of U), and that $P = UP_0U^\top = VV^\top$ as observed before. The embedded Riemannian metric on $\mathscr{G}(p, d) \simeq \mathscr{PG}(d, p)$ is then given by

$$\langle U \begin{bmatrix} 0 & \Omega_1 \\ \Omega_1^\top & 0 \end{bmatrix} U^\top, U \begin{bmatrix} 0 & \Omega_2 \\ \Omega_2^\top & 0 \end{bmatrix} U^\top \rangle_{UP_0U^\top} = \operatorname{Tr} \Omega_1 \Omega_2^\top + \operatorname{Tr} \Omega_1^\top \Omega_2$$

in terms of this representation. It is not difficult to see that this Riemannian metric, in fact, is the same as the normal Riemannian metric constructed above [24, Proposition 2.3].

The unique geodesic starting at a point $U_0 P_0 U_0^\top \in \mathscr{PG}(d, p)$ in direction

$$U_0 \begin{bmatrix} 0 & \Omega \\ \Omega^\top & 0 \end{bmatrix} U_0^\top$$

is given by

$$P(t) = U_0 \operatorname{expm} \left(t \begin{bmatrix} 0 & -\Omega \\ \Omega^\top & 0 \end{bmatrix} \right) \begin{bmatrix} I_p & 0 \\ 0 & 0 \end{bmatrix} \operatorname{expm} \left(-t \begin{bmatrix} 0 & -\Omega \\ \Omega^\top & 0 \end{bmatrix} \right) U_0^\top,$$

where expm denotes the matrix exponential [24, Theorem 2.2]. Alternatively, it is given by

$$U(t)H = U_0 \operatorname{expm} \left(t \begin{bmatrix} 0 & -\Omega \\ \Omega^\top & 0 \end{bmatrix} \right) H$$

in the quotient representation. In particular, the *Riemannian exponential map* on the Grassmannian is given by

$$\exp_{U_0P_0U_0^\top} \left(U_0 \begin{bmatrix} 0 & \Omega \\ \Omega^\top & 0 \end{bmatrix} U_0^\top \right) = U_0 \operatorname{expm} \begin{bmatrix} 0 & -\Omega \\ \Omega^\top & 0 \end{bmatrix} \begin{bmatrix} I_p & 0 \\ 0 & 0 \end{bmatrix} \operatorname{expm} \begin{bmatrix} 0 & \Omega \\ -\Omega^\top & 0 \end{bmatrix} U_0^\top$$

in the embedded representation and by

$$\exp_{U_0H} \left(U_0 \left(\begin{bmatrix} 0 & -\Omega \\ \Omega^\top & 0 \end{bmatrix} + T_I H \right) \right) = U_0 \operatorname{expm} \begin{bmatrix} 0 & -\Omega \\ \Omega^\top & 0 \end{bmatrix} H$$

in the quotient representation.

Geodesics

The Grassmannian with the above Riemannian metric is a complete Riemannian manifold, hence any pair of points $\mathscr{X}_1$ and $\mathscr{X}_2$ on the Grassmannian can be connected by a length-minimizing geodesic. The *geodesic distance* $d_{\text{geod}}(\mathscr{X}_1, \mathscr{X}_2)$ is then defined as the length of this minimizing geodesic. Since points on the Grassmannian can be moved around arbitrarily by application of an orthogonal transformation $U \in O(d)$, and since the above Riemannian metric is invariant under such transformations, it is sufficient to compute the length of a minimizing geodesic connecting the special point $\mathscr{X}_0 = \operatorname{colspan}(P_0)$ to any other point $\mathscr{X} = \operatorname{colspan}(P)$. By the above formula for the exponential map

$$
\begin{aligned}
P &= \operatorname{expm}\begin{bmatrix} 0 & -\Omega \\ \Omega^\top & 0 \end{bmatrix} \begin{bmatrix} I_p & 0 \\ 0 & 0 \end{bmatrix} \operatorname{expm}\begin{bmatrix} 0 & \Omega \\ -\Omega^\top & 0 \end{bmatrix} \\
&= \begin{bmatrix} \cos^2 \sqrt{\Omega\Omega^\top} & \operatorname{sinc}\left(2\sqrt{\Omega\Omega^\top}\right)\Omega \\ \Omega^\top \operatorname{sinc}\left(2\sqrt{\Omega\Omega^\top}\right) & \sin^2 \sqrt{\Omega^\top\Omega} \end{bmatrix}
\end{aligned}
$$

for some $\Omega \in \mathbf{R}^{p\times(d-p)}$, where we have used [24, Eq. (2.68)] in the second line. The geodesic distance from $\mathscr{X}_0$ to $\mathscr{X}$ is then $d_{\text{geod}}(\mathscr{X}_0, \mathscr{X}) = \sqrt{2 \operatorname{Tr} \Omega\Omega^\top}$, that is the length of the tangent vector

$$
\begin{bmatrix} 0 & \Omega \\ \Omega^\top & 0 \end{bmatrix} \in T_{P_0}\mathscr{PG}(d, p)
$$

under the Riemannian metric at P_0. In more explicit terms, starting with a block representation

$$
P = \begin{bmatrix} P_1 & P_2 \\ P_2^\top & P_3 \end{bmatrix},
$$

where $P_1 \in \operatorname{Sym}(p)$, compute the eigenvalue decomposition $P_1 = U_1 \operatorname{diag}(\lambda_1, \ldots, \lambda_p)U_1^\top$ with $U_1 \in O(p)$, then $U_1\operatorname{diag}(\lambda_1, \ldots, \lambda_p)U_1^\top = P_1 = \cos^2\sqrt{\Omega\Omega^\top}$ is equivalent to $\Omega\Omega^\top = U_1\operatorname{diag}(\arccos^2(\sqrt{\lambda_1}), \ldots, \arccos^2(\sqrt{\lambda_p}))U_1^\top$ and hence

$$
d_{\text{geod}}(\mathscr{X}_0, \mathscr{X}) = \sqrt{2 \operatorname{Tr} \Omega\Omega^\top} = \sqrt{2\sum_{i=1}^{p} \arccos^2(\sqrt{\lambda_i})},
$$

cf. [24, Corollary 2.1].

A geometric interpretation of the above distance formula can be obtained as follows. Swapping back to the quotient representation and using [24, Eq. (2.66)], it follows that

$$
\mathscr{X} = \operatorname{colspan}\begin{bmatrix} \cos\sqrt{\Omega\Omega^\top} \\ \frac{\sin\sqrt{\Omega^\top\Omega}}{\sqrt{\Omega^\top\Omega}}\Omega^\top \end{bmatrix},
$$

where the columns of this matrix have unit length in the 2-norm since they are the first p columns of an orthogonal matrix. The *first principal angle* θ_1 between the subspaces $\mathscr{X}_0$ and $\mathscr{X}$ is given by

$$
\begin{aligned}
\cos\theta_1 &= \max_{u\in\mathscr{X}_0, v\in\mathscr{X}} \frac{u^\top v}{\|u\|_2\|v\|_2} \\
&= \max_{\|x\|_2=1, \|y\|_2=1} \begin{bmatrix} x^\top & 0 \end{bmatrix} \begin{bmatrix} \cos\sqrt{\Omega\Omega^\top} \\ \frac{\sin\sqrt{\Omega^\top\Omega}}{\sqrt{\Omega^\top\Omega}}\Omega^\top \end{bmatrix} y \\
&= \max_{\|x\|_2=1, \|y\|_2=1} x^\top U_1 \mathrm{diag}(\sqrt{\lambda_1}, \ldots, \sqrt{\lambda_p}) U_1^\top y \\
&= \sqrt{\lambda_1},
\end{aligned}
$$

assuming that the eigenvalues λ_i are ordered in nonincreasing order. Similarly, it can be shown that the ith principal angle $\theta_i = \sqrt{\lambda_i}$ for $i = 2, \ldots, p$. It follows that, in general,

$$d_{\text{geod}}(\mathscr{X}_1, \mathscr{X}_2) = \sqrt{2}\|\Theta\|_2,$$

where $\Theta = \begin{bmatrix}\theta_1 \ldots \theta_p\end{bmatrix}^\top$ is the vector of principal angles between $\mathscr{X}_1$ and $\mathscr{X}_2$. Note that some authors use a different scaling of the Frobenius inner product (usually an additional factor of $\frac{1}{2}$) to arrive at a formula for the geodesic distance without the factor of $\sqrt{2}$. Obviously, this does not change the geometry.

Principal angles. The geodesic distance has an interpretation as the magnitude of the smallest rotation that takes one subspace to the other. If $\Theta = [\theta_1, \theta_2, \ldots, \theta_p]$ is the sequence of principal angles between two subspaces $\mathscr{X}_1 \in \mathscr{G}(p, d)$ and $\mathscr{X}_2 \in \mathscr{G}(p, d)$, then $d_{\text{geod}}(\mathscr{X}_1, \mathscr{X}_2) = \|\Theta\|_2$.

Definition 1 (*Principal Angles*) Let X_1 and X_2 be two matrices of size $d \times p$ with orthonormal columns. The principal angles $0 \leq \theta_1 \leq \theta_2 \leq \cdots \leq \theta_p \leq \pi/2$ between two subspaces $\mathrm{Span}(X_1)$ and $\mathrm{Span}(X_2)$, are defined recursively by

$$
\begin{aligned}
\cos(\theta_i) &= \max_{u_i\in\mathrm{Span}(X_1)} \max_{v_i\in\mathrm{Span}(X_2)} u_i^T v_i \qquad (6.11)\\
\text{s.t.:} \quad & \|u_i\|_2 = \|v_i\|_2 = 1 \\
& u_i^T u_j = 0;\ j = 1, 2, \ldots, i-1 \\
& v_i^T v_j = 0;\ j = 1, 2, \ldots, i-1
\end{aligned}
$$

In other words, the first principal angle θ_1 is the smallest angle between all pairs of unit vectors in the first and the second subspaces. The rest of the principal angles are defined similarly.

Two operators, namely the logarithm map $\log_x(\cdot) : \mathscr{M} \to T_x(\mathscr{M})$ and its inverse, the exponential map $\exp_x(\cdot) : T_x(\mathscr{M}) \to \mathscr{M}$ are defined over Riemannian manifolds to switch between the manifold and the tangent space at x. A key point here is the fact that both the logarithm map and its inverse do not have closed-form solutions for Grassmann manifolds. Efficient numerical approaches for computing both maps were proposed by [5, 15]. In this paper, however, the exponential and logarithm maps will only be used when describing previous work of other authors.

6.4 Dictionary Learning on Grassmannian

In this part, we propose to make use of the projective representation of Grassmannian to perform coding and dictionary learning on Grassmannian. We recall that working with $\mathcal{PG}(p, d)$ instead of $\mathcal{G}(p, d)$ has the advantage that each element of $\mathcal{PG}(p, d)$ is a single matrix, whereas elements of $\mathcal{G}(p, d)$ are equivalence classes of matrices. Hereinafter, we shall denote XX^T by $\widehat{X}$, the hat representing the action of the projection embedding. Furthermore, $\langle \cdot, \cdot \rangle$ represents the Frobenius inner product: thus $\langle \widehat{X}, \widehat{Y} \rangle = \mathrm{Tr}(\widehat{X}\widehat{Y})$. Note that in computing $\langle \widehat{X}, \widehat{Y} \rangle$ it is not necessary to compute $\widehat{X}$ and $\widehat{Y}$ explicitly (they may be large matrices). Instead, note that $\langle \widehat{X}, \widehat{Y} \rangle = \mathrm{Tr}(\widehat{X}\widehat{Y}) = \mathrm{Tr}(XX^T YY^T) = \mathrm{Tr}(Y^T XX^T Y) = \|Y^T X\|_F^2$. This is advantageous, since $Y^T X$ may be a substantially smaller matrix.

Apart from the geodesic distance metric, an important metric used in this paper is the *chordal metric*, defined by

$$d_{\mathrm{chord}}(\widehat{X}, \widehat{Y}) = \|\widehat{X} - \widehat{Y}\|_F \ , \tag{6.12}$$

This metric will be used in the context of (6.4) to recast the coding and consequently dictionary-learning problem in terms of chordal distance. Before presenting our proposed methods, we establish an interesting link between coding and the notion of weighted mean in a metric space.

6.4.1 Weighted Karcher Mean

The underlying concept of coding using a dictionary is to represent in some way a point in a space of interest as a combination of other elements in that space. In the usual method of coding in $\mathbf{R}^d$ given by (6.1), each x is represented by a linear combination of dictionary elements d_j, where the first term represents the coding error. For coding in a manifold, the problem to address is that linear combinations do not make sense. We wish to find some way in which an element $\mathcal{X}$ may be represented in terms of other dictionary elements $\mathcal{D}_j$ as suggested in (6.4). For a proposed method to generalize the $\mathbf{R}^d$ case, one may prefer a method that is a direct generalization of the Euclidean case in some way.

In $\mathbf{R}^d$, a different way to consider the expression $\sum_{j=1}^N y_j d_j$ in (6.1) is as a weighted mean of the points d_j This observation relies on the following fact, which is verified using a Lagrange multiplier method.

Lemma 1 *Given coefficients y with $\sum_{i=1}^N y_i = 1$, and dictionary elements $\{d_1, \ldots, d_N\}$ in $\mathbf{R}^d$, the point $x^* \in \mathbf{R}^d$ that minimizes $\sum_{i=1}^N y_i \, \|x - d_i\|_F^2$ is given by $x^* = \sum_{i=1}^N y_i \, d_i$.*

In other words, the affine combination of dictionary elements is equal to their weighted mean. Although linear combinations are not defined for points on manifolds or metric spaces, a weighted mean is.

Definition 2 Given points $\mathscr{D}_i$ on a Riemannian manifold $\mathscr{M}$, and weights y_i, the point $\mathscr{X}^*$ that minimizes $\sum_{i=1}^{N} y_i\, d_g(\mathscr{X}, \mathscr{D}_i)^2$, is called the weighted Karcher mean of the points $\mathscr{D}_i$ with weights y_i. Here, $d_g(\cdot, \cdot)$ is the geodesic distance on $\mathscr{M}$.

Generally, finding the Karcher mean [28] on a manifold involves an iterative procedure, which may converge to a local minimum, even on a simple manifold, such as $SO(3)$ [23, 32]. However, one may replace the geodesic metric with a different metric in order to simplify the calculation. To this end, we propose the *chordal metric* on a Grassman manifold, defined for matrices $\widehat{X}$ and $\widehat{Y}$ in $\mathscr{PG}(p, d)$ in Eq. (6.12). The corresponding mean, as in Definition 2 (but using the chordal metric) is called the *weighted chordal mean* of the points. In contrast to the Karcher mean, the weighted chordal mean on a Grassman manifold has a simple closed form.

Theorem 1 *The weighted chordal mean of a set of points $\widehat{D}_i \in \mathscr{PG}(p, d)$ with weights y_i is equal to* $\mathrm{Proj}(\sum_{i=1}^{m} y_i \widehat{D}_i)$, *where* $\mathrm{Proj}(\cdot)$ *represents the closest point on $\mathscr{PG}(p, d)$ [20].*

The function $\mathrm{Proj}(\cdot)$ has a closed-form solution in terms of the singular value decomposition. More specifically,

Lemma 2 *Let X be an $d \times d$ symmetric matrix with eigenvalue decomposition $X = UDU^T$, where D contains the eigenvalues λ_i of X in descending order. Let U_p be the $d \times p$ matrix consisting of the first p columns of U. Then $\widehat{U}_p = U_p U_p^T$ is the closest matrix in $\mathscr{PG}(p, d)$ to X (under the Frobenius norm) [20].*

The chordal metric on a Grassman manifold is not a geodesic metric (that is it is not equal to the length of a shortest geodesic under the Riemannian metric). However, it is closely related. In fact, one may easily show that for $\mathscr{G}(p, d) \ni \mathscr{X} = \mathrm{span}(X)$ and $\mathscr{G}(p, d) \ni \mathscr{Y} = \mathrm{span}(Y)$

$$\frac{2}{\pi}\, d_{\mathrm{geod}}(\mathscr{X}, \mathscr{Y}) \le d_{\mathrm{chord}}(\widehat{X}, \widehat{Y}) \le d_{\mathrm{geod}}(\mathscr{X}, \mathscr{Y}) .$$

Furthermore, the path-metric [23] induced by $d_{\mathrm{chord}}(\cdot, \cdot)$ is equal to the geodesic distance.

Sparse Coding

Given a dictionary $\mathbb{D}$ with atoms $\widehat{D}_j \in \mathscr{PG}(p, d)$ and a query sample $\widehat{X}$ the problem of sparse coding can be recast extrinsically as:

$$\ell(\mathscr{X}, \mathbb{D}) \triangleq \min_{y} \Big\| \widehat{X} - \sum_{j=1}^{N} y_j \widehat{D}_j \Big\|_F^2 + \lambda \|y\|_1 . \tag{6.13}$$

The formulation here varies slightly from the general form given in (6.4), in that the point $\sum_{j=1}^{N} y_j\widehat{D}_j$ does not lie exactly on the manifold $\mathcal{PG}(p,d)$, since it is not idempotent nor its rank is necessarily p. We call this solution an *extrinsic solution*; the point coded by the dictionary is allowed to step out of the manifold.

Expanding the Frobenius norm term in (6.13) results in a convex function in y:

$$\left\|\widehat{X} - \sum_{j=1}^{N} y_j\widehat{D}_j\right\|_F^2 = \|\widehat{X}\|_F^2 + \left\|\sum_{j=1}^{N} y_j\widehat{D}_j\right\|_F^2 - 2\langle\sum_{j=1}^{N} y_j\widehat{D}_j, \widehat{X}\rangle \ .$$

The sparse codes can be obtained without explicitly embedding the manifold in $\mathcal{PG}(p,d)$ using $\Pi(\mathcal{X})$. This can be seen by defining $[\mathcal{K}(X,\mathbb{D})]_i = \langle\widehat{X},\widehat{D}_i\rangle$ as an N dimensional vector storing the similarity between signal X and dictionary atoms in the induced space and $[\mathbb{K}(\mathbb{D})]_{i,j} = \langle\widehat{D}_i,\widehat{D}_j\rangle$ as an $N \times N$ symmetric matrix encoding the similarities between dictionary atoms (which can be computed offline). Then, the sparse coding in (6.13) can be written as:

$$\ell(\mathcal{X},\mathbb{D}) = \min_y y^T\mathbb{K}(\mathbb{D})y - 2y^T\mathcal{K}(X,\mathbb{D}) + \lambda\|y\|_1 \ . \tag{6.14}$$

The symmetric matrix $\mathbb{K}(\mathbb{D})$ is positive semidefinite since for all $v \in \mathbf{R}^N$:

$$\begin{aligned} v^T\mathbb{K}(\mathbb{D})v &= \sum_{i=1}^{N}\sum_{j=1}^{N} v_i v_j\,\langle\widehat{D}_i,\widehat{D}_j\rangle = \left\langle\sum_{i=1}^{N} v_i\widehat{D}_i,\ \sum_{j=1}^{N} v_j\widehat{D}_j\right\rangle \\ &= \left\|\sum_{i=1}^{N} v_i\widehat{D}_i\right\|_F^2 \geq 0. \end{aligned}$$

Therefore, the problem is convex and can be efficiently solved. The problem in (6.14) can be transposed into a vectorized sparse coding problem. More specifically, let $U\Sigma U^T$ be the SVD of $\mathbb{K}(\mathbb{D})$. Then (6.14) is equivalent to

$$\ell(\mathcal{X},\mathbb{D}) = \min_y \|x^* - Ay\|^2 + \lambda\|y\|_1, \tag{6.15}$$

where $A = \Sigma^{1/2}U^T$ and $x^* = \Sigma^{-1/2}U^T\mathcal{K}(X,\mathbb{D})$. This can be easily verified by plugging A and x^* into (6.15). Algorithm 2 provides the pseudo-code for performing Grassmann Sparse Coding (gSC).

A special case is sparse coding on the Grassmann manifold $\mathcal{G}(1,d)$, which can be seen as a problem on $d-1$ dimensional unit sphere, albeit with a subtle difference. More specifically, unlike conventional sparse coding in vector spaces, $x \sim -x, \forall x \in \mathcal{G}(1,d)$, which results in having antipodals points being equivalent. For this special case, the solution proposed in (6.13) can be understood as sparse coding in the higher dimensional quadratic space, i.e, $f : \mathbf{R}^d \to \mathbf{R}^{d^2}, f(x) = [x_1^2, x_1x_2, \ldots, x_d^2]^T$. We note that in the quadratic space, $\|f(x)\| = 1$ and $f(x) = f(-x)$.

Algorithm 2: Sparse coding on Grassmann manifolds (gSC).

Input: Grassmann dictionary $\{\mathcal{D}_i\}_{i=1}^N$, $\mathcal{D}_i \in \mathcal{G}(p,d)$ with $\mathcal{D}_i = \mathrm{span}(D_i)$; the query $\mathcal{G}(p,d) \ni \mathcal{X} = \mathrm{span}(X)$

Output: The sparse code y^*

Initialization.

```
for i, j ← 1 to N do
    [K(D)]_{i,j} ← ||D_i^T D_j||_F^2
end
K(D) = U Σ U^T /* compute SVD of K(D) */
A ← Σ^{1/2} U^T
```

$$[\mathbb{K}(\mathbb{D})]_{i,j} \leftarrow \|D_i^T D_j\|_F^2, \quad \mathbb{K}(\mathbb{D}) = U\Sigma U^T, \quad A \leftarrow \Sigma^{1/2}U^T$$

Processing.

```
for i ← 1 to N do
    [K(X, D)]_i ← ||X^T D_i||_F^2
end
x* ← Σ^{-1/2} U^T K(X, D)
y* ← arg min_y ||x* − Ay||^2 + λ||y||_1
```

$$[\mathcal{K}(X,\mathbb{D})]_i \leftarrow \|X^T D_i\|_F^2, \quad x^* \leftarrow \Sigma^{-1/2}U^T\mathcal{K}(X,\mathbb{D}), \quad y^* \leftarrow \arg\min_y \|x^* - Ay\|^2 + \lambda\|y\|_1$$

Classification Based on Coding

If the atoms in the dictionary are not labeled (e.g, if $\mathbb{D}$ is a generic dictionary not tied to any particular class), the generated sparse codes (vectors) for both training and query data can be fed to Euclidean-based classifiers like support vector machines [36] for classification. Inspired by the Sparse Representation Classifier (SRC) [46], when the atoms in sparse dictionary $\mathbb{D}$ are labeled, the generated codes of the query sample can be directly used for classification. In doing so, let

$$y_c = \begin{pmatrix} y_0\delta(l_0 - c) \\ y_1\delta(l_1 - c) \\ \vdots \\ y_N\delta(l_N - c) \end{pmatrix}$$

be the class-specific sparse codes, where l_j is the class label of atom $\mathcal{G}(p,d) \ni \mathcal{D}_j = \mathrm{span}(D_j)$ and $\delta(x)$ is the discrete Dirac function. An efficient way of utilizing class-specific sparse codes is through computing residual errors. In this case, the residual error of query sample $\mathcal{G}(p,d) \ni \mathcal{X} = \mathrm{span}(X)$ for class c is defined as:

$$\varepsilon_c(\mathcal{X}) = \left\| \widehat{X} - \sum_{j=1}^{N} y_j \widehat{D}_j \delta(l_j - c) \right\|_F^2 . \tag{6.16}$$

Alternatively, the similarity between query sample $\mathcal{X}$ to class c can be defined as $s(\mathcal{X}, c) = h(y_c)$. The function $h(\cdot)$ could be a linear function like $\sum_{j=1}^{N}(\cdot)$ or even a nonlinear one like max $(\cdot)$. Preliminary experiments suggest that Eq. (6.16) leads to higher classification accuracies when compared to the aforementioned alternatives.

6.4.2 Dictionary Learning

Given a finite set of observations $\mathbb{X} = \{\mathscr{X}_i\}_{i=1}^m$, $\mathscr{G}(p,d) \ni \mathscr{X}_i = \mathrm{span}(X_i)$, the problem of dictionary learning on Grassmann manifolds is defined as minimizing the following cost function:

$$h(\mathbb{D}) \triangleq \sum_{i=1}^{m} \ell_{\mathscr{G}}(\mathscr{X}_i, \mathbb{D}), \tag{6.17}$$

with $\mathbb{D} = \left\{\mathscr{D}_j\right\}_{j=1}^N$, $\mathscr{G}(p,d) \ni \mathscr{D}_j = \mathrm{span}(D_j)$ being a dictionary of size N. Here, $\ell_{\mathscr{G}}(\mathscr{X}, \mathbb{D})$ is a loss function and should be small if $\mathbb{D}$ is "good" at representing $\mathscr{X}$. In the following text, we elaborate on how a Grassmann dictionary can be learned.

Aiming for sparsity, the ℓ_1-norm regularization is usually employed to obtain the most common form of $l_{\mathscr{G}}(\mathscr{X}, \mathbb{D})$ as depicted in Eq. (6.13). With this choice, the problem of dictionary learning on Grassmann manifolds can be written as:

$$\min_{\{y_i\}_{i=1}^m, \mathbb{D}} \sum_{i=1}^{m} \left\| \widehat{X}_i - \sum_{j=1}^{N} y_{i_j} \widehat{D}_j \right\|_F^2 + \lambda \sum_{i=1}^{m} \|y_i\|_1. \tag{6.18}$$

Due to the non-convexity of (6.18) and inspired by the solutions in Euclidean spaces, we propose to solve (6.18) by alternating between the two sets of variables, $\mathbb{D}$ and $\{y_i\}_{i=1}^m$. More specifically, minimizing (6.18) over sparse codes y while dictionary $\mathbb{D}$ is fixed is a convex problem. Similarly, minimizing the overall problem over $\mathbb{D}$ with fixed $\{y_i\}_{i=1}^m$ is convex as well.

Therefore, to update dictionary atoms we break the minimization problem into N sub-minimization problems by independently updating each atom, $\widehat{D}_r$, in line with general practice in dictionary learning [12]. To update $\widehat{D}_r$, we write

$$\sum_{i=1}^{m} \left\| \widehat{X}_i - \sum_{j=1}^{N} y_{i_j} \widehat{D}_j \right\|_F^2 = \sum_{i=1}^{m} \left\| \left(\widehat{X}_i - \sum_{j \neq r} y_{i_j} \widehat{D}_j \right) - y_{i_r} \widehat{D}_r \right\|_F^2. \tag{6.19}$$

All other terms in (6.18) being independent of $\widehat{D}_r$, and since $\|\widehat{D}_r\|_F^2 = p$ is fixed, minimizing this with respect to $\widehat{D}_r$ is equivalent to minimizing $\mathscr{J}_r = -2\left\langle S_r, \widehat{D}_r \right\rangle$ where

$$S_r = \sum_{i=1}^{m} y_{i_r} \left(\widehat{X}_i - \sum_{j \neq r} y_{i_j} \widehat{D}_j \right). \tag{6.20}$$

Fig. 6.2 **a** Examples of actions performed by a ballerina. **b** The dominant eigenvectors for four atoms learned by the proposed Grassmann Dictionary Learning (gDL) method (grayscale images were used in gDL)

Finally, minimizing $\mathcal{J}_r = -2\langle S_r, \widehat{D}_r\rangle$ is the same as minimizing $\|S_r - \widehat{D}_r\|$ over $\widehat{D}_r$ in $\mathcal{PG}(n, p)$. The solution to this problem is given by the p-leading eigenvectors of S_r according to the Lemma 2. Algorithm 3 details the pseudocode for learning a dictionary on Grassmann manifolds. Figure 6.2 shows examples of a ballet dance.

To perform coding, we have relaxed the idempotent and rank constraints of the mapping $\Pi(\cdot)$ since matrix addition and subtraction do not preserve these constraints. However, for dictionary learning, the orthogonality constraint ensures the dictionary atoms have the required structure.

6.5 Kernel Coding

In this section, we propose to perform coding and dictionary learning in a reproducing Kernel Hilbert space (RKHS). This has the twofold advantage of yielding simple solutions to several popular coding techniques and of resulting in a potentially better representation than standard coding techniques due to the nonlinearity of the approach. Before formulating our kernel solutions, we need to make sure that positive definite kernels on Grassmann manifolds are at our disposal. Formally,

Definition 3 (*Real-valued Positive Definite Kernels*) Let $\mathcal{X}$ be a nonempty set. A symmetric function $k : \mathcal{X} \times \mathcal{X} \to \mathbb{R}$ is a positive definite (***pd***) kernel on $\mathcal{X}$ if and only if $\sum_{i,j=1}^{n} c_i c_j k(x_i, x_j) \geq 0$ for any $n \in \mathbb{N}$, $x_i \in \mathcal{X}$ and $c_i \in \mathbb{R}$.

Definition 4 (*Grassmannian Kernel*) A function $k : \mathcal{G}(p, d) \times \mathcal{G}(p, d) \to \mathbb{R}$ is a Grassmannian kernel, if it is well defined and *pd*. In our context, a function is well defined if it is invariant to the choice of basis, i.e, $k(XR_1, YR_2) = k(X, Y)$, for all $X, Y \in \mathcal{G}(p, d)$ and $R_1, R_2 \in \mathrm{SO}(p)$, where $\mathrm{SO}(p)$ denotes the special orthogonal group.

Algorithm 3: Grassmann Dictionary Learning (gDL)

Input: training set $\mathbb{X}=\{\mathcal{X}_i\}_{i=1}^{m}$, where each $\mathcal{G}(p,d) \ni \mathcal{X}_i = \text{span}(X_i)$; *nIter*: number of iterations

Output: Grassmann dictionary $\mathbb{D} = \{\mathcal{D}_i\}_{i=1}^{N}$, where $\mathcal{G}(p,d) \ni \mathcal{D}_i = \text{span}(D_i)$

Initialization.

Initialize the dictionary $\mathbb{D}$ by selecting N samples from $\mathbb{X}$ randomly

Processing.

```
for t = 1 to nIter do
    // Sparse Coding Step using Algorithm 2
    for i = 1 to m do
        y_i ← min_y ‖X̂_i − Σ_{j=1}^{N} [y]_j D̂_j‖_F^2 + λ‖y‖_1
    end
    // Dictionary update step
    for r = 1 to N do
        Compute S_r according to Eq. (6.20).
        {λ_k, v_k} ← eigenvalues and eigenvectors of S_r
        S_r v = λv; λ_1 ≥ λ_2 ≥ ··· ≥ λ_d
        D_r^* ← [v_1|v_2| ··· |v_p]
    end
end
```

The most widely used kernel is arguably the Gaussian or radial basis function (RBF) kernel. It is therefore tempting to define a Radial Basis Grassmannian kernel by replacing the Euclidean distance with the geodesic distance. Unfortunately, although symmetric and well defined, the function $\exp(-\beta d_{\text{geod}}^2(\cdot,\cdot))$ is not *pd* [21]. Nevertheless, two Grassmannian kernels, i.e, the Binet–Cauchy kernel [45] and the projection kernel [18], have been proposed to embed Grassmann manifolds into RKHS. In this work, we are only interested in the projection kernels[5]

$$k_p(X,Y) = \left\|X^T Y\right\|_F^2 . \tag{6.21}$$

From the previous discussions, k_p, defined in Eq. (6.21) can be seen as a linear kernel in the space induced by the projection embedding. However, the inner products defined by the projection embedding can actually be exploited to derive many new Grassmannian kernels, including universal kernels.

Universal Grassmannian Kernels

Although often used in practice, linear kernels are known not to be universal [38]. This can have a crucial impact on their representation power for a specific task. Indeed, from the *Representer Theorem* [35], we have that, for a given set of training data $\{x_j\}$, $j \in \mathbb{N}_n$, $\mathbb{N}_n = \{1, 2, \ldots, n\}$ and a *pd* kernel k, the function learned by any

[5]In our experiments, we observed that the projection kernel almost always outperforms the Binet–Cauchy kernel.

algorithm can be expressed as

$$\hat{f}(x_*) = \sum_{j \in \mathbb{N}_n} c_j k(x_*, x_j) \; . \tag{6.22}$$

Importantly, only *universal kernels* have the property of being able to approximate any target function f_t arbitrarily well given sufficiently many training samples. Therefore, k_p may not generalize sufficiently well for certain problems. Below, we develop several universal Grassmannian kernels. To this end, we make use of negative definite kernels and of their relation to *pd* ones. Let us first formally define negative definite kernels.

Definition 5 (*Real-valued Negative Definite Kernels*) Let $\mathscr{X}$ be a nonempty set. A symmetric function $\psi : \mathscr{X} \times \mathscr{X} \to \mathbb{R}$ is a negative definite (***nd***) kernel on $\mathscr{X}$ if and only if $\sum_{i,j=1}^{n} c_i c_j k(x_i, x_j) \leq 0$ for any $n \in \mathbb{N}$, $x_i \in \mathscr{X}$ and $c_i \in \mathbb{R}$ with $\sum_{i=1}^{n} c_i = 0$.

Note that, in contrast to positive definite kernels, an additional constraint of the form $\sum c_i = 0$ is required in the negative definite case. The most important example of *nd* kernels is the distance function defined on a Hilbert space. More specifically,

Theorem 2 ([27]) *Let $\mathscr{X}$ be a nonempty set, $\mathscr{H}$ be an inner product space, and $\psi : \mathscr{X} \to \mathscr{H}$ be a function. Then $f : (\mathscr{X} \times \mathscr{X}) \to \mathbb{R}$ defined by $f(x_i, x_j) = \|\psi(x_i) - \psi(x_j)\|_{\mathscr{H}}^2$ is negative definite.*

Therefore, being distances in Hilbert spaces, d_{chord}^2 is a *nd* kernel. We now state an important theorem which establishes the relation between *pd* and *nd* kernels.

Theorem 3 (Theorem 2.3 in Chap. 3 of [6]) *Let μ be a probability measure on the half line $\mathbb{R}_+$ and $0 < \int_0^{\infty} t\mathrm{d}\mu(t) < \infty$. Let $\mathscr{L}_\mu$ be the Laplace transform of μ, i.e, $\mathscr{L}_\mu(s) = \int_0^{\infty} e^{-ts}\mathrm{d}\mu(t)$, $s \in \mathbb{C}_+$. Then, $\mathscr{L}_\mu(\beta f)$ is positive definite for all $\beta > 0$ if and only if $f : \mathscr{X} \times \mathscr{X} \to \mathbb{R}_+$ is negative definite.*

The problem of designing a *pd* kernel on the Grassmannian can now be cast as that of finding an appropriate probability measure μ. Below, we show that this lets us reformulate popular kernels in Euclidean space as Grassmannian kernels.

RBF Kernels.

Grassmannian RBF kernels can be obtained by choosing $\mu(t) = \delta(t-1)$ in Theorem 3, where $\delta(t)$ is the Dirac delta function. This choice yields the Grassmannian RBF kernels (after discarding scalar constants)

$$k_{r,p}(X, Y) = \exp\left(\beta \left\|X^T Y\right\|_F^2\right), \quad \beta > 0 \; . \tag{6.23}$$

Table 6.1 The proposed Grassmannian kernels and their properties

Kernel	Equation	Properties
Linear	$k_p(X, Y) = \|X^T Y\|_F^2$	*pd*
RBF	$k_{r,p}(X, Y) = \exp\left(\beta \|X^T Y\|_F^2\right), \ \beta > 0$	*pd*, universal
Laplace	$k_{l,p}(X, Y) = \exp\left(- \beta\sqrt{p - \|X^T Y\|_F^2}\right), \beta > 0$	*pd*, universal

Laplace Kernels.

The Laplace kernel is another widely used Euclidean kernel, defined as $k(x, y) = \exp(-\beta\|x - y\|)$. To obtain heat kernels on the Grassmannian, we make use of the following theorem for *nd* kernels.

Theorem 4 (Corollary 2.10 in Chap. 3 of [6]) *If* $\psi : \mathscr{X} \times \mathscr{X} \to \mathbb{R}$ *is negative definite and satisfies* $\psi(x, x) \geqq 0$ *then so is* ψ^α *for* $0 < \alpha < 1$.

As a result $d_{\text{chord}}(\cdot, \cdot)$ is *nd* by choosing $\alpha = 1/2$ in Theorem 4. By employing $d^2_{\text{chord}}(\cdot, \cdot)$ along with $\mu(t) = \delta(t - 1)$ in Theorem 3, we obtain the Grassmannian heat kernels

$$k_{l,p}(X, Y) = \exp\left(- \beta\sqrt{p - \|X^T Y\|_F^2}\right), \quad \beta > 0\,. \tag{6.24}$$

As shown in [38], the RBF and heat kernels are universal for $\mathbb{R}^d, d > 0$. The kernels described above are summarized in Table 6.1. Note that many other kernels can be derived by, e.g, exploiting different measures in Theorem 3. However, the kernels derived here correspond to the most popular ones in Euclidean space, and we therefore leave the study of additional kernels as future work.

6.5.1 Kernel-Based Riemannian Coding

Let $\phi : \mathscr{M} \to \mathscr{H}$ be a mapping to an RKHS induced by the kernel $k(x, y) = \phi(x)^T\phi(y)$. Sparse coding in $\mathscr{H}$ can then be formulated by rewriting (6.1) as

$$\ell_\phi(x, \mathbb{D}) \triangleq \min_y \left\| \phi(x) - \sum\nolimits_{j=1}^{N} [y]_j \phi(d_j) \right\|_2^2 + \lambda\|y\|_1. \tag{6.25}$$

Expanding the reconstruction term in (6.25) yields

$$\left\|\phi(x) - \sum_{j=1}^{N}[y]_j\phi(d_j)\right\|_2^2 = \phi(x)^T\phi(x)$$
$$- 2\sum_{j=1}^{N}[y]_j\phi(d_j)^T\phi(x) + \sum_{i,j=1}^{N}[y]_i[y]_j\phi(d_i)^T\phi(d_j)$$
$$= k(x, x) - 2y^T k(x, \mathscr{D}) + y^T K(\mathscr{D}, \mathscr{D})y, \tag{6.26}$$

where $k(x, \mathscr{D}) \in \mathbb{R}^N$ is the kernel vector evaluated between x and the dictionary atoms, and $K(\mathscr{D}, \mathscr{D}) \in \mathbb{R}^{N\times N}$ is the kernel matrix evaluated between the dictionary atoms.

This shows that the reconstruction term in (6.25) can be kernelized. More importantly, after kernelization, this term remains quadratic, convex, and similar to its counterpart in Euclidean space. To derive an efficient solution to kernel sparse coding, we introduce the following theorem.

Theorem 5 ([19]) *Consider the least-squares problem in an RKHS* $\mathscr{H}$

$$\min_y \left\|\phi(x) - \sum_{j=1}^{N}[y]_j\phi(d_j)\right\|_2^2 \Leftrightarrow$$
$$\min_y \; y^T K(\mathscr{D}, \mathscr{D})y - 2y^T k(x, \mathscr{D}) + f(x)\ , \tag{6.27}$$

where $f(x)$ is a constant function (i.e, independent of α). Let $U\Sigma U^T$ be the SVD of the symmetric positive definite matrix $K(\mathscr{D}, \mathscr{D})$. Then (6.27) *is equivalent to the least-squares problem in* $\mathbb{R}^N$

$$\min_\alpha \left\|\tilde{x} - \tilde{D}y\right\|_2^2\ , \tag{6.28}$$

with $\tilde{D} = \Sigma^{1/2}U^T$ and $\tilde{x} = \Sigma^{-1/2}U^T k(x, \mathscr{D})$.

This theorem lets us write kernel sparse coding as

$$\min_y \left\|\tilde{x} - \tilde{D}y\right\|_2^2 + \lambda\|y\|_1\ , \tag{6.29}$$

which is a standard linear sparse coding problem. Algorithm 4 provides the pseudocode for performing kernel Sparse Coding (kSC).

6.5.2 Kernel Dictionary Learning

To obtain a dictionary in $\mathscr{H}$, we follow an alternating optimization strategy to update the codes and the dictionary. Since obtaining the codes with a given dictionary was discussed in the previous part, here we focus on the dictionary update.

Algorithm 4: Kernel sparse coding (kSC).

Input: Dictionary $\mathscr{D} = \{d_i\}_{i=1}^N,\ d_i \in \mathscr{M}$; the query $x \in \mathscr{M}$, a positive definite kernel $k : \mathscr{M} \times \mathscr{M} \to \mathbb{R}$.
Output: The sparse codes y^*

Initialization.

for $i, j \leftarrow 1$ **to** N **do**
  $[K(\mathscr{D}, \mathscr{D})]_{i,j} \leftarrow k(d_i, d_j)$
end
$K(\mathscr{D}, \mathscr{D}) = U \Sigma U^T$ `/* apply SVD */`
$A \leftarrow \Sigma^{1/2} U^T$

Processing.

for $i \leftarrow 1$ **to** N **do**
  $[k(x, \mathscr{D})]_i \leftarrow k(x, d_i)$
end
$x^* \leftarrow \Sigma^{-1/2} U^T k(x, \mathscr{D})$
$y^* \leftarrow \underset{y}{\arg\min}\ \|x^* - Ay\|^2 + \lambda \|y\|_1$

Algorithm 5: Learning a generic dictionary.

Input: Training data $\{x_i\}_{i=1}^M,\ x_i \in \mathscr{M}$; kernel function $k(\cdot, \cdot) : \mathscr{M} \times \mathscr{M} \to \mathbb{R}$; size of dictionary N.
Output: Dictionary $\phi(\mathscr{D})$ in the RKHS $\mathscr{H}$ described as $\phi(\mathscr{X})V$

Processing.

`/* Initialize` $\phi(\mathscr{D})$ `either randomly or through kernel k-means algorithm. */`
for iter $\leftarrow 1$ **to** nIter **do**
  Compute kernel codes $y_i,\ i \in [1, \ldots, M]$ using Algorithm 4.
  `/* fix kernel codes` y_i `and update dictionary. */`
  $\phi(\mathscr{D}) = \phi(\mathscr{X}) A^\dagger$
  $K(\mathscr{D}, \mathscr{D}) \leftarrow (A^\dagger)^T K(\mathscr{X}, \mathscr{X}) A^\dagger$
  $k(x_i, \mathscr{D}) \leftarrow (A^\dagger)^T k(x_i, \mathscr{X})$
end

With fixed codes for the training data (and a fixed kernel parameter), learning the dictionary can be expressed as solving the optimization problem

$$\min_{\mathscr{D}} \frac{1}{M} \sum\nolimits_{i=1}^{M} \ell_\phi(\mathscr{D}; x_i). \tag{6.30}$$

Here, we make use of the *Representer theorem* [35] which enables us to express the dictionary as a linear combination of the training samples in RKHS. That is

$$\phi(d_j) = \sum_{i=1}^{M} v_{i,j} \phi(x_i), \tag{6.31}$$

where $\{v_{i,j}\}$ is the set of weights, now corresponding to our new unknowns. By stacking these weights for the M samples and the N dictionary elements in a matrix $V_{M\times N}$, we have

$$\phi(\mathscr{D}) = \phi(\mathscr{X})V \,. \tag{6.32}$$

The only term that depends on the dictionary is the reconstruction error (i.e, the first term in the objective of (6.25)). Given the matrix of sparse codes $A_{N\times M} = [y_1|y_2|\cdots|y_M]$, this term can be expressed as

$$\begin{aligned} R(V) &= \left\|\phi(\mathscr{X}) - \phi(\mathscr{X})VA\right\|_F^2 \\ &= \mathrm{Tr}\left(\phi(\mathscr{X})(\mathbf{I}_M - VA)(\mathbf{I}_M - VA)^T\phi(\mathscr{X})^T\right) \\ &= \mathrm{Tr}\left(K(\mathscr{X},\mathscr{X})(\mathbf{I}_M - VA - A^TV^T + VAA^TV^T)\right) \,. \end{aligned} \tag{6.33}$$

The new dictionary, fully defined by V, can then be obtained by zeroing out the gradient of $R(V)$ w.r.t. V. This yields

$$\nabla R(V) = 0 \Leftrightarrow V = (AA^T)^{-1}A = A^{\dagger} \,. \tag{6.34}$$

6.6 Experiments

To compare and contrast the proposed techniques against state-of-the-art methods, we used the Ballet dataset [44] to classify actions from videos. The Ballet dataset contains 44 videos collected from an instructional ballet DVD [44]. The dataset consists of eight complex motion patterns performed by three subjects, The actions include: *'left-to-right hand opening'*, *'right-to-left hand opening'*, *'standing hand opening'*, *'leg swinging'*, *'jumping'*, *'turning'*, *'hopping'*, and *'standing still'*. Figure 6.3 shows examples. The dataset is challenging due to the significant intra-class variations in terms of speed, spatial and temporal scale, clothing, and movement.

We extracted 2200 image sets by grouping 6 frames that exhibited the same action into one image set. We described each image set by a subspace of order 6 with histogram of oriented gradients (HOG) as frame descriptor [10] using SVD. To this

Fig. 6.3 Examples from the Ballet dataset [44]

Table 6.2 Average recognition rate on the Ballet dataset.

Method	gSC	kSC-RBF	kSC-Laplace	kSC-Poly
Accuracy	64.5	**69.7**	67.9	68.5

end, frame were first resized to 128×128 and HoG descriptor from four 64×64 nonoverlapping blocks were extracted. The HoG descriptors were concatenated to form the 124 dimensional frame descriptor.

Extracted subspaces were randomly split into training and testing sets (the number of image sets in both sets was even). The process of random splitting was repeated ten times and the average classification accuracy is reported.

Table 6.2 reports the average accuracies along their standard deviations for the studied methods. All the results were obtained by training a dictionary of size 128. To classify the sparse codes, we used a linear SVM. For the kSC algorithm, we used three different kernels, namely RBF, Laplace and a polynomial kernel of degree 2 as described in Sect. 6.5.

The highest accuracy is obtained by the universal RBF kernel. Interestingly, the polynomial kernel performs better than the Laplace kernel. All the kernel methods outperform the gSC algorithm, implying that the data is highly nonlinear.

References

1. P.A. Absil, R. Mahony, R. Sepulchre, *Optimization Algorithms on Matrix Manifolds* (Princeton University Press, Princeton, 2008)
2. M. Aharon, M. Elad, A. Bruckstein, K-SVD: an algorithm for designing overcomplete dictionaries for sparse representation. IEEE Trans. Signal Process. **54**(11), 4311–4322 (2006)
3. V. Arsigny, P. Fillard, X. Pennec, N. Ayache, Log-Euclidean metrics for fast and simple calculus on diffusion tensors. Magn. Reson. Med. **56**(2), 411–421 (2006)
4. R. Basri, D.W. Jacobs, Lambertian reflectance and linear subspaces. IEEE Trans. Pattern Anal. Mach. Intell. **25**(2), 218–233 (2003)
5. E. Begelfor, M. Werman, Affine invariance revisited, in *Proceedings of the IEEE Conference on Computer Vision and Pattern Recognition (CVPR)* (2006), pp. 2087–2094
6. C. Berg, J.P.R. Christensen, P. Ressel, *Harmonic Analysis on Semigroups* (Springer, New York, 1984)
7. E.J. Candès, J. Romberg, T. Tao, Robust uncertainty principles: exact signal reconstruction from highly incomplete frequency information. IEEE Trans. Inf. Theory **52**(2), 489–509 (2006)
8. H.E. Cetingul, M.J. Wright, P.M. Thompson, R. Vidal, Segmentation of high angular resolution diffusion MRI using sparse Riemannian manifold clustering. IEEE Trans. Med. Imaging **33**(2), 301–317 (2014)
9. S. Chen, C. Sanderson, M. Harandi, B.C. Lovell, Improved image set classification via joint sparse approximated nearest subspaces, in *Proceedings of the IEEE Conference on Computer Vision and Pattern Recognition (CVPR)* (2013), pp. 452–459
10. N. Dalal, B. Triggs, Histograms of oriented gradients for human detection, in *Proceedings of the IEEE Conference on Computer Vision and Pattern Recognition (CVPR)* (2005), pp. 886–893
11. D.L. Donoho, Compressed sensing. IEEE Trans. Inf. Theory **52**(4), 1289–1306 (2006)

12. M. Elad, *Sparse and Redundant Representations - From Theory to Applications in Signal and Image Processing* (Springer, New York, 2010)
13. E. Elhamifar, R. Vidal, Sparse subspace clustering: algorithm, theory, and applications. IEEE Trans. Pattern Anal. Mach. Intell. **35**(11), 2765–2781 (2013)
14. M. Faraki, M. Harandi, F. Porikli, More about VLAD: a leap from Euclidean to Riemannian manifolds, in *Proceedings of the IEEE Conference on Computer Vision and Pattern Recognition (CVPR)* (2015), pp. 4951–4960
15. K.A. Gallivan, A. Srivastava, X. Liu, P. Van Dooren, Efficient algorithms for inferences on Grassmann manifolds, in *IEEE Workshop on Statistical Signal Processing* (2003), pp. 315–318
16. B. Gong, Y. Shi, F. Sha, K. Grauman, Geodesic flow kernel for unsupervised domain adaptation, in *Proceedings of the IEEE Conference on Computer Vision and Pattern Recognition (CVPR)* (2012), pp. 2066–2073
17. R. Gopalan, R. Li, R. Chellappa, Unsupervised adaptation across domain shifts by generating intermediate data representations. IEEE Trans. Pattern Anal. Mach. Intell. **36**(11), 2288–2302 (2014). doi:10.1109/TPAMI.2013.249
18. J. Hamm, D.D. Lee, Grassmann discriminant analysis: a unifying view on subspace-based learning, in *Proceedings of the International Conference on Machine Learning (ICML)* (2008), pp. 376–383
19. M. Harandi, M. Salzmann, Riemannian coding and dictionary learning: kernels to the rescue, in *Proceedings of the IEEE Conference on Computer Vision and Pattern Recognition (CVPR)* (2015), pp. 3926–3935
20. M. Harandi, R. Hartley, C. Shen, B. Lovell, C. Sanderson, Extrinsic methods for coding and dictionary learning on Grassmann manifolds. Int. J. Comput. Vis. **114**(2–3), 113–136 (2015)
21. M.T. Harandi, M. Salzmann, S. Jayasumana, R. Hartley, H. Li, Expanding the family of Grassmannian kernels: an embedding perspective, in *Proceedings of the European Conference on Computer Vision (ECCV)*, vol. 8695, Lecture Notes in Computer Science, ed. by D. Fleet, T. Pajdla, B. Schiele, T. Tuytelaars (Springer International Publishing, Cham, 2014), pp. 408–423. doi:10.1007/978-3-319-10584-0_27
22. M.T. Harandi, R. Hartley, B.C. Lovell, C. Sanderson, Sparse coding on symmetric positive definite manifolds using Bregman divergences. IEEE Trans. Neural Netw. Learn. Syst. (TNNLS) **PP**(99), 1–1 (2015)
23. R. Hartley, J. Trumpf, Y. Dai, H. Li, Rotation averaging. Int. J. Comput. Vis. **103**(3), 267–305 (2013)
24. U. Helmke, K. Hper, J. Trumpf, Newton's method on Gramann manifolds (2007)
25. U. Helmke, K. Hüper, P.Y. Lee, J.B. Moore, Essential matrix estimation using Gauss-Newton iterations on a manifold. Int. J. Comput. Vis. **74**(2), 117–136 (2007). doi:10.1007/s11263-006-0005-0
26. J. Ho, Y. Xie, B. Vemuri, On a nonlinear generalization of sparse coding and dictionary learning, in *Proceedings of the International Conference on Machine Learning (ICML)* (2013), pp. 1480–1488
27. S. Jayasumana, R. Hartley, M. Salzmann, H. Li, M. Harandi, Kernel methods on Riemannian manifolds with Gaussian RBF kernels. IEEE Trans. Pattern Anal. Mach. Intell. **37**(12), 2464–2477 (2015). doi:10.1109/TPAMI.2015.2414422
28. H. Karcher, Riemannian center of mass and mollifier smoothing. Commun. Pure Appl. Math. **30**(5), 509–541 (1977)
29. J.M. Lee, *Introduction to Smooth Manifolds*, vol. 218 (Springer, New York, 2012)
30. J. Mairal, F. Bach, J. Ponce, G. Sapiro, A. Zisserman, Discriminative learned dictionaries for local image analysis, in *Proceedings of the IEEE Conference on Computer Vision and Pattern Recognition (CVPR)* (IEEE, 2008), pp. 1–8
31. J. Mairal, M. Elad, G. Sapiro, Sparse representation for color image restoration. IEEE Trans. Image Process. (TIP) **17**(1), 53–69 (2008)
32. J.H. Manton, A globally convergent numerical algorithm for computing the centre of mass on compact lie groups. Int. Conf. Control Autom. Robot. Vis. **3**, 2211–2216 (2004)

33. B.A. Olshausen, D.J. Field, Emergence of simple-cell receptive field properties by learning a sparse code for natural images. Nature **381**(6583), 607–609 (1996)
34. R. Ramamoorthi, Analytic PCA construction for theoretical analysis of lighting variability in images of a Lambertian object. IEEE Trans. Pattern Anal. Mach. Intell. **24**(10), 1322–1333 (2002)
35. B. Schölkopf, R. Herbrich, A.J. Smola, A generalized representer theorem, *Computational Learning Theory* (Springer, New York, 2001), pp. 416–426
36. J. Shawe-Taylor, N. Cristianini, *Kernel Methods for Pattern Analysis* (Cambridge University Press, Cambridge, 2004)
37. S. Shirazi, M. Harandi, B. Lovell, C. Sanderson, Object tracking via non-Euclidean geometry: a Grassmann approach, in *IEEE Winter Conference on Applications of Computer Vision (WACV)* (2014), pp. 901–908. doi:10.1109/WACV.2014.6836008
38. I. Steinwart, A. Christmann, *Support Vector Machines* (Springer, Berlin, 2008)
39. R. Subbarao, P. Meer, Nonlinear mean shift over Riemannian manifolds. Int. J. Comput. Vis. **84**(1), 1–20 (2009)
40. R. Tibshirani, Regression shrinkage and selection via the lasso. J. R. Stat. Soc. Ser. B (Methodol.) **58**, 267–288 (1996)
41. P. Turaga, A. Veeraraghavan, A. Srivastava, R. Chellappa, Statistical computations on Grassmann and Stiefel manifolds for image and video-based recognition. IEEE Trans. Pattern Anal. Mach. Intell. **33**(11), 2273–2286 (2011)
42. R. Vemulapalli, J.K. Pillai, R. Chellappa, Kernel learning for extrinsic classification of manifold features, in *Proceedings of the IEEE Conference on Computer Vision and Pattern Recognition (CVPR)* (2013), pp. 1782–1789
43. J. Wang, J. Yang, K. Yu, F. Lv, T. Huang, Y. Gong, Locality-constrained linear coding for image classification, in *Proceedings of the IEEE Conference on Computer Vision and Pattern Recognition (CVPR)* (2010), pp. 3360–3367
44. Y. Wang, G. Mori, Human action recognition by semilatent topic models. IEEE Trans. Pattern Anal. Mach. Intell. **31**(10), 1762–1774 (2009)
45. L. Wolf, A. Shashua, Learning over sets using kernel principal angles. J. Mach. Learn. Res. **4**, 913–931 (2003)
46. J. Wright, A.Y. Yang, A. Ganesh, S.S. Sastry, Y. Ma, Robust face recognition via sparse representation. IEEE Trans. Pattern Anal. Mach. Intell. **31**(2), 210–227 (2009)
47. J. Wright, Y. Ma, J. Mairal, G. Sapiro, T.S. Huang, S. Yan, Sparse representation for computer vision and pattern recognition. Proc. IEEE **98**(6), 1031–1044 (2010)
48. J. Yang, K. Yu, Y. Gong, T. Huang, Linear spatial pyramid matching using sparse coding for image classification, in *Proceedings of the IEEE Conference on Computer Vision and Pattern Recognition (CVPR)* (2009), pp. 1794–1801

Chapter 7
Regression on Lie Groups and Its Application to Affine Motion Tracking

Fatih Porikli

Abstract In this chapter, we present how to learn regression models on Lie groups and apply our formulation to visual object tracking tasks. Many transformations used in computer vision, for example orthogonal group and rotations, have matrix Lie group structure. Unlike conventional methods that proceed by directly linearizing these transformations, thus, making an implicit Euclidean space assumption, we formulate a regression model on the corresponding Lie algebra that minimizes a first order approximation to the geodesic error. We demonstrate our method on affine motions, however, it generalizes to any matrix Lie group transformations.

7.1 Introduction

Suppose we are given with a set of pairs $\{(\mathrm{M}_i, f_i)\}$ where M_i's are on an n-dimensional Lie group G and f_i's are their associated field vectors in $\mathbb{R}^d$. Our goal is to derive a regression function $\beta : \mathbb{R}^d \mapsto G$ that approximates the corresponding point M on the Lie group for a vector f

$$\mathrm{M} = \beta(f). \tag{7.1}$$

We take advantage of the Lie algebra $\mathfrak{g}$ and solve the corresponding linear regression problem instead

$$\log \mathrm{M} = f^T \Omega. \tag{7.2}$$

After a brief overview of Lie groups in Sect. 7.2, we define an approximate solution of (7.1) for matrix Lie groups in Sect. 7.3, and apply it to 2D affine motion tracking in Sect. 7.4. Part of the discussion can also be found in [24, 30].

F. Porikli (✉)
Australian National Univeristy, Canberra, Australia
e-mail: fatih.porikli@anu.edu.au

F. Porikli
Data61/CSIRO, Eveleigh, Australia

H.Q. Minh and V. Murino (eds.), *Algorithmic Advances in Riemannian Geometry and Applications*, Advances in Computer Vision and Pattern Recognition, DOI 10.1007/978-3-319-45026-1_7

7.2 Lie Group

A Lie group is a set G that is a group with the topology of an n-dimensional smooth differentiable manifold, in which the group operations multiplication $G \times G \mapsto G$: $(\mathrm{X}, \mathrm{Y}) \mapsto \mathrm{XY}$ and inversion $G \mapsto G : \mathrm{X} \mapsto \mathrm{X}^{-1}$ are smooth maps. In other words, the mapping $(\mathrm{X}, \mathrm{Y}) \mapsto \mathrm{X}^{-1}\mathrm{Y}$ is a smooth mapping of the product manifold $G \times G$ into G.

Some simple examples of Lie groups are the non-zero real numbers, the circle, the torus, the set of rotations of 3-dimensional space, the 3-sphere, and the set of square matrices that have nonzero determinant. Consider the sphere $S^2 \subset \mathbb{R}^3$ under rotations. The group property means that any two consecutive rotations of the sphere can also be done by rotating it over a single angle, and any rotation has an inverse, i.e. rotating the sphere over an opposite angle. This shows the sphere has rotational symmetries. Since these rotations can be arbitrarily small and many small rotations adds up for a big rotation, these operations are smooth maps (it is indistinguishable from ordinary Euclidean space at small scales), therefore the rotation group $SO(3)$ acting on S^2 is a Lie group. This can be observed for the set of square matrices that have non-zero determinant. Such a matrix corresponds to a transformation of the space. The set of such transformations for a group: the matrices can be multiplied, each has an inverse, the multiplication is associative, and the identity transformation fixes each point of space. From these examples, we abstract the concept of a Lie group as a set of transformations or symmetries that has the structure of a smooth manifold, i.e. continuous symmetries.

Any Lie group gives rise to a Lie algebra. There is a corresponding connected Lie group unique up to covering to any finite-dimensional Lie algebra over real numbers. This correspondence between Lie groups and Lie algebras allows one to study Lie groups in terms of Lie algebras then transfer results from algebras back to groups. The tangent space to the identity element I of the group forms a Lie algebra $\mathfrak{g}$, which is a vector space together with a non-associative multiplication called Lie bracket $[\mathrm{x}, \mathrm{y}]$. Lie bracket is a binary operator over $\mathfrak{g} \times \mathfrak{g} \mapsto \mathfrak{g}$ defined as $[\mathrm{x}, \mathrm{y}] := \mathrm{xy} - \mathrm{yx}$ and satisfies the bilinearity, alternativity, and Jacobi identity axioms, i.e. $[a\mathrm{x} + b\mathrm{y}, \mathrm{z}] = a[\mathrm{x}, \mathrm{z}] + b[\mathrm{y}, \mathrm{z}]$, $[\mathrm{x}, \mathrm{x}] = 0$, $[\mathrm{x}, [\mathrm{y}, \mathrm{z}]] + [\mathrm{z}, [\mathrm{x}, \mathrm{y}]] + [\mathrm{y}, [\mathrm{z}, \mathrm{x}]] = 0$ for all scalars a, b and all elements $\mathrm{x}, \mathrm{y}, \mathrm{z} \in \mathfrak{g}$. We can reinterpret most of the properties of a Lie group into properties of the bracket on the Lie algebra.

The distances on a manifold are measured by the lengths of the curves connecting the points, and the minimum length curve between two points is called the geodesic. There exists a unique geodesic starting with vector $\mathrm{m} \in \mathfrak{g}$ at the group element I. The exponential map $\exp : \mathfrak{g} \to G$ maps the vector m to the point reached by this geodesic. Let $\exp(\mathrm{m}) = \mathrm{M}$, then the length of the geodesic is given by $\rho(\mathrm{I}, \mathrm{M}) = \|\mathrm{m}\|$. In general, the exponential map is onto but not one-to-one. Therefore, the inverse mapping $\log : G \to \mathfrak{g}$ is uniquely defined only around the neighborhood of I. If for any $\mathrm{M} \in G$, there exist several $\mathrm{m} \in \mathfrak{g}$ such that $\mathrm{M} = \exp(\mathrm{m})$, then $\log(\mathrm{M})$ is selected as the vector with the smallest norm. Left multiplication by the inverse of a group element $\mathrm{M}^{-1} : G \to G$ gives way to map the point M to I. The tangent space at I is the Lie algebra. The action of M^{-1} on the tangent space is through the adjoint action map. See [22] for more explanation.

Using the logarithm map and the group operation, the geodesic distance between two group elements is measured by

$$\rho(\mathrm{M}_1, \mathrm{M}_2) = \| \log(\mathrm{M}_1^{-1}\mathrm{M}_2) \|. \tag{7.3}$$

The norm above for the Euclidean space $\mathbb{R}^d$ with ordinary vector addition as the group operation is the Euclidean norm. How basis elements in the Lie algebra map to natural basis elements in R^d is not unique, which amounts to a choice of weighting, as explained in [7].

The exponential and logarithm maps for matrix Lie groups are given by the matrix exponential and logarithm operators

$$\exp(\mathrm{m}) = \sum_{k=0}^{\infty} \frac{1}{k!}\mathrm{m}^k \quad , \quad \log(\mathrm{M}) = \sum_{k=1}^{\infty} \frac{(-1)^{k-1}}{k}(\mathrm{M} - \mathrm{I})^k. \tag{7.4}$$

A comprehensive discussion on matrix manifolds and higher-order optimization methods on manifolds can be found in [17].

7.3 Linear Regression on Matrix Lie Groups

The regression function $\beta : \mathbb{R}^d \mapsto G$ estimates the element M on the matrix Lie group G for a given d-dimensional feature vector f as $\mathrm{M} = \beta(f)$. This concepts is illustrated in Fig. 7.1.

The parameters of the regression function are learned from a set of N training pairs $\{(\mathrm{M}_i, f_i)\}$. Since these matrices are on a differentiable manifold, the sum of the

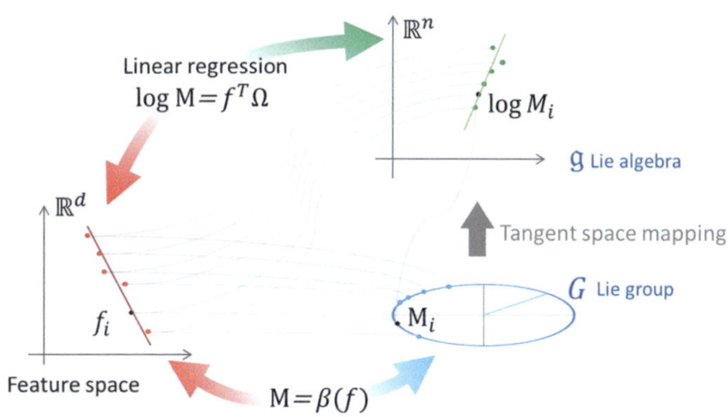

Fig. 7.1 Conceptual illustration of linear regression on Lie group

squared geodesic distances between the estimations $\beta(f_i)$ and the given matrices M_i can be used as the loss function

$$L = \sum_{i=1}^{N} \rho^2 \left(\beta(f_i), M_i\right). \tag{7.5}$$

In general, the exponential map does not satisfy the identity $\exp(m_1)\exp(m_2) = \exp(m_1 + m_2)$. The Baker–Campbell–Hausdorff (BCH) formula [21] expresses the logarithm $\log(\exp(M_1)\exp(M_2))$ of the product of two Lie group elements as a Lie algebra element using only Lie algebraic operations for noncommutative Lie groups. A first order approximation to the BCH is

$$\log(\exp(M_1)\exp(M_2)) = M_1 + M_2 + \frac{1}{2}[M_1, M_2] + O(M_1^2, M_2^2) \tag{7.6}$$

using the Lie bracelet operator. Since the corresponding Lie algebra elements for M_1 and M_2 are $m_1 = \log(M_1)$ and $m_2 = \log(M_2)$, the geodesic distance can be approximated by

$$\begin{aligned}\rho(M_1, M_2) &= \| \log(M_1^{-1}M_2)\| \\ &= \left\|\log\left[\exp(-m_1)\exp(m_2)\right]\right\| \\ &= \left\|m_2 - m_1 + 0.5[-m_1, m_2] + O(m_1^2, m_2^2)\right\| \\ &\approx \|m_2 - m_1\|. \end{aligned} \tag{7.7}$$

Using (7.7), the loss function (7.5) can be approximated as

$$L \approx \sum_{i=1}^{N} \left\|\log(M_i) - \log\left(\beta(f_i)\right)\right\|^2. \tag{7.8}$$

up to the first-order terms. The approximation is good enough as long as the training samples are in a small neighborhood of the identity.

To formulate the the loss function in terms of a linear regression in a vector space, a matrix $\Omega : \mathbb{R}^d \mapsto \mathbb{R}^n$ that estimates the tangent vectors $\log(M_i)$ on Lie algebra is defined

$$\beta(f) = \exp\left(f^T \Omega\right) \tag{7.9}$$

where Ω is a $d \times n$ matrix of linear regression coefficients. Selecting d orthonormal bases on the Lie algebra, the matrix norm can be computed as the Euclidean distance between two vectors.

By taking the advantage of the Lie algebra, the tangent vectors at the identity $\log(M_i)$ can be rearranged from matrix to n-dimensional vector form. Let $\mathbf{X}$ be a $N \times d$ matrix of row-wise arranged feature vectors, and $\mathbf{Y}$ be the corresponding $N \times n$ matrix of vector form mappings of the tangent vectors

$$\mathbf{X} = \begin{bmatrix} f_1 \\ \vdots \\ f_N \end{bmatrix} \quad \mathbf{Y} = \begin{bmatrix} \log(\mathrm{M}_1) \\ \vdots \\ \log(\mathrm{M}_N) \end{bmatrix}. \tag{7.10}$$

Then, the loss function (7.8) can be written as

$$L \approx \mathrm{tr}\big((\mathbf{Y} - \mathbf{X}\Omega)^T(\mathbf{Y} - \mathbf{X}\Omega)\big) \tag{7.11}$$

where the trace tr replaces the summation in (7.8). Differentiating the loss function with respect to Ω, the minimum is achieved at

$$\Omega = (\mathbf{X}^T\mathbf{X})^{-1}\mathbf{X}^T\mathbf{Y}. \tag{7.12}$$

For rank deficient cases where the number of training samples is smaller than the dimension of the feature space $N < d$, the least squares estimate becomes inaccurate since $\mathbf{X}^T\mathbf{X}$ has determinant zero. To avoid overfitting, a penalty on the magnitude of the regression coefficients in the loss function is introduced

$$L \approx \mathrm{tr}\big((\mathbf{Y} - \mathbf{X}\Omega)^T(\mathbf{Y} - \mathbf{X}\Omega)\big) + \lambda\|\Omega\|^2 \tag{7.13}$$

which is also known as the *ridge regression* [11]. The minimizer of the loss function is given by

$$\Omega = (\mathbf{X}^T\mathbf{X} + \lambda\mathrm{I})^{-1}\mathbf{X}^T\mathbf{Y} \tag{7.14}$$

where I is an $d \times d$ identity matrix. The regularization coefficient λ determines the degree of shrinkage on the regression coefficients.

7.4 Application to Affine Motion Tracking

Locating an image region that undergoes 2D affine transformations is an essential task for camera motion estimation, pose invariant object recognition, and object tracking. In addition to challenging problems such as appearance changes, lighting variations, background clutters, and temporary occlusions, affine motion tracking confronts with computational issues due to the high dimensionality of the motion parameter space that induces an intractable number of hypotheses to be tested.

7.4.1 Related Work

Conventional methods often attempt to solve affine motion tracking in a vector space by state-space estimation [1, 6, 13, 25], template alignment [4, 8, 18] and feature correspondence [12, 19] approaches. State-space estimators assume affine tracking as a Markovian process and construct a probability density function of object parameters, which is supposed to be a normal distribution in case of Kalman filtering [6]. Due to this assumption, Kalman filters fail to describe multi-modal distributions, thus, Monte Carlo integration methods such as particle filters [13] are utilized. In theory, particle filter can track any parametric variation including affine motion. However, its dependency to random sampling induces degenerate likelihood estimations especially for the higher dimensional parameter spaces. Moreover, its computational requirements exponentially grow with the number of the state variables. In template alignment, the parametrized motion models -often more complex than affine motion- is estimated using appearance and shape models that are usually fitted by nonlinear optimization, e.g. iteratively solving for incremental additive updates to the shape parameters [8] or compositional updates to the warped models [3]. Alternatively, affine tracking can be formulated as a minimization on a cost function that consists of the sum of squares differences between the model instance obtained with a linear transformation and input image. However, rarely the relationship between the image intensity values and the model variation can be expressed in a linear form. To accommodate nonlinear transformations, stochastic gradient descent [25], relevance vector machine [28], Tikhonov regularization [1] are employed. One shortcoming of these algorithms is that they require computation of partial derivatives, Jacobian, and Hessian for each iteration, which makes them impractical. Several methods utilize feature point correspondences. Feature point based methods mainly differ in the type of features and descriptors, e.g. using SIFT [26], SURF [12], a combination of primitive features like simple differences between intensity values at randomly chosen locations [19], used for matching the object model to the current frame. The feature-point based trackers are highly sensitive to the available texture information on the object.

Tracking in general can also be regarded as a detection and model fitting problem. A typical tracking-by-detection framework is composed mainly of motion model, observation model and model updater [23, 27, 29]. Motion model generates a set of candidates which might contain the target in the current frame based on the estimation from the previous frame. Observation model judges whether a candidate is the target based on the features extracted from it. Model updater online updates the observation model to adapt the change of the object appearance. Conventional models range from histograms, templates, classifier ensembles, to more intricate appearance models such as region covariance matrices [20] where the matrix is updated on a manifold. Model fitting is considered as a classification problem in [2] by training an ensemble of classifiers with object and background pixels and integrating classifiers over time. More recently, [10] proposed directly predicting the change in object location between frames by an online structured output support vector machine. This method uniformly samples the state space to generate positive and negative support vectors. Such a brute force approach on a larger search window, however, is computationally intractable.

7.4.2 *Tracking as a Regression Problem on Lie Group*

We interpret object tracking task as a supervised learning problem and solve it using a regression function on the Lie algebra. We focus on 2D region motions that establish a matrix Lie group structure. The transformations that we are interested (affine motion $\text{Aff}(2, \mathbb{R})$, similarity transform $S(2)$, Euclidean motion $SE(2)$, etc.) are closed subgroups of general linear group $GL(3, \mathbb{R})$, which is the group of 3×3 nonsingular square matrices. We develop formulation for 2D affine motion group, however the tracking method is applicable to any matrix Lie group structured motion transformation. A two-dimensional affine transformation $\text{Aff}(2, \mathbb{R})$ is given by a 3×3 matrix M as

$$\mathrm{M} = \begin{pmatrix} \theta & \mathbf{t} \\ 0 & 1 \end{pmatrix} \tag{7.15}$$

where θ is a nonsingular 2×2 rotation matrix and $\mathbf{t} \in \mathbb{R}^2$ is a translation vector. The set of all affine transformations forms a matrix Lie group. The structure of affine matrices in (7.15) is a $d = 6$ dimensional manifold. The associated Lie algebra is the set of matrices

$$\mathrm{m} = \begin{pmatrix} \mathrm{U} & \mathbf{v} \\ 0 & 0 \end{pmatrix} \tag{7.16}$$

where, U is a 2×2 matrix and $\mathbf{v} \in \mathbb{R}^2$. The matrix m can be formed into a $d = 6$ dimensional vector by selecting the entries of U and $\mathbf{v}$ as an orthonormal basis.

The set of 2D affine transformations $\text{Aff}(2, \mathbb{R})$ do not constitute a vector space, but rather a manifold that has the structure of a Lie group. Existing methods for the most part disregard this manifold structure and flatten the topology in a vector space. Vector forms cannot globally parameterize the intrinsic topology on the manifold in a homogeneous fashion, thus fail to accurately evaluate the distance between affine motion matrices causing unreliable tracking performance. There are only a few relevant work for parameter estimation on Lie groups, e.g. [9] for tracking an affine snake and [5, 15, 24] for tracking a template. However, [5] fails to account for the noncommutativity of the matrix multiplications thus the estimations are valid only around the initial transformation. [24] learned the correlation between affine motions and the observed descriptors using a regression model on Lie algebra. Inherent topology is considered by [15] where a conventional particle filter based tracker where the state dynamics are defined on a manifold using a log-Euclidean metric. However, none of these methods incorporate an efficient mechanism to incorporate object appearance changes.

Our formulation has several advantages. After learning the regression function, the tracking reduces to evaluating the function at the previous location, therefore it can be performed very fast. In addition, the framework gives flexibility to use any region descriptor.

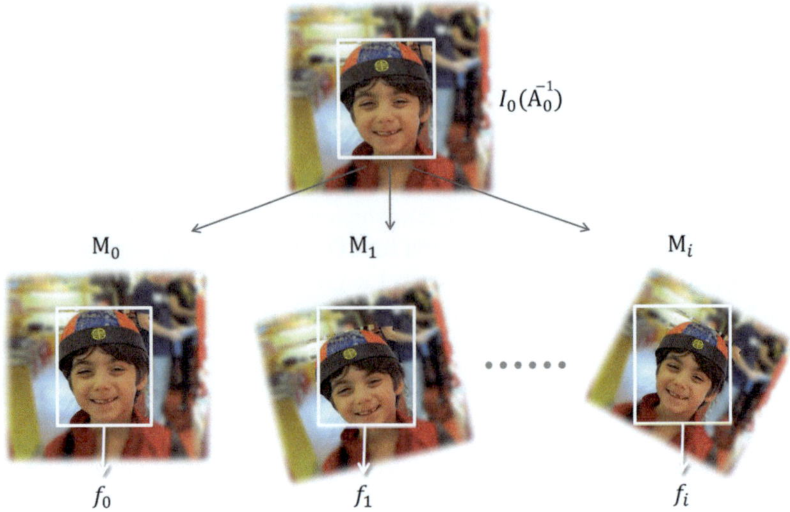

Fig. 7.2 Training samples are generated by applying N affine motions M_i at the object coordinates

Learning Regression Function:

During the initialization of the tracking at frame I_0, we generate a training set of N random affine transformation matrices $\{M_i\}$ around the identity matrix and compute their corresponding observed descriptors f_i within the initial object region to obtain the training set samples $\{f_i, M_i\}$. The process is illustrated in Fig. 7.2.

Specifically, the object coordinates are transformed by multiplying on the right with M_i^{-1} and the corresponding descriptor $f_i = f(I_0(A_0^{-1}.M_i^{-1}))$ is computed Using the initial location of the object A_0. The motion matrix A transforms a unit rectangle at the origin to the affine region enclosing the target object

$$[x_{im} \ y_{im} \ 1]^T = A[x_{ob} \ y_{ob} \ 1]^T \tag{7.17}$$

where, the subscripts indicate the object coordinates and image coordinates respectively. The inverse transform A^{-1} is also an affine motion matrix and transforms the image coordinates to the object coordinates as illustrated in Fig. 7.3. Notice that, the transformation A_0^{-1} moves the object region back to the unit rectangle and the image in the object coordinates is denoted as $I(A_0^{-1})$.

The appearance of an object is described with an feature vector. We use only the pixel values inside the unit rectangle. Since we expect the feature vector to be an indicator of affine motion, we use a motion sensitive region feature. The target region is represented with a concatenated set of orientation histograms computed at a regular grid inside the unit rectangle in object coordinates (see Fig. 7.3). With this, a d-dimensional vector $f\left(I(A^{-1})\right) \in \mathbb{R}^d$ is obtained. The unit rectangle is divided into $6 \times 6 = 36$ cells and a cell histogram is computed in each of them. Each his-

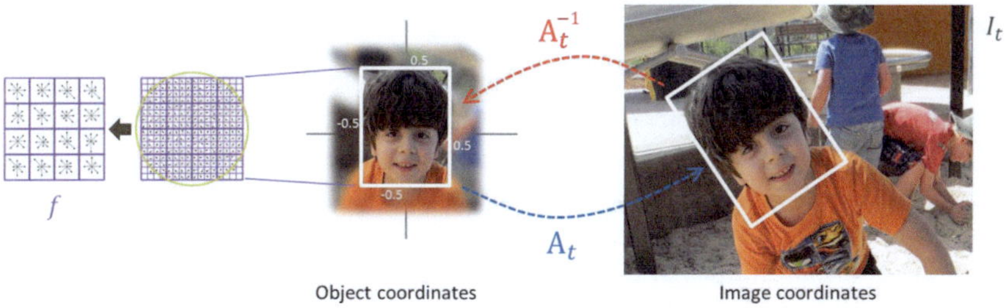

Fig. 7.3 The mapping and its inverse, between the object and image coordinates. The gradient weighted orientation histograms are utilized as region descriptors

togram is quantized at $\pi/4$ degrees between 0 and 2π. The size of each histogram is eight dimensional and the descriptor is $d = 288$ dimensional. Similar to SIFT descriptors [16], the contribution of each pixel to the histogram is proportional to its gradient magnitude. During tracking the peripheral pixels are frequently contaminated by the background, hence we leave a 10 % boundary outside the unit rectangle and construct the descriptor inside the inner rectangle.

After obtaining the training pairs, we form the data matrices as in (7.10) and apply (7.14) to learn the linear regression function Ω to model the correlation between the tangent space projection of the affine motion matrices and their corresponding observed descriptors.

Tracking Region in the Next Frame:

Tracking process estimates the transformation matrix A_t, given the observations $I_{0...t}$ up to time t, and the initial location A_0. We model the transformations incrementally

$$A_t = M_t . A_{t-1} \tag{7.18}$$

and estimate the increments M_t at each time. The transformation M_t corresponds to motion of target from time $t-1$ to t in the object coordinates. Given the previous location of the object A_{t-1} and the current image I_t, we estimate the incremental motion M_t by the regression function

$$M_t = \exp\left(\left(f\left(I_t(A_{t-1}^{-1}) \right)\right)^T \Omega \right). \tag{7.19}$$

After learning the regression function Ω, the tracking problem reduces to estimating the motion via (7.19) using current observation I_t and updating the target location via (7.18). To better localize the target, at each frame we repeat the motion estimation using Ω a maximum of $K = 10$ times or the estimated incremental motion M_t becomes equal to identity.

Model Update:

Since objects can undergo appearance changes in time, it is necessary to adapt to these variations. In our case, we update the regression function Ω.

During tracking, we generate a set of random observations at each frame. The observations stored for last $p = 100$ frames constitute the update training set. Let $\mathbf{X}_p$ and $\mathbf{Y}_p$ be the new training set stored in the matrix form, and Ω_p be the previous model. After each p frames of tracking, we update the coefficients of the regression function by minimizing the loss

$$L \approx \text{tr}\big((\mathbf{Y}_p - \mathbf{X}_p\Omega)^T(\mathbf{Y}_p - \mathbf{X}_p\Omega)\big) + \lambda\|\Omega\|^2 + \gamma\|\Omega - \Omega_p\|^2.$$

The error function is similar to (7.13), but another constraint is introduced on the difference of regression coefficients. The minimum is achieved at

$$\Omega = (\mathbf{X}_p^T\mathbf{X}_p + (\lambda+\gamma)\mathrm{I})^{-1}(\mathbf{X}_p^T\mathbf{Y}_p + \gamma\Omega_p) \tag{7.20}$$

where the parameter γ controls how much change on the regression parameters are allowed from the last estimation. To take into account the bias terms all the function estimations are performed using centered data.

A pseudo-code of the tracking algorithm is given in Algorithm 1.

Algorithm 1 Affine motion tracking

Require: Initial location A_0, images I_t, λ, γ, update frequency p, max iteration K
  procedure TRAINING($t = 0$)
    Generate N motion matrices M_i, $i = 1 \ldots N$
    Extract features $f_i = f\left(I_0(\mathrm{A}_0^{-1}.\mathrm{M}_i^{-1})\right)$
    Form $\mathbf{X}$, $\mathbf{Y}$
    Learn Ω by Eq. 7.14
  procedure TRACKING($t > 0$)
    repeat
      $\mathrm{M}_t = \Omega f\left(I_t(\mathrm{A}_{t-1}^{-1})\right)$
      $\mathrm{A}_t \leftarrow \mathrm{M}_t.\mathrm{A}_{t-1}$
      $k \leftarrow k+1$
    until $\mathrm{M}_t = \mathrm{I}$ or $k \leq K$
    if mod$(t, p) = 0$ **then**
      Update Ω by Eq. 7.20
    $t \leftarrow t+1$

Experiments:

We compare the Lie algebra based parametrization with the linearization (7.21) around the identity matrix [8, 14, 28]

$$\mathrm{M}(x_0 + \delta x) \approx \mathrm{M}(x_0) + \frac{\partial \mathrm{M}}{\partial x}\delta x \tag{7.21}$$

where $\mathrm{M}(x_0) = \mathrm{I}$, by measuring the estimation errors. We also compare orientation histograms with the intensity difference features used in optical flow estimation and tracking.

We generate a training set of $N = 200$ samples by random affine transformations of a single object. The motions are generated on the Lie algebra, by giving random values between -0.2 and 0.2 to each of the six parameters, and mapped to affine matrices via exponentiation. Since the size of the training set is large enough, there is no rank deficiency problem. The function Ω is estimated by ridge regression with $\lambda = 2.10^{-3}$ for orientation histograms and $\lambda = 5.0$ for intensity features, determined by cross validation. Each test set consists of $N_T = 1000$ samples. The samples inside a set have fixed norm. The norms $\| \log(\mathrm{M}) \|$ vary from 0.025 to 0.35. We perform a single tracking iteration by each method, and measure the mean squared geodesic error

$$\frac{1}{N_T} \sum_{i=1}^{N_T} \rho^2 \left(\Omega f_i, \mathrm{M}_i \right) \tag{7.22}$$

between the estimations and the true values.

As shown in Fig. 7.4, the estimation based on the Lie algebra is better than the linearization for transformation of all norms. The ratio is almost constant and on the average the linearization have 12 % larger error. This is expected since our approach minimizes the sum of squared geodesic error. The estimations with orientation histograms are significantly better than the intensity based features.

We show sample tracking examples in Fig. 7.5. In the experiments, the parameters of the ridge regression were $\lambda = \gamma = 2.10^{-3}$, which were learned offline via cross validation. The training dataset is generated on the Lie algebra, by giving random values between -0.1 and 0.1 to each of the six parameters. Although we track the targets with an affine model, these targets are not planar. Therefore, an affine model cannot perfectly fit the target but produces the best affine approximation. Since nonplanar objects undergo significant appearance variations due to pose changes,

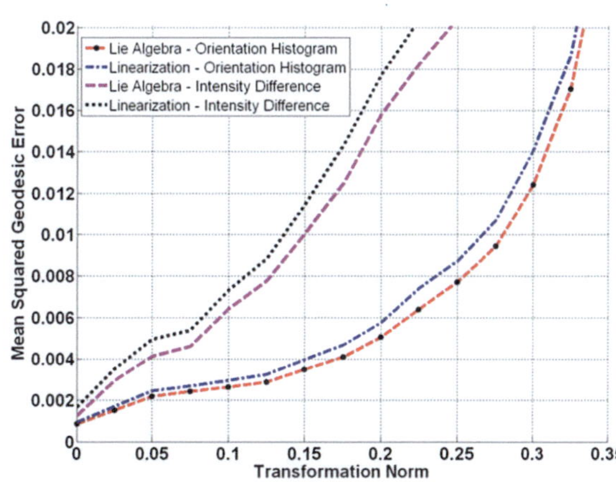

Fig. 7.4 Estimation errors of the Lie algebra and the linearization methods using orientation histograms and intensity features

Fig. 7.5 Sample affine tracking results. Target region boundaries are color-coded

the model update becomes important. The targets have large in-plane and off-plane rotations, translations, scale changes and occlusions. The estimations are accurate, which shows the robustness of the tracking approach.

References

1. T. Albrecht, M. Luthi, T. Vetter, A statistical deformation prior for non-rigid image and shape registration, in *Proceedings of the IEEE Conference on Computer Vision and Pattern Recognition (CVPR)*, 2008
2. S. Avidan. Ensemble tracking, in *Proceedings of the IEEE International Conference on Computer Vision and Pattern Recognition (CVPR)*, 2005
3. S. Baker, I. Matthews, Equivalence and efficiency of image alignment algorithms. Proc. IEEE Conf. Comput. Vis. Pattern Recogn. Kauai, HI, **1**, 1090–1097 (2001)
4. S. Baker, I. Matthews, Lucas-Kanade 20 years on: a unifying framework. Int. J. Comput. Vis. **56**(3), 221–255 (2004)
5. E. Bayro-Corrochano, J. Ortegon-Aguilar. Lie algebra template tracking, in *Proceedings of the 17th International Conference on Pattern Recognition*, vol. 2, pp. 56–59, 2004
6. Y. Boykov, D. Huttenlocher, *Adaptive Bayesian recognition in tracking rigid objects* (In *Proceedings of the IEEE Conf. on Computer Vision and Pattern Recognition*, Hilton Head, SC, II, 2000), pp. 697–704
7. G. Chirikjian, *Stochastic Models, Information Theory, and Lie Groups* (Birkhauser, Boston, 2011)
8. T. Cootes, G. Edwards, C. Taylor, *Active appearance models* (In Proc. European Conf. on Computer Vision, Freiburg, Germany, 1998), pp. 484–498
9. T. Drummond, R. Cipolla, Application of Lie algebras to visual servoing. Int. J. Comp. Vis. **37**, 21–41 (2000)
10. S. Hare, A. Saffari, P.H.S. Torr, Struck: Structured output tracking with kernels, in *Proceedings of the IEEE International Conference on Computer Vision (ICCV)*, 2011
11. T. Hastie, R. Tibshirani, J. Freidman. *The Elements of Statistical Learning*. (Springer, Berlin, 2001)

12. W. He, T. Yamashita, H. Lu, S. Lao, SURF tracking, in *Proceedings of the International Conference on Computer Vision (ICCV)*, 2009
13. M. Isard, I. Blake, Condensation-conditional density propagation for visual tracking. Int. J. Comp. Vis. **29**, 5–28 (1998)
14. F. Jurie, M. Dhome, Hyperplane approximation for template matching. IEEE Trans. Pattern Anal. Mach. Intell. **24**, 996–1000 (2002)
15. J. Kwon, K. M. Lee, F. Park. Visual tracking via geometric particle filtering on the affine group with optimal importance functions, in *Proceedings of the IEEE Conference on Computer Vision and Pattern Recognition (CVPR)*, 2009
16. D. Lowe, Distinctive image features from scale-invariant keypoints. Int. J. Comp. Vis. **60**(2), 91–110 (2004)
17. R. Mahony, P.A. Absil, R. Sepulchre. *Optimization algorithms on matrix manifolds*. (Princeton University Press, Princeton, 2009)
18. I. Matthews, S. Baker, Active appearance models revisited. Int. J. Comp. Vis. **60**, 135–164 (2004)
19. M. Ozuysal, P. Fua, V. Lepetit. Fast keypoint recognition in ten lines of code, in *Proceedings of the IEEE Conference on Computer Vision and Pattern Recognition (CVPR)*, 2007
20. F. Porikli, O. Tuzel, P. Meer. Covariance tracking using model update based on Lie algebra, in *Proceedings of the IEEE International Conference on Computer Vision and Pattern Recognition (CVPR)*, 2006
21. W. Rossmann. *Lie Groups: An Introduction Through Linear Groups*. (Oxford Press, Oxford, 2002)
22. S.S. Sastry, J. Kosecka, Y. Ma, S. Sastry, *An invitation to 3-d vision: from images to geometric models*. (Springer, Berlin, 2003)
23. A.W.M. Smeulders, D.M. Chu, R. Cucchiara, S. Calderara, A. Dehghan, M. Shah. Visual tracking: an experimental survey. IEEE Trans. Pattern Anal. Mach. Intell. **36**(7), 1442–1468 (2014)
24. O. Tuzel, F. Porikli, P. Meer. Learning on Lie groups for invariant detection and tracking, in *Proceedings of the IEEE Conference on Computer Vision and Pattern Recognition (CVPR)*, 2008
25. T. Vetter, T. Poggio, Linear object classes and image synthesis from a single example image. IEEE Trans. Pattern Anal. Machine Intell. **19**, 733–742 (1997)
26. D. Wagner, T. Langlotz, D. Schmalstieg. Robust and unobtrusive marker tracking on mobile phones, in *Proceedings of the ACM International Symposium on Mixed and Augmented Reality (ISMAR)*, 2008
27. N. Wang, S. Li, A. Gupta, and D. Yeung. Transferring rich feature hierarchies for robust visual tracking. *CoRR*, 2015
28. O. Williams, A. Blake, R. Cipolla, Sparse Bayesian learning for efficient visual tracking. IEEE Trans. Pattern Anal. Machine Intell. **27**, 1292–1304 (2005)
29. Y. Wu, J. Lim, M.H. Yang. Online object tracking: a benchmark, in *Proceedings of the IEEE Conference on Computer Vision and Pattern Recognition (CVPR)*, 2013
30. G. Zhu, F. Porikli, H. Li. Lie-Struck: affine tracking on Lie groups using structured SVM, in *Proceedings of the IEEE Winter Conference on Applications of Computer Vision (WACV)*, 2015

Chapter 8
An Elastic Riemannian Framework for Shape Analysis of Curves and Tree-Like Structures

Adam Duncan, Zhengwu Zhang and Anuj Srivastava

Abstract Shape analysis of complex structures formed by Euclidean curves and trees are of interest in many scientific domains. The difficulty in this analysis comes from: (1) Manifold representations of curves and trees, and (2) need for the analysis to be invariant to certain shape-preserving transformations. Additionally, one is faced with the difficult task of registering points and parts across objects during shape comparisons. We present a Riemannian framework to solve this problem as follows: we select an elastic Riemannian metric that is invariant to the action of re-parameterization group and use a square-root velocity function to transform this metric into the $\mathbb{L}^2$ norm. Re-parameterization of objects is considered to control registrations across objects, and an inherited distance on the quotient space of shape representations modulo shape-preserving transformations forms the shape metric. The resulting framework is used to compute geodesic paths and sample means, and to discover modes of variability in sample shapes. We demonstrate these ideas using some simple examples involving planar shapes and neuron morphology.

8.1 Introduction

Many important scientific endeavors of our time seek to **correlate forms with functionality** in biological systems. For instance, one asks, how well does the structure of an anatomical part or a biomolecule facilitate or predict its role or functionality in a biological system? This current era is also characterized by a remarkable **abundance of digital data** to be able to potentially hypothesize and validate such

A. Duncan (✉) · A. Srivastava
Department of Statistics, Florida State University, Tallahassee, FL, USA
e-mail: a.duncan@stat.fsu.edu

A. Srivastava
e-mail: anuj@stat.fsu.edu

Z. Zhang
SAMSI, Research Triangle Park, Durham, NC, USA
e-mail: zzhang@samsi.info

H.Q. Minh and V. Murino (eds.), *Algorithmic Advances in Riemannian Geometry and Applications*, Advances in Computer Vision and Pattern Recognition, DOI 10.1007/978-3-319-45026-1_8

connections. Take, for example, the protein database (PdB) or neuromorpho [1], the database of neurons, or the human connectome database. These databases represent tens or even hundreds of thousands of relevant structures that can be analyzed using formal statistical tools. Consequently, the focus is on **data driven** research, where high-dimensional observations of forms and associated functionalities are to be analyzed jointly. While there is a free availability of data to infer functionality from structures, the relevant mathematical and statistical tools to discover connections between forms and functions are severely lacking. This lack of tools forms a bottleneck in building broad and ambitious pipelines for validating biological hypotheses.

This chapter will develop computational tools, based on novel mathematical theories, for statistical comparisons of certain complex structures – curves and trees – that are encountered in different scientific problems. Instances of these structures include backbones of proteins in protein structure analysis, shape classification of leaf boundaries for determining species, morphological analysis of neurons for characterizing mutations, shape analysis of skeleton videos for recognition human actions and activities, and so on. A common thread in these problems is: (1) High level of **complexity** and high-dimensionality of the structures of interest. These structures may differ not only in their geometries and topologies, but also in nuisance variables such as position, rotation, scale, and parameterization. (2) Tremendous **statistical variability** that is exhibited within and across functional classes associated with these structures. For statistical inferences involving such structures, for instance the labeling of healthy versus diseased subjects, one needs to develop statistical models that can efficiently capture the observed variability in those two classes. A direct consequence of complexity of these structures is that one cannot use tools from traditional multivariate statistics directly. The representation spaces of these structures form nonlinear manifolds that are endowed with appropriate Riemannian structures to in order to define the basic notions of distance, averages, and covariances. The choice of a metric in a problem is not straightforward either. For instance, the most commonly used metric in function data analysis, i.e., the Hilbert structure based on the $\mathbb{L}^2$ norm, is problematic when it comes to analyzing function data under registration variability (unaligned data). Thus, the task of developing a principled and a comprehensive framework for statistical analysis and modeling of such complex structures is a difficult one.

8.1.1 From Discrete to Continuous and Elastic

A large majority of past statistical analyses of shapes uses discrete, point-set representations, while the more recent trend is to study continuous objects. Since continuous objects, such as parameterized curves and trajectories, are represented by coordinate functions, and functional spaces are typically infinite-dimensional, this change introduces an additional complexity of infinite-dimensionality. So, the question arises: Are we making the problem unnecessarily complicated by going to functional representations? Let us study the choices more carefully. Say we are given two sets containing finite number of unregistered points, and our goal is to register them and to compare their shapes. The problem of registration is a combinatorial one and adds considerable computational complexity to the solution. On the other hand, let us assume the original objects are parameterized curves: $t \mapsto f_1(t), f_2(t)$, for $t \in [0, 1]$. The interesting part in this approach is the following. For each t, the pair of points, $f_1(t)$ and $f_2(t)$ are considered registered. In order to change registration, one simply has to re-parameterize one of the objects. In other words, find a re-parameterization γ of f_2 such that $f_1(t)$ is now registered to $f_2(\gamma(t))$. Thus, we can find optimal registration, or alignment, of curves by optimizing over the variable γ under a proper objective function. If this objective function is a metric that is invariant of all shape-preserving transformations, then we simultaneously achieve a joint solution for registration and shape comparison. Thus, parameterization controls registration between curves and an optimal registration can be found using algorithms with complexity much smaller than those encountered in combinatorial solutions. Similar arguments can be made for more complex objects such as trees. The optimization over parameterization in shape analysis of objects, under a metric with proper invariance properties, leads to a framework called *elastic shape analysis*.

8.1.2 General Elastic Framework

From here onwards we will focus exclusively on parameterized objects – curves, trajectories, and trees – and use parameterizations to control registrations. The re-parameterization set Γ is chosen to be the set of all positive diffeomorphisms such that $\gamma(0) = 0$ and $\gamma(1) = 1$. An interesting property of Γ is that it forms a group action under composition, with the identity element given by the function $\gamma_{id}(t) = t$. Therefore, for any two γ_1, γ_2, the composition $\gamma_1 \circ \gamma_2$ is also a valid re-parameterization, and so is the inverse γ^{-1} for any γ. The next issue is to decide the objective function so that the optimal re-parameterization can be found in a variational framework. A seemingly natural idea of performing alignment is using the criterion $\inf_\gamma \|f_1 - f_2 \circ \gamma\|$, where $\|\cdot\|$ denotes the $\mathbb{L}^2$ norm, but it turns out to be problematic. The main issue is that it allows degeneracy, that is, one can reduce this cost arbitrarily close to zero even when the two functions are quite different. This is commonly referred to as the *pinching problem* in the literature [7]. Pinching implies that a severely distorted γ is used to eliminate (or minimize) those parts of f_2 that do not match with f_1; this can be done even when f_2 is mostly different from f_1. Another way to state the problem is that one can easily manipulate $\|f \circ \gamma\|$ into a broad range of values, by choosing an appropriate γ. Of course, once can avoid the pinching problem by imposing a roughness penalty on γ, thus avoiding a severe distortion of γs but that leads to other issues including asymmetry. A related problem from the registration perspective is that $\|f_1 - f_2\| \neq \|f_1 \circ \gamma - f_2 \circ \gamma\|$ in general. Note that if we warp two functions by the same γ, their registration remains unchanged but their $\mathbb{L}^2$ norm changes. Hence, the $\mathbb{L}^2$ norm is not a proper objective function to help solve the registration problem.

The solution comes from deriving an elastic metric-based objective function that is better suited for registration and shape analysis. While the discussion of the underlying elastic Riemannian metric is complicated, we directly move on to a simplification which is based on certain square-root transforms of data objects. Denoted by q, these objects take different mathematical forms in different contexts, as explained in later sections. The important mathematical property of these representations is that $\|q_1 - q_2\| = \|(q_1, \gamma) - (q_2, \gamma)\|$, for all γ, where each q_i represents the object f_i, and each (q_i, γ) represents the warped object $(f_i \circ \gamma)$. This property allows us to define a solution for all important problems

$$\inf_{\gamma} \|q_1 - (q_2, \gamma)\|. \tag{8.1}$$

Not only does the optimal γ help register the object f_2 to f_1, but also the infimum value of the objective function is a proper metric for shape comparison of the two objects. (In the case of shape analysis of curves and surfaces one needs to perform an additional rotation alignment for shape comparisons.) This metric enables statistical analysis of shapes. One can compute mean shapes and the dominant modes of variations in a shape sample, develop statistical models to capture observed shape variability, and use these models to perform hypothesis tests. While we focus on static shapes in this paper, these ideas can also be naturally extended to dynamic shapes.

In the next few sections we take different examples of this elastic framework in the contexts of shape analysis of Euclidean curves, curves in infinite-dimensional Hilbert spaces, and shape analysis of trees with arbitrary number of branches.

8.2 Shape Analysis of Euclidean Curves

Here the objects of interest are curves of the type $f : [0, 1] \to \mathbb{R}^n$. (Note that in case of closed curves it is natural to use $\mathbb{S}^1$ as the parameterization domain, rather than an interval.) The $\mathbb{L}^2$ metric is given by $\langle f_1, f_2 \rangle = \int_0^1 \langle f_1(t), f_2(t) \rangle \, dt$ and the resulting norm $\|f_1 - f_2\| = \int_0^1 |f_1(t) - f_2(t)|^2 dt$, where $|\cdot|$ denotes the vector norm. The mathematical representation of curves is in form of the *square-root velocity function* (SRVF) given by [6, 10]

$$q(t) = \frac{\dot{f}(t)}{\sqrt{|\dot{f}(t)|}}.$$

The re-parameterization group here is the set of all positive diffeomorphisms of $[0, 1]$. If q is the SRVF of a curve f, then the SRVF of the re-parameterized curve $f \circ \gamma$ is given by $(q \circ \gamma)\sqrt{\dot{\gamma}}$; we will denote this by (q, γ).

From the perspective of shape analysis, a re-parameterization of a curve does not alter its shape. An illustration of different parameterizations of a curve is shown in Fig. 8.1. The shape of f is exactly same as the shape of $f \circ \gamma$, for any γ. The same

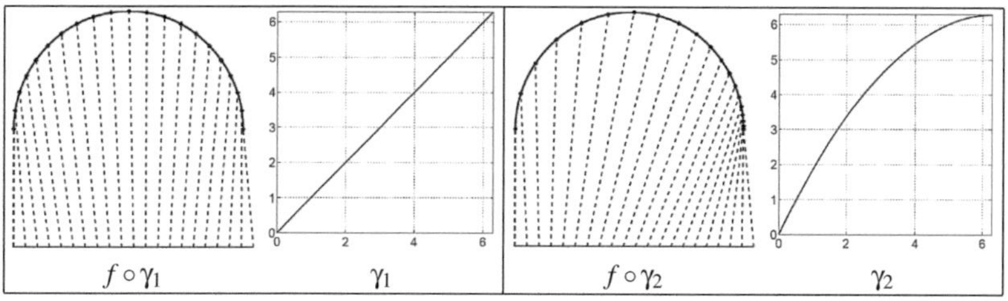

Fig. 8.1 An illustration of re-parameterization of a curve on domain $D = [0, 2\pi]$

holds for the rigid rotation of a curve. For any $O \in SO(n)$, the rotated curve $Of(t)$ has the same shape as the original curve. This leads to formulation of equivalence classes, or orbits, of representations that all correspond to the same shape. Let $[f]$ denote all possible rotations and re-parameterizations of a curve f. The corresponding set in SRVF representation is given by $[q] = \{O(q, \gamma) | O \in SO(n), \gamma \in \Gamma\}$. Each such class represents a shape uniquely and shapes are compared by computing a distance between the corresponding orbits.

As mentioned earlier, the SRVF representation satisfies the property that $\|q\| = \|(q, \gamma)\|$, and $\|q_1 - q_2\| = \|(q_1, \gamma) - (q_2, \gamma)\|$ for all $\gamma \in \Gamma$ and all q, q_1, q_2. Using this property, the shape distance between any two shapes is given by

$$d([q_1], [q_2]) = \inf_{\gamma \in \Gamma, O \in SO(n)} \|q_1 - O(q_2 \circ \gamma)\sqrt{\dot{\gamma}}\|. \tag{8.2}$$

This optimization emphasizes the joint nature of our analysis – on one hand we optimally register points across two curves using re-parameterization and rotation and other hand we obtain a metric for comparing shapes of the two curves. The optimization over $SO(n)$ and Γ is performed using coordinate relaxation – optimizing over one variable while fixing the other. The optimization over $SO(n)$ uses the Procrustes method while the optimization over Γ uses the dynamic programming algorithm (DPA) [10]. In the absence of any other constraints on the curves, a straight line between q_1 and the registered q_2, i.e., $O^*(q_2, \gamma^*)$, with these quantities being the minimizers in the equation above, forms the desired geodesic. However, if we rescale the curves to be of unit length or restrict ourselves to only the closed curves, then the underlying space becomes nonlinear and requires additional techniques for computing geodesics. We have developed a path-straightening for computing geodesics in the shape space of closed curves under the elastic metric, as described in [10]. Figure 8.2 shows some examples of geodesic paths between several pairs of closed curves taken from the MPEG7 dataset [4]. One can see that under this joint framework, we deform one shape to another in a natural way – the features are preserved across shapes and deformations are smooth.

Mean Shape and Modes of Variations This framework is amenable to the development of tools for statistical analysis of shapes. For example, given a set of observations of curves, we may want to calculate the sample mean and modes of variations.

Fig. 8.2 Examples of geodesic paths between shapes under the elastic shape metric

Furthermore, we are interested in capturing the variability associated with the shape samples using probability models. The notion of a sample mean on a nonlinear manifold is typically defined using the Karcher mean [5]. Let $f_1, f_2, \ldots, f_n$ be the observed sample shapes, and their SRVFs are denoted by $q_1, q_2, \ldots, q_n$, the Karcher mean is defined as a quantity that satisfies: $[\mu] = \text{argmin}_{[q]} \sum_{i=1}^{n} d([q], [q_i])^2$, where $d([q], [q_i])$ is calculated using Eq. 8.2, and μ is the SRVF representation of the mean shape $\bar{f}$. The search for the optimal mean shape $\bar{f}$ can be solved using iterative gradient-based algorithm [5, 6, 9]. Figure 8.3 shows some sample mean shapes calculated using this approach.

In addition to the Karcher mean, the Karcher covariance and modes of variation can be calculated to summarize the given sample shapes. Since the shape space is a nonlinear manifold, we use the tangent space at the mean shape μ, which is a

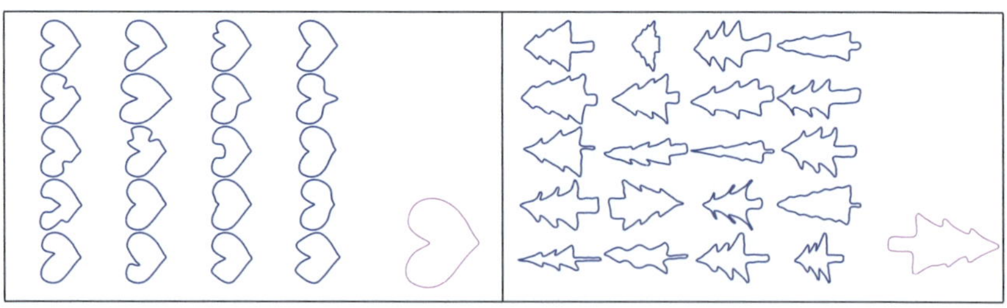

Fig. 8.3 Mean shapes of two different classes of shapes. Each mean shape (shown in *magenta color*) is calculated from shapes on its *left*

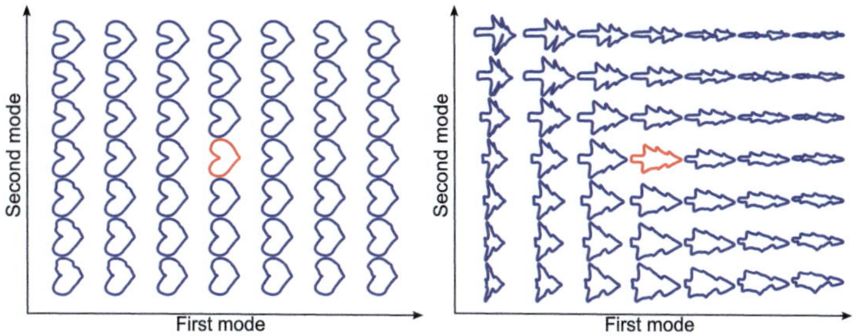

Fig. 8.4 Modes of variations: for each class of shapes in mean shape example, we show the variation along the first and second principle modes. Shape in the *center* with *red color* is the mean shape

standard vector space, to perform the statistical analysis. We first map each sample shape onto the tangent space using inverse exponential map: $v_i = \log_\mu(q_i)$, then we define the covariance matrix to be: $C = \frac{1}{n-1}\sum_{i=1}^{n} v_i v_i^t$. Using Principle Component Analysis (PCA) of C, one can get the modes of shape variation. Let PC_k denote the k-th principal direction, exponential map $\exp_\mu(t\text{PC}_k s_k)$ shows the shape variation in PC_k principal direction with standard deviation s_k. Figure 8.4 shows the modes of variations for different classes of shapes in Fig. 8.3.

Statistical Shape Models After obtaining the mean and covariance matrix, the further step is to develop probability models to capture the distribution of given sample shapes. It is challenging to directly impose a probability density on the nonlinear manifold shape space. A common solution is to impose a distribution on the tangent vector space. For example, the tangent space at mean μ can be estimated by a principal subspace, and each inverse exponential mapped shape in this space can be expressed as a linear combination of the orthonormal basis obtained from PCA. Then, we can impose a multivariate Gaussian distribution on the principal subspace with zero mean and covariance matrix obtained from the sample shapes. Figure 8.5 shows the examples of random samples using mean and covariance matrices estimated from shapes shown in Fig. 8.3.

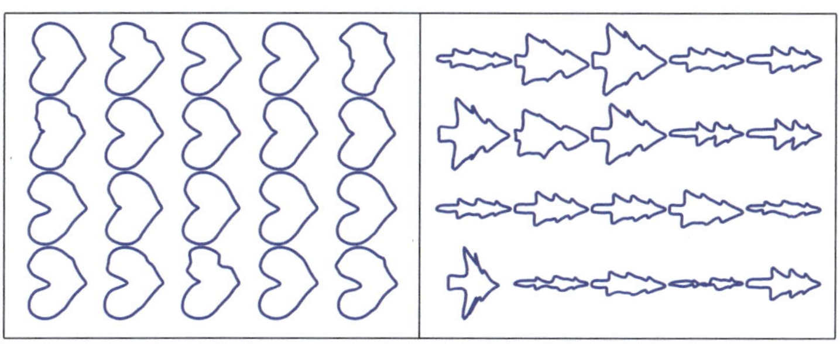

Fig. 8.5 Random samples from the Gaussian shape distribution of different classes of shapes

Traditional shape analysis removes the transformations resulting from rigid motions and global scaling in shape considerations, while in elastic shape analysis we additionally remove the effects of re-parameterizations. In some situations, however, there is a need for removing other groups such as the affine and projective groups. For a discussion on the resulting affine-elastic shape analysis of planar curves, we refer the reader to the paper [2]. The paper also describes a framework for projective-invariant shape analysis of planar objects but using point-set representations rather than continuous curves.

8.3 Shape Analysis of Trajectories in Hilbert Spaces

The framework for comparing Euclidcan curves (curves in $\mathbb{R}^n$) can be extended to compare and analyze trajectories in function spaces, e.g., $\mathbb{L}^2([0,1],\mathbb{R})$, in a natural fashion. Even the computer implementations for Euclidean curves can be adapted to Hilbert spaces (with $\mathbb{L}^2$ norm) with minimal changes. This is because the $\mathbb{L}^2$ norm on the function space, $\|f\|_{\mathbb{L}^2}^2 = \int_0^1 f(t)^2 dt \approx \delta \sum_{i=1}^n f(t_i)^2 = \delta \|v\|_{\mathbb{R}^n}^2$, where $\{t_i\}$ is a uniform partition of $[0,1]$ with bin size δ and $v \in \mathbb{R}^n$ is a vector of values $\{f(t_i)\}$. A comparison of any two functions under the $\mathbb{L}^2$ norm is performed numerically using the 2-norm of vectors formed by their samples. Thus, in practice, one can treat the given functions as elements of $\mathbb{R}^n$, for a large n, and use the elastic shape analysis setup for curves in this space.

8.3.1 *Elastic Comparison of Trajectories in* $\mathbb{L}^2([0,1],\mathbb{R})$

Let $X : [0,1] \to \mathbb{L}^2([0,1],\mathbb{R})$ be a trajectory in the space of square-integrable functions. We can also view $X : [0,1] \times [0,1] \to \mathbb{R}$ as an indexed family of curves in $\mathbb{R}$. For each $t \in [0,1]$, the mapping $X(t,\tau)$ denotes a curve in $\mathbb{R}$ with the parameter τ; t forms a continuous indexing of these curves. Define the SRVF of trajectory X according to

$$q_X(t,\tau) = \frac{\frac{dX(t,\tau)}{dt}}{\sqrt{\left\|\frac{dX(t,\cdot)}{dt}\right\|_{\mathbb{L}^2([0,1],\mathbb{R})}}}\,,\ \text{where}\ \left\|\frac{dX(t,\cdot)}{dt}\right\|_{\mathbb{L}^2([0,1],\mathbb{R})} = \sqrt{\int_0^1 \left|\frac{dX(t,\tau)}{dt}\right|^2 d\tau}.$$

We can reconstruct X back from q_X (unto a different translation for each t) using

$$X(t,\tau) = \int_0^t q(s,\tau)\|q(s,\cdot)\|_{\mathbb{L}^2([0,1],\mathbb{R})} ds\ .$$

Under appropriate smoothness assumptions on the original X, q_X is an element of $\mathbb{L}^2([0, 1], \mathbb{L}^2([0, 1], \mathbb{R}))$. Since they are square-integrable maps, we can extend the standard $\mathbb{L}^2$ on curve space to measure the differences between these maps

$$\|q_{X_1} - q_{X_2}\|^2 = \int_0^1 \|q_{X_1}(t, \cdot) - q_{X_2}(t, \cdot)\|^2 dt = \int_0^1 \left(\int_0^1 (q_{X_1}(t, \tau) - q_{X_2}(t, \tau))^2 d\tau \right) dt \, .$$

One can define a geodesic path between any two such SRVFs as a straight line

$$\alpha : [0, 1] \to \mathbb{L}^2([0, 1], \mathbb{L}^2([0, 1], \mathbb{R})), \;\; \alpha(s) = (1 - s)q_{X_1} + s q_{X_2} \, .$$

In order to perform elastic shape analysis of trajectories in $\mathbb{L}^2([0, 1], \mathbb{R})$, define the action of the re-parameterization group Γ on a trajectory according to $(X, \gamma)(t) = X(\gamma(t))$ (that is, $(X(t, \tau), \gamma(t)) \equiv X(\gamma(t), \tau)$). The corresponding action on the SRVF of X is given by: $(q_X, \gamma)(t) = q_X(\gamma(t))\sqrt{\dot{\gamma}(t)}$. In order to perform temporal registration between any two trajectories X_1 and X_2, we take the associated SRVFs q_{X_1} and q_{X_2}, and solve the following optimization problem:

$$\inf_{\gamma \in \Gamma} \|q_{X_1} - (q_{X_2} \circ \gamma)\sqrt{\dot{\gamma}}\|^2 = \inf_{\gamma \in \Gamma} \left[\int_0^1 \left(\int_0^1 (q_{X_1}(t, \tau) - \sqrt{\dot{\gamma}(t)} q_{X_2}(\gamma(t)), \tau)^2 d\tau \right) dt \right] . \tag{8.3}$$

The infimum of this objective function provides a distance between the re-parameterization orbits of q_{X_1} and q_{X_2}, and the geodesic equation mentioned above applies with q_{X_2} replaced by the optimal element of its orbit.

Figure 8.6 shows an example of this idea. The top left panel shows a path X_1 in $\mathbb{L}^2$. For each value of t, we have a unimodal function (as a function of τ) and the location of peak changes as t changes. We use a time-warping function γ to create a new path $X_2 = X_1 \circ \gamma$ shown in the bottom left panel. Then, we use the corresponding

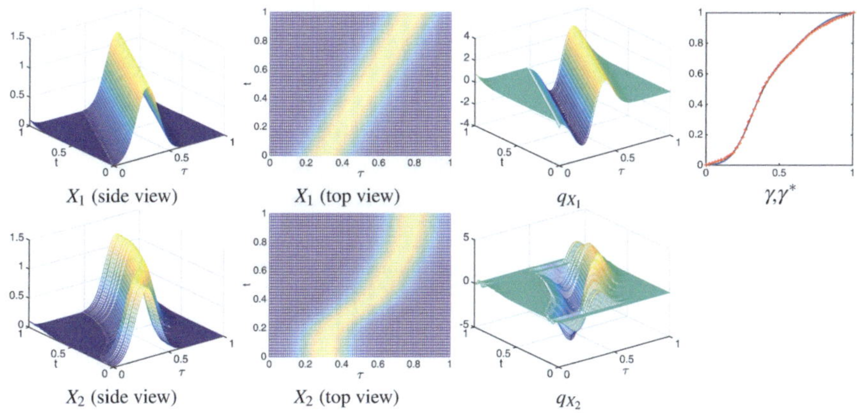

Fig. 8.6 A simulated example of registration of two trajectories in $\mathbb{L}^2$: here $f_2 = f_1 \circ \gamma$, a warped version of f_1. The original γ and the estimated γ^* are shown in the *top right panel*

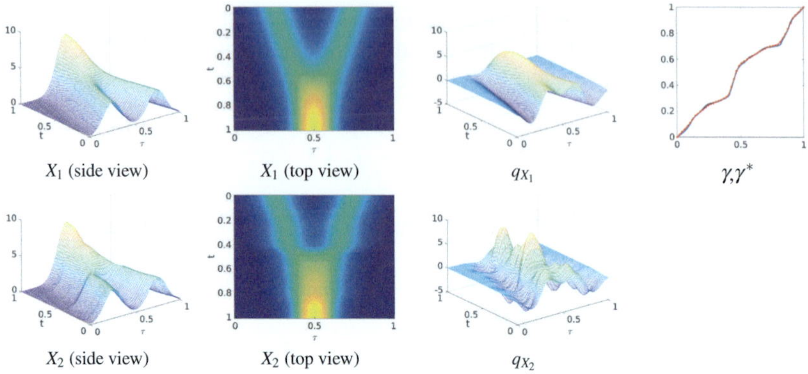

Fig. 8.7 Same as Fig. 8.6

SRVFs, shown in the third column, and Eq. 8.3 to perform alignment via DPA. The estimated optimal γ^* is shown in the top right panel, drawn over the original γ. The high degree of overlap implies a high accuracy in alignment of trajectories. Figure 8.7 shows a similar example in which a bimodal function at $t = 0$ smoothly turns into a unimodal function at $t = 1$.

8.3.2 Elastic Comparison of Trajectories in $\mathbb{L}^2([0,1], \mathbb{R})/\Gamma$

Now we consider a situation where we are given trajectories of the type $X : [0, 1] \to \mathbb{L}^2([0, 1], \mathbb{R})/\Gamma$ and our goal is align any two such trajectories. What this means is that for each $t \in [0, 1]$, $X(t, \cdot) \in \mathbb{L}^2([0, 1], \mathbb{R})/\Gamma$ is not just a real-valued function on $[0, 1]$, it is an *elastic function*. Additionally, we want to compute geodesics between them under the elastic framework. In other words, not only we have to match $X_1(t)$ and $X_2(\gamma(t))$ using a certain $\gamma \in \Gamma$, but also we will have to match $X_1(t, \tau)$ with $X_2(t, \gamma_t(\tau))$ using some $\gamma_t \in \Gamma$.

The definition of X and its SRVF q_X remains same as above. The only change is in the metric used and the mechanism for registration of two trajectories. The norm in the space $\mathbb{L}^2([0, 1], \mathbb{R})/\Gamma$ is given by $\|f_1 - f_2\|^2_{\mathbb{L}^2([0,1],\mathbb{R})/\Gamma} = \inf_{\gamma \in \Gamma} \|f_1 - \sqrt{\dot{\gamma}}(f_2 \circ \gamma)\|^2$, and the corresponding norm between the two SRVFs get modified to

$$\|q_{X_1} - q_{X_2}\|^2 = \int_0^1 \|q_{X_1}(t, \cdot) - q_{X_2}(t, \cdot)\|^2_{\mathbb{L}^2([0,1],\mathbb{R})/\Gamma} dt$$
$$= \int_0^1 \inf_{\gamma_t \in \Gamma} \left(\int_0^1 (q_{X_1}(t, \tau) - \sqrt{\dot{\gamma}_t(\tau)} q_{X_2}(t, \gamma_t(\tau)))^2 d\tau \right) dt \,.$$

Now, the alignment of trajectories is given by

$$\inf_{\gamma} \|q_{X_1} - (q_{X_2} \circ \gamma)\sqrt{\dot{\gamma}}\|^2 = \inf_{\gamma} \int_0^1 \|q_{X_1}(t, \cdot) - \sqrt{\dot{\gamma}} q_{X_2}(\gamma(t), \cdot)\|^2_{\mathbb{L}^2([0,1],\mathbb{R})/\Gamma} dt$$
$$= \inf_{\gamma} \int_0^1 \inf_{\gamma_t \in \Gamma} \left(\int_0^1 (q_{X_1}(t, \tau) - \sqrt{\dot{\gamma}(t)}\sqrt{\dot{\gamma}_t(\tau)} q_{X_2}(\gamma(t), \gamma_t(\tau)))^2 d\tau \right) dt \, .$$

In Sect. 8.3.1 we noted that a trajectory $X : [0, 1] \to \mathbb{L}^2([0, 1], \mathbb{R})$ can be alternately notated as a real-valued function on the unit square, $X : [0, 1]^2 \to \mathbb{R}$. Similarly, the re-parameterizations of trajectories in $\mathbb{L}^2([0, 1], \mathbb{R})/\Gamma$ can be written as a special class of re-parameterizations of the unit square. Let $\xi : [0, 1]^2 \to [0, 1]^2$ be an orientation and boundary-preserving diffeomorphism. We can express ξ in terms of its two components, $\xi(t, \tau) = (\xi_1(t, \tau), \xi_2(t, \tau))$ but for our needs ξ is somewhat restricted. First, ξ_1 is constant with respect to τ for any fixed t and is a function of t only. In other words, $\xi_1(t)$ is a re-parameterization of the interval $[0, 1]$ with t as the parameter. Second, the second component ξ_2 satisfies the following property: for each t, $\xi_2(t, \tau)$ is a re-parameterization of $[0, 1]$ with the parameter τ. This ξ can be written more concisely as $\xi(t, \tau) = (\xi_1(t), \xi_2(t, \tau))$. These two components correspond to the re-parameterizations mentioned in the last equation $\xi_1(t)$ is $\gamma(t)$, the trajectory re-parameterization, and $\xi_2(t, \tau)$ is $\gamma_t(\tau)$.

For any two trajectories, X_1 and X_2, and the corresponding SRVFs q_{X_1} and q_{X_2}, once the optimal registration ξ has been estimated, the geodesic paths and the geodesic distances can be computed rather easily in the corresponding $\mathbb{L}^2$ space between the registered trajectories.

8.4 Elastic Shape Analysis of Axonal Trees

Now that we have tools for representing and comparing individual curves, both in $\mathbb{R}^3$ and $\mathbb{L}^2([0, 1], \mathbb{R}^3)$, we consider problem of comparing shapes of trees. In this section, we specify a mathematical representation of axonal trees as *composite trajectories* and develop a framework for comparing the shapes of these composite trajectories as described in [3]. Our goal is to develop a metric that can be used to: (1) compare the geometries of any two axonal trees, i.e., quantify differences in their shapes, (2) find a geodesic path for morphing one tree into the other, and (3) generate statistical summaries for any finite collection of trees.

8.4.1 *Representing Trees as Composite Trajectories*

We start with a definition of a composite trajectory that has two types of objects – the main branch and an indexed family of side branches. We take the viewpoint that attached to the main branch is a continuum of branches – at each point $t \in [0, 1]$, the corresponding branch is a curve with a certain shape, scale, and orientation attributes.

Definition 1 (*Composite Trajectory*)
Define $\beta : [0, 1] \times [0, 1] \rightarrow \mathbb{R}^3$ to be a composite trajectory with the following components:

1. The base curve $\beta_0 : [0, 1] \rightarrow \mathbb{R}^3$, defined by $\beta_0(t) \equiv \beta(t, 0)$ is a parameterized curve in $\mathbb{R}^3$ and denotes the main branch.
2. For each t, $\beta^{(t)}(\tau) : [0, 1] \rightarrow \mathbb{R}^3$ defined by $\beta^{(t)}(\tau) = \beta(t, \tau)$ is a parameterized curve in $\mathbb{R}^3$ representing a side branch attached to the main branch at the point $\beta_0(t)$. In other words, $\beta^{(t)}(0) = \beta_0(t) = \beta(t, 0)$.

For each t, we consider the side branch $\beta^{(t)}$ and represent it by its SRVF

$$q^{(t)}(\tau) = \frac{\frac{d\beta^{(t)}(\tau)}{d\tau}}{\sqrt{\left\| \frac{d\beta^{(t)}(\tau)}{d\tau} \right\|_{\mathbb{R}^3}}}.$$

We will make enough assumptions on β to ensure that $q^{(t)} \in \mathbb{L}^2([0, 1], \mathbb{R}^3)$. Note that given the SRVF $q^{(t)}$ and the starting point $\beta_0(t)$, we can reconstruct the original side branch $\beta^{(t)}$ exactly. Therefore, there is no loss of information in representing the original tree by the set

$$\left\{ \left(\beta_0(t), q^{(t)}\right) \middle| t \in [0, 1] \right\}.$$

Setting $\mathscr{P} \equiv (\mathbb{R}^3 \times \mathbb{L}^2([0, 1], \mathbb{R}^3))$, the product space of all base points in $\mathbb{R}^3$ and SRVFs of all 3D parameterized curves, we can impose a weighted product norm. For any $P_1, P_2 \in \mathscr{P}$, where $P_i = (P_i^{(1)}, P_i^{(2)})$ and $P_i^{(1)} \in \mathbb{R}^3$, $P_i^{(2)} \in \mathbb{L}^2([0, 1], \mathbb{R}^3)$, for $i = 1, 2$, we have

$$\|P_1 - P_2\|_{\mathscr{P}}^2 = (1 - \lambda) \left\| P_1^{(1)} - P_2^{(1)} \right\|_{\mathbb{R}^3}^2 + \lambda \left\| P_1^{(2)} - P_2^{(2)} \right\|_{\mathbb{L}^2([0,1],\mathbb{R}^3)}^2 .$$

where $\lambda \in [0, 1]$ is a parameter which controls the relative importance of the base point versus the curve.

We will denote the new representation of the tree as $Y : [0, 1] \rightarrow \mathscr{P}$, $Y(t) = (\beta_0(t), q^{(t)})$; Y is a parameterized curve, or a trajectory, in $\mathscr{P}$ and we are interested in studying its shape. Towards this goal, we will use the same idea as in Sect. 8.3.1 except the trajectories have additional information in the form of the base curve β_0. We can form SRVFs of these trajectories according to

$$Q_Y(t) = \frac{\frac{dY(t)}{dt}}{\sqrt{\| \frac{dY(t)}{dt} \|_{\mathscr{P}}}} = \frac{\left(\frac{d\beta_0(t)}{dt}, \frac{dq^{(t)}}{dt}\right)}{\sqrt{\| \left(\frac{d\beta_0(t)}{dt}, \frac{dq^{(t)}}{dt}\right) \|_{\mathscr{P}}}}.$$

We will use $\mathscr{Q}$ to denote the space $\mathbb{L}^2([0, 1], \mathscr{P})$, so that $Q_Y \in \mathscr{Q}$.

The $\mathbb{L}^2$ norm on $\mathscr{P}$ can be extended to impose a norm on space $\mathscr{Q}$ as follows: For any $Q_1, Q_2 \in \mathscr{Q}$, where $Q_i(t) \equiv \left(Q_i^{(1)}(t), Q_i^{(2)}(t)\right) \in \mathscr{P} \equiv \mathbb{R}^3 \times \mathbb{L}^2([0,1], \mathbb{R}^3)$, define the norm

$$\begin{aligned}
\|Q_1 - Q_2\|_{\mathscr{Q}}^2 &= \int_0^1 \|Q_1(t) - Q_2(t)\|_{\mathscr{P}}^2 dt \\
&= \int_0^1 \left((1-\lambda)\|Q_1^{(1)}(t) - Q_2^{(1)}(t)\|_{\mathbb{R}^3}^2 + \lambda\|Q_1^{(2)}(t) - Q_2^{(2)}(t)\|_{\mathbb{L}^2}^2\right) dt \\
&= \int_0^1 \left((1-\lambda)\|Q_1^{(1)}(t) - Q_2^{(1)}(t)\|_{\mathbb{R}^3}^2 + \lambda\left(\int_0^1 \|Q_1^{(2)}(t)(\tau) - Q_2^{(2)}(t)(\tau)\|_{\mathbb{R}^3}^2 d\tau\right)\right) dt\,.
\end{aligned}$$

The geodesic path between any two trees, represented by $Q_1, Q_2 \in \mathscr{Q}$ is given by a straight line

$$\alpha : [0,1] \to \mathscr{Q}, \ \alpha(s) = (1-s)Q_1 + sQ_2,$$

For each time point s, $\alpha(s) \in \mathscr{Q}$ is a trajectory in $\mathscr{P}$; it can be written as $\alpha(s)(t)$ where t is the parameter of the trajectory in $\mathscr{P}$. Furthermore, for each s and t, $\alpha(s)(t)$ has two components $\left(\alpha^{(1)}(s)(t), \alpha^{(2)}(s)(t)\right)$ where the first component is a point in $\mathbb{R}^3$ denoting the starting point of a side chain and the second component denotes the SRVF of that side chain.

Figures 8.8 and 8.9 show examples of the geodesic paths between two trees viewed as elements of $\mathscr{Q}$. In case when the corresponding points across the two trees – $Q_1(t)$ and $Q_2(t)$ – have nontrivial branches, the geodesic will depict one branch deforming into the other. When only one of these branches is trivial (length zero), and other is

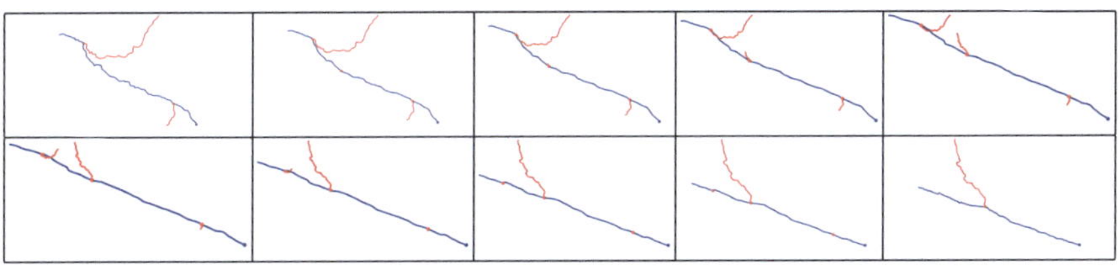

Fig. 8.8 An example of a geodesic path between two axonal trees, shown in *top left* and *bottom-right*, represented as composite trajectories, in the pre-space $\mathscr{Q}$

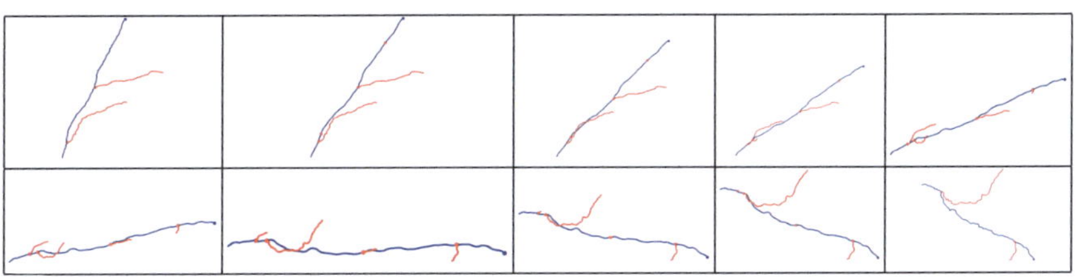

Fig. 8.9 An example of a geodesic path between two axonal trees, represented as composite trajectories, in the pre-space $\mathscr{Q}$

not, we see a growth from a point into a branch. Finally, when both the branches are zero, then there is no change in the structure across geodesics.

8.4.2 Shape Space of Axonal Trees and Geodesic Paths

So far we have not performed any registration across trees but now we look at this problem using the two-dimensional re-parameterization $\xi(t, \tau)$ mentioned in the last section. Recall that ξ takes a special form so that $\xi(t, \tau) = (\xi_1(t), \xi_2(t, \tau))$. For an ξ of the type discussed previously, $\tilde{Q} \equiv (Q, \xi)$ denotes a re-parameterization of components of Q as follows. Since $Q^{(1)}(t)$ denotes the main branch of that tree and $\xi_1(t)$ is a re-parameterization [0, 1], then $\tilde{Q}^{(1)}(t) = (Q^{(1)}, \xi_1)(t)$ is the re-parameterization of the main branch. Similarly, for each t, $\tilde{Q}^{(2)}(t, \tau) = Q^{(2)}(\xi_1(t), \xi_2(t, \tau))$ denotes the re-parameterization of the size branches. This sets up the large registration problem across trees

$$\xi^* = \operatorname*{argmin}_{\xi \in \Xi} \| Q_1 - (Q_2, \xi) \|^2 . \tag{8.4}$$

The solution of this minimization problem is difficult to reach analytically and we use a purely numerical approach for solving it. This approach requires nested calls to DPA as follows: We discretize the parameter domain for the main axis [0, 1] into a dense, finite partition $T \equiv \{0, t_1, t_2, \ldots, 1\}$ with a mild assumption that the parameter values corresponding locations of non-degenerate side branches on both the trees are included in this set. In general, the size of this partition $|T|$ is much larger than the number of side branches on either tree. Then, we compute optimal curve by curve registrations/metric across the two sets of side branches associated with each partition point, using DPA. In principle, this implies $|T| \times |T|$ calls to the DPA, for computing all potential pairings of the side branches across trees, keeping in mind that many of the side branches are null curves. In case both the curves being matched as null, the call to DPA is avoided, and the actual DPA calls is only $n \times m$, where n, m are the number of non-degenerate side branches. Given these pairwise registrations of the side branches, we use a final DPA call to register points along the main branch according to the cost function in Eq. 8.4.

8.4.3 Experimental Results

In this section we present some results on matching and deforming of simple trees with a small number of branches. To focus on the problem of matching the main branch, we further reduce the optimization problem stated in Eq. 8.4 by assuming that $\xi_2(t, \tau) = \tau$ for all t. In other words, the points along the side branches are matched linearly, and only the main branch is matched elastically. This reduces the computational cost to a single DPA algorithm.

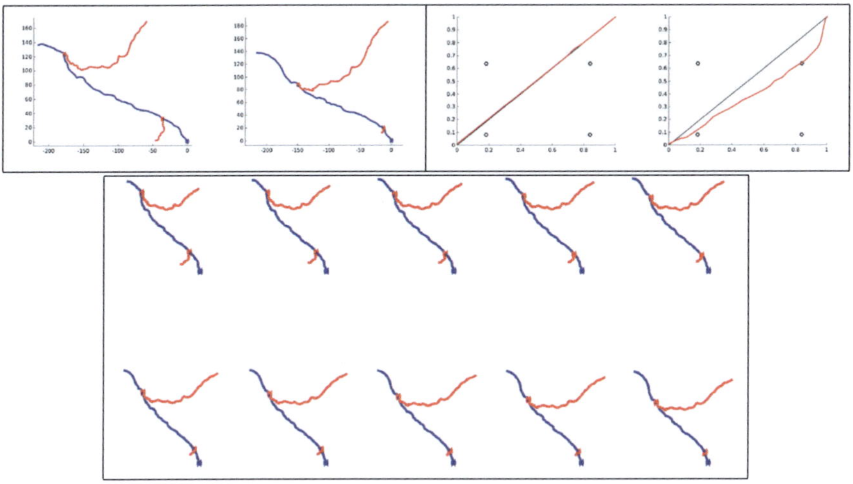

Fig. 8.10 *Upper left* Two trees which are same except for the locations of side branches. *Upper right* Alignment using $\lambda = 0.9$ and $\lambda = 0.999$. *Bottom* Geodesic path between them for $\lambda = 0.999$

To start with, we consider two simple artificially created trees. In this example, the two trees come from identical data. The two trees shown in Fig. 8.10 have the same main branches and same side branches – the only difference is that side branches of the second tree have been artificially moved to a new location. The side branches have the same shape and orientation but are translated to a new position so that they each start at a different location on the main branch. Using smoothing parameter, $h = 0.5$, we examine alignment/geodesic between the two trees with weight $\lambda = 0.9$ and $\lambda = 0.999$. The top right two panels show the re-parameterizations that optimally align the two trees under $\lambda = 0.9$ and $\lambda = 0.999$ respectively. Although $\lambda = 0.9$ gives more weight to the side branches, it is low enough in this case that the optimal re-parameterization does not significantly deviate from identity. With $\lambda = 0.999$, the optimal γ aligns the pairs of identical branches exactly or nearly so.

In the second example we consider two trees that have the same shape side branches occurring at the same locations along the main branch. The difference between the trees lies in their main branches, that are completely different curves. The two trees are depicted in Fig. 8.11, along with their optimal re-parameterizations under $\lambda = 0.5$ and $\lambda = 0.9999$ and the geodesic corresponding to $\lambda = 0.5$.

Now we show examples of geodesic paths between some simple trees taken from the neuromorpho [1] database. It is a centrally curated inventory of digitally reconstructed neurons associated with peer-reviewed publications. It contains contributions from over 100 laboratories worldwide and is continuously updated as new morphological reconstructions are collected, published, and shared. To date, NeuroMorpho.Org [1] is the largest collection of publicly accessible 3D neuronal reconstructions and associated metadata, containing data from human, rat, mice, and drosophila, between other species. Figure 8.12 shows the two trees to be aligned and

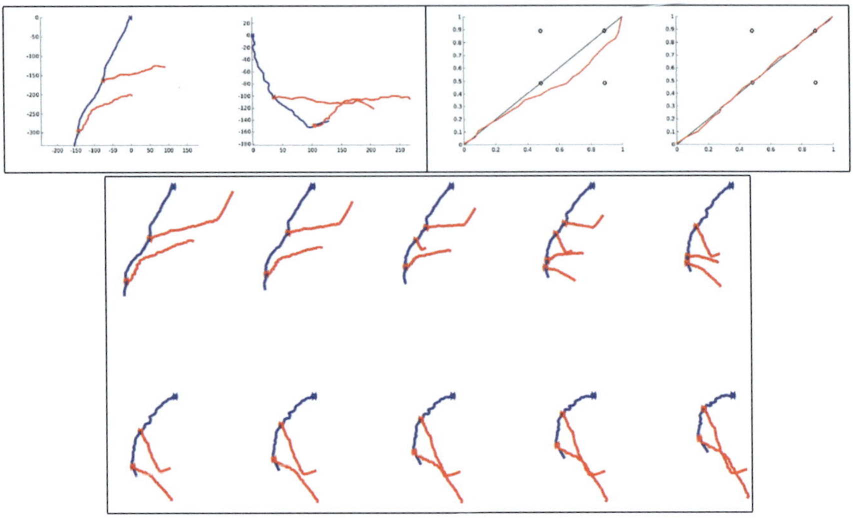

Fig. 8.11 *Top left* Two trees which have the same side branches in the same positions along the main branch. Their main branches are completely different curves. *Top right* Optimal re-parameterizations using $\lambda = 0.5$ (*left*) and $\lambda = 0.9999$ (*right*). *Bottom* Geodesic between the two trees using $\lambda = 0.5$

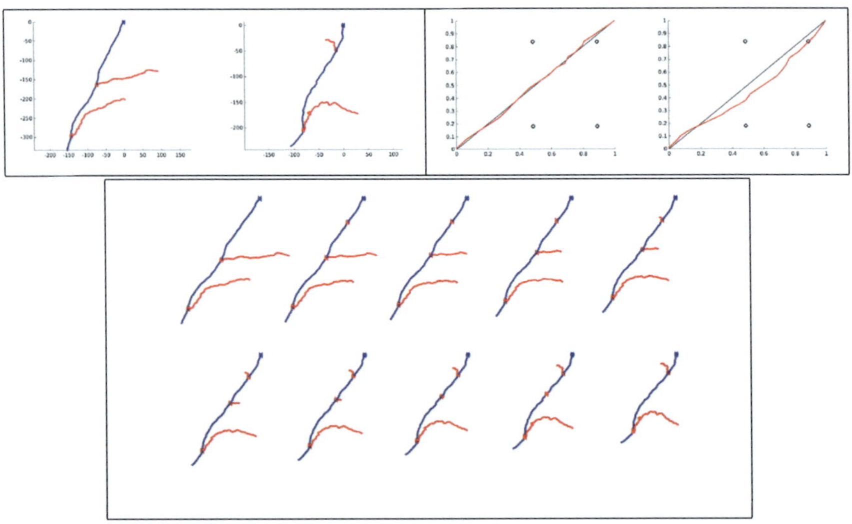

Fig. 8.12 *Upper left* Two trees to be aligned. *Upper right* re-parameterizations to align the two trees. On the *left* is the alignment using $\lambda = 0.999$ and on the *right* is the alignment using $\lambda = 0.5$. In each, the *black line* is the identity re-parameterization, the *red line* is the computed re-parameterization, and the *black circles* show points of alignment which would match side branches to each other at the base

optimal re-parameterizations using $\lambda = 0.5$ and $\lambda = 0.999$. For the $\lambda = 0.999$ case the geodesic path is also shown in Fig. 8.12.

8.5 Karcher Mean of Tree-Like Structures

Given the method of computing distances between trees developed so far, we can define and compute a Karcher mean the same way that we did for elastic curves. Given a set of trees $Q_1, Q_2, \dots, Q_n \in \mathscr{Q}$, we seek a composite curve $Q \in \mathscr{Q}$ that minimizes the sum of square of distances to the given trees

$$\mu_Q = \underset{Q \in \mathscr{Q}}{\operatorname{argmin}} \sum_{i=1}^{n} \left(\min_{\xi_i \in \Xi} \|Q - (Q_i, \xi_i)\|^2 \right) .$$

To compute μ_Q, we use an iterative algorithm similar to the one given in Algorithm 2 of [11], which computes the Karcher mean amplitude of a set of functions. For the case of $\mu_Q \in \mathscr{Q}$ we give an iterative procedure in Algorithm 1.

Algorithm 1 Karcher Mean of $\{[Q_i]\} \in \mathscr{Q}/\Gamma$

Require: $Q_1, Q_2, \dots, Q_n \in \mathscr{Q}$
Ensure: μ_Q as defined in Eq. (8.5)
1: Initialize $\mu_Q = Q_j$, where $j = \operatorname{argmin}_{i \in \{1,\dots,n\}} \left\| Q_i - \frac{1}{n}\sum_{k=1}^{n} Q_k \right\|_{\mathscr{Q}}^2$
2: For each $i = 1, \dots, n$, find $\gamma_i^* \leftarrow \operatorname{argmin}_{\gamma \in \Gamma} \left\| \mu_Q - (Q_i, \gamma) \right\|_{\mathscr{Q}}^2$ using DPA.
3: For each $i = 1, \dots, n$, compute the aligned trajectory $\tilde{Q}_i \leftarrow (Q_i, \gamma_i^*) = (Q_i \circ \gamma_i^*)\sqrt{\dot{\gamma}}$
4: Assign a new mean estimate from the aligned trajectories: $\mu_Q \leftarrow \frac{1}{n}\sum_{i=1}^{n} \tilde{Q}_i$
5: Repeat steps 2–4 until the change in μ_Q is below some small threshold.

Figure 8.13 shows an example result of this algorithm: seven sample trees selected from [8] and their Karcher mean.

Summary and Discussion

In this paper we describe a Riemannian framework to handle shape analysis for three different types of objects. We start with the past work on shape analysis of Euclidean curves using elastic Riemannian metric and square-root velocity functions. The efficiency and utility of this framework in generating statistical summaries and modes of variability of shape data is demonstrated using several examples. This framework is then naturally extended to include analysis of curves in Hilbert spaces. These curves denote a continuous sequences of real-valued functions on [0, 1], and we can compare their *shapes* by temporally registering points across curves. With this tool in mind, we present tree-like structures as composite curves in a Hilbert space, imagining a continuum of side branches along the main branch. Using

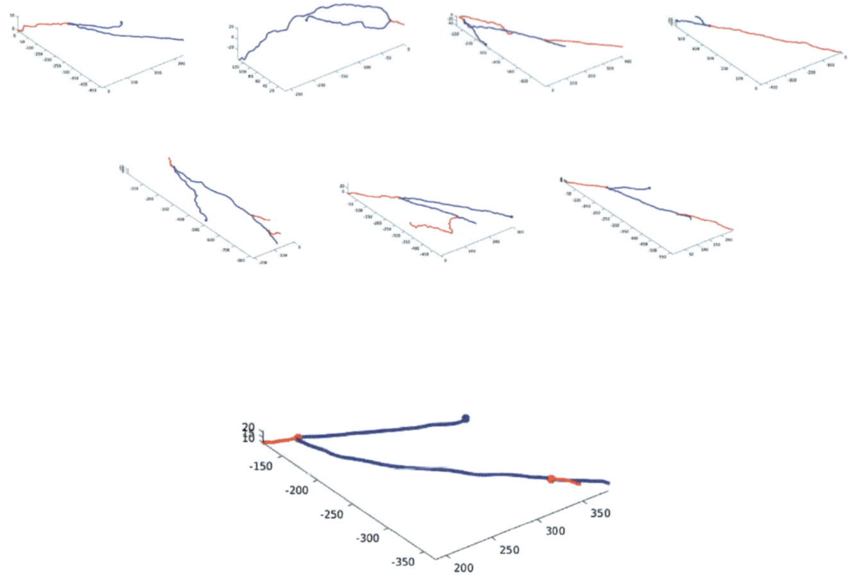

Fig. 8.13 *Top* seven trees from [8]. *Bottom* Their Karcher mean tree

square-root representations for the curves involved – main branch, side branches, and the composite curves, we elastically register points across trees and compare their shapes using the elastic metric.

References

1. G.A. Ascoli, D.E. Donohue, M. Halavi, Neuromorpho.org: a central resource for neuronal morphologies. J. Neurosci. **27**(35), 9247–9251 (2007)
2. D. Bryner, E. Klassen, H. Le, A. Srivastava, 2D affine and projective shape analysis. IEEE Trans. Pattern Anal. Mach. Intell. **36**(5), 998–1011 (2014)
3. A. Duncan, E. Klassen, X. Descombes, A. Srivastava, Geometric analysis of axonal tree structures, in *Proceedings of the 1st International Workshop on DIFFerential Geometry in Computer Vision for Analysis of Shapes, Images and Trajectories (DIFF-CV 2015)* (2015)
4. S. Jeannin, M. Bober, Shape data for the MPEG-7 core experiment ce-shape-1 @ONLINE (1999)
5. H. Karcher, Riemannian center of mass and mollifier smoothing. Commun. Pure Appl. Math. **30**(5), 509–541 (1977)
6. S. Kurtek, A. Srivastava, E. Klassen, Z. Ding, Statistical modeling of curves using shapes and related features. J. Am. Stat. Assoc. **107**(499), 1152–1165 (2012)
7. J.O. Ramsay, B.W. Silverman, *Functional Data Analysis*, 2nd edn., Springer Series in Statistics (Springer, New York, 2005)
8. L.F. Santiago, E.G. Rocha, C.L. Santos, A. Pereira Jr., C.W. Picanço-Diniz, J.G. Franca, S1 to S2 hind- and forelimb projections in the agouti somatosensory cortex: axon fragments morphological analysis. J. Chem. Neuroanat. **40**(4), 339–345 (2010)

9. A. Srivastava, S.H. Joshi, W. Mio, X. Liu, Statistical shape analysis: clustering, learning, and testing. IEEE Trans. Pattern Anal. Mach. Intell. **27**(4), 590–602 (2005)
10. A. Srivastava, E. Klassen, S.H. Joshi, I.H. Jermyn, Shape analysis of elastic curves in Euclidean spaces. IEEE Trans. Pattern Anal. Mach. Intell. **33**(7), 1415–1428 (2011a)
11. A. Srivastava, W. Wu, S. Kurtek, E. Klassen, J.S. Marron, Registration of functional data using Fisher-Rao metric (2011b). arXiv:11033817v2

Index

H.Q. Minh and V. Murino (eds.), *Algorithmic Advances in Riemannian Geometry and Applications*, Advances in Computer Vision and Pattern Recognition, DOI 10.1007/978-3-319-45026-1

MIX
Papier aus verantwortungsvollen Quellen
Paper from responsible sources
FSC® C105338

If you have any concerns about our products,
you can contact us on
ProductSafety@springernature.com

In case Publisher is established outside the EU,
the EU authorized representative is:
Springer Nature Customer Service Center GmbH
Europaplatz 3, 69115 Heidelberg, Germany

Printed by Libri Plureos GmbH
in Hamburg, Germany